Introduction to Agroecology
Principles and Practices

Introduction to Agroecology
Principles and Practices

Paul A. Wojtkowski, PhD

Routledge
Taylor & Francis Group
NEW YORK AND LONDON

First published 2006 by Food Products Press®

Published 2021 by Routledge
605 Third Avenue, New York, NY 10017
2 Park Square, Milton Park, Abingdon, Oxon OX14 4RN

Routledge is an imprint of the Taylor & Francis Group, an informa business

ISBN 13: 978-1-56022-316-0 (hbk)

Cover design by Kerry Mack.

Library of Congress Cataloging-in-Publication Data

Wojtkowski, Paul A. (Paul Anthony), 1947-
Introduction to agroecology : principles and practices / Paul A. Wojtkowski.
p. cm.
Includes bibliographical references and index.
ISBN-13: 978-1-56022-316-0 (hard : alk. paper)
ISBN-10: 1-56022-316-2 (hard : alk. paper)
ISBN-13: 978-1-56022-317-7 (soft : alk. paper)
ISBN-10: 1-56022-317-0 (soft : alk. paper)
1. Agricultural ecology. I. Title.

S589.7.W647 2006
577.5'5—dc22

2005027702

We must tackle and grasp the larger encompassing themes of our universe, but we make our best approach through small curiosities that rivet our attention. . . .

S.J. Gould (1989)
Wonderful Life: The Burghers Shade

and the nature of history
a gardyn saw, I, ful of blosmy bowes,
upon a river, in a grene mede,
ther as swetnesse evermore y-now is,
with floures white, blewe, yelowe, and rede;
and colde welle-stremes, no-thing dede,
that swommen ful of smale fishes lighte,
with fines rede and scales silver-brighte.

Chaucer (1382)
Parliament of Foules

ABOUT THE AUTHOR

Paul A. Wojtkowski, PhD, brings over 25 years of research, teaching, and outreach experience from many corners of the world. Dr. Wojtkowski has lived in eight countries and has conducted professional observations in another 70. His applied efforts span diverse ecological zones and a wide range of natural and human-impacted ecosystems. Some of his residences were in environmental trouble spots or in regions where food and fiber are grown only through nature-accommodating measures. His travel includes Africa (almost four years), Australia (one year), the South Pacific (two years), Latin America (three years), Europe, and North America.

From this vast experience, Dr. Wojtkowski is well-versed in both commercial agriculture and subsistence farming. From experience he knows how those confronted with poor soils and harsh climates coax yields from the land. With equal importance, he also knows how those operating on good soils and favorable climates can utilize nature-assisted farming techniques to maximize profits. Dr. Wojtkowski is the author of four books on advanced agroecology, including *Landscape Agroecology* from Haworth.

CONTENTS

Preface

Agricultural ecology (or agroecology) is a compilation of views on how food, fuel, and fiber are produced from the land. Useful yields can be produced directly, as with a grain harvest, or indirectly, as when chickens consume this grain. Perspectives vary on how these outputs, from plants and/or animals, are best obtained.

Agroecology

Foremost in interpreting agroecology is the driving vision that agriculture should be nature friendly. Anchoring this drive is a continually expanding body of concepts and theories. These explain the underlying rationales for why things are done, and provide paths for future progress. Since agroecology is a close cousin to ecology, with shared values and ideas, much can be accomplished, through their commonalities, to put agriculture on a solid environmental footing.

The outward manifestation of agroecology, farm practice, can range from improvements on what can be called conventional usage to unique forms of organic agriculture. In considering agroecology, most subscribe to the view of a more harmonious interface with nature. For a few, agroecology encompasses traditional means to produce crops, at times under trying circumstances.

Prior to the advent of agrochemicals, most farming incorporated some degree of natural harmony. Only in recent times have farmers come out from under the rather broad agroecological umbrella. With advantages of science and interest in ecology, what has been lost can be reintroduced, only with a sharper focus.

Introduction to Agroecology: Principles and Practices

doi:10.1300/5553_a

Agriculture

Since agronomy and agroecology partake in a vast store of information and research output, agroecology may seem a subset of agronomy. This is not the case. Agroecology and agronomy are both separate but interrelated, tracks toward separate but overlapping objectives. For the textual presentation, agronomy and mainstream or conventional agriculture are synonymous terms: agronomy and agroecology are considered directional partitions of agriculture.

Natural Associations

For explanation purposes, it helps that agriculture is an offshoot of nature. Some practices reside close to the source; others have moved farther afield. As a result, agroecology has commonalities and differences with its natural counterparts. These occur on a number of levels.

Natural and agroecosystems are internally governed by a host of plant-on-plant relationships. The interaction of unlike plant species exists in nature, but the explanations are more advanced in agroecology.

The interaction of unlike ecosystems is yet another operational level. This can be in an agriculturally pure form, as with a farm composed entirely of agricultural ecosystems, or some nature-farm mix, as when agriculture shares space with native flora and fauna.

A host of factors set agroecology apart from anything found in nature. Dwellings, barns, roads, along with land ownership patterns are all ex-nature elements. Also, agroecosystems are judged through economic criteria, whereas natural ecosystems are exempt or should be exempt from this type of scrutiny. Despite these differences, the first step in addressing an agroecological problem is the question "What would nature do?"

Starting Points

One way to approach agroecology is by looking at the plant-plant interface. How plants interact together to achieve good growth and yields, sans unwanted competition, accentuates a wide swath of agriculture. This carries forth and explains topics such as field layout, threat countermeasures, etc.

To keep the book on course and to extract as much as possible from available sources, the starting approach is the one plot–one agrotechnology model. There are reasons for what may seem a more obscure view, that is, that of agrotechnology. Rather than attempting to bite off large chunks of complex agroecology and economy, the agrotechnologies deliver ecological and economic bite-sized pieces. Also, there is a certain unpretentiousness as the agrotechnologies are shorn of cultural and philosophical preconceptions, shrouding beliefs, and encumbering images.

In contrast to agrotechnologies, classification based upon the so-called organic practices offer far fewer definable points upon which to focus and far more extraneous preconceptions to overcome. Discussion of the organic alternatives is therefore deferred until some degree of comprehension is in place.

Agrotechnologies, as definable agricultural units, are comparatively recent. Development stems from the mainstreaming of agroforestry and the need to label and define the many agroforestry options associated with tree-crop mixes. This is unnecessary in conventional agronomy as the mainstay, the monoculture, offers comparatively little ecological variation.

In essence, the agrotechnologies demarcate points along what might be termed the agroecological mosaic. This has the added advantage in that the points are more or less evenly distributed with plenty of ecological and technical diversity. The monoculture, as a small group of agrotechnologies, represents only a few points in this vast mosaic.

Sources

In a major deviation from what one might expect from an introductory text, many seemingly superficial topics are fair game for inclusion. Historic, geographic, cultural, popular, and philosophical inputs help complete the agroecological picture.

Without becoming subservient to current trends, popular sources are appropriate as biodiversity and sustainability, presently prominent in the environmental movement, are also watchwords for agroecological ideals.

Many historical sources, first person and otherwise, have been looked at. This is because

1. all agriculture once fell under the heading of agroecology, and therefore lessons long forgotten or techniques discarded because of ill-advised convention or false direction have future application;
2. past events, as a continuation of ideas, help shape present and future agricultural activities; and
3. science has been remiss in that many highly appropriate practices lack any formal study, when formal research is missing, thousands of years of use constitutes proof as to effectiveness.

Agriculture, as practiced in remote regions or by farmers operating away from the mainstream, is also fair game. This is for many of the same reasons that historic accounts find reference; the lessons are revealing, especially when these practices employ innovative methods to overcome challenging climates. It is these exceptions, as much as the mundane, that contribute to our understanding of the nature-agroecology interface.

Entwined with history and immediate necessity is culture. This is far from an inert influence. Defined broadly, culture puts an unequivocal stamp on what is acceptable and what is outside societal bounds. The eating of savory bugs is solely a cultural dictate: some societies relish the opportunity, while others forgo a nutritious snack. The dictates of culture, with or without religious significance, influence field practice. To this end, some mention is made of the opportunities presented (bugs included).

Agriculture, as a philosophical expression, is seldom overtly discussed. Although not a subject ripe for discussion here, insights do help in understanding why certain practices achieve prominence. Along these lines, the human-nature interface ranges from a total domination of nature (unfortunately, a view often associated with conventional agronomy) to those few that farm in complete natural harmony (the traditional agroforests of the South Pacific appear highly nature balanced).

Given the scope described, reading everything is impossible. The only question is that the sample taken is wide and inclusive enough. The historic and cultural sources, along with the publications of organic adherents, help ensure that no major topics have slipped through the cracks.

Of course, not all is recorded. Many farmers and gardeners, past and present, toil in obscurity, some utilizing unique and viable methods that, if brought into the mainstream, could enrich the whole. This author is fortunate in having observed uncharted practices. These have made their way into this text; many more remain to be categorized and documented.

Limits

Despite the stated intent to embrace all of agroecology, there are limits. Agronomic convention holds that forests and forest trees, those managed for only wood output, are not part of agriculture. This is not a good indicator of farm practice, as forest trees are plentiful in and part of the ecology of most agricultural landscapes.

In a shift away from agronomic convention, agroforestry has been included. Trees are an integral part of raising agricultural crops and, therefore, the wide array of agroforestry options are more than admissible in agroecology. As an added benefit, tree-crop mixes do incorporate some very diverse (agro)ecological principles and, in understanding the mechanisms of advantage, the picture is made more complete. The danger is that, since many adhere to the view of agriculture without trees, this will be regarded as an agroforestry rather than an agroecology text.

One can go only so far into agroforestry without becoming involved in forestry. Bowing to conventional thinking, forest trees, as in farm forestry and as a nonagricultural peripheral activity, are discussed only when they directly contribute to the agricultural whole.

Domestic animals are a lage part of the agricultural scene. Despite this, plants are the core of husbandry, as animals are dependent upon plants for food. Keeping things simple, animals are discussed only when they ecologically intrude beyond a simple fauna-feed association.

Endnotes

In an introductory text, topics should be offered in a near-effortless manner. This is problematic, as much comes from firsthand, citation-requiring sources. Endnotes keep the text whole and flowing, uninterrupted by numerous cites.

As textual unintrusive excursions, these notes contain, in addition to source references, explanations or remarks that normally accompany new materials. Taking full advantage of this feature, the notes also contain comments, asides, cross linkages, dissenting opinions, and other augmentations not normally found in an introductory or even an advanced text.

Goals

The goal is to present enough of agroecology so that a broad picture emerges. The problem is that biodiversity is not a simple topic; rather than sidestepping the complexity, the book confronts it head on.

There are dangers in taking this stance. Unless the ideas are well ordered, they can come across as a discombobulated aggregate of practices and approaches, entangling the reader in what may seem like overwhelming intricacy. To avoid this, the flow pattern employed starts with a theoretical and conceptional framework. Once some degree of understanding is in place, a host of secondary, more practical topics are introduced. As the flow accelerates, various supporting concepts are introduced.

Along the way, digressions show the linkages to other subjects. Digressions foster the idea that agroecology is not a single, beginning-to-end, topic, but radiates in many directions.

If digressions cause readers to think about or to increase their sensitivity on how others express agriculture, this is for the good. The danger lies in the opposite direction, the assumption that one size fits all. The ultimate agroecological crime, one all too common, is where experts counsel locals to abandon ecologically sound practices in favor of those that are not sustainable, more risky, too costly for the resource-poor farmer, and/or out of sync with the climate and/or culture.

Rather than being past relics or out of step with modern ways, farmers in remote areas or gardeners in their backyards can create some very interesting nature-agricultural associations. This is part of an unfolding of ideas where, upon understanding the basics of agroecology, the more inquisitive will be in a position to ecologically interpret what is or has happened. Steering readers around pitfalls and toward innovative agroecology is an important but often overlooked goal.

In defining agroecology, some far-reaching and lofty objectives are outlined. These may seem idealistic and out of reach, but given the influence and exalted ideals of environmental science, they are entirely appropriate. An overall goal of agroecology and of this text is, by visualizing the environmental and societal whole, to make agroecology a more effective vehicle in bettering the human experience. The ideal is to do this with little intrusion on the earth's natural ecosystems.

Acknowledgments

This text is built on direct observations, inputs from countless individuals, and long hours in libraries. The libraries visited include the Berkshire Athenaeum, Pittsfield, Massachusetts; Berkshire Community College, Pittsfield, Massachusetts; the Mann Library at Cornell University; and the libraries at William's College, Smith College, and the University of Massachusetts at Amherst. Without their help, this text would not have been possible.

Thanks goes to the technical staff at Aldea Global, Jenotega, Nicaragua. Their work advancing regional agroecology provided many good photographs. Also enriching this text are those who, over the years, have shared their ideas or have pointed out interesting features in farm landscapes. As there have been many, they cannot be singled out. They still deserve thanks.

Introduction to Agroecology: Principles and Practices

doi:10.1300/5553_b

Chapter 1

Introduction

Agriculture, in its beginning, was a clear offshoot of nature. People, through their close-to-the-land lifestyle, interfaced with nature, gathered, modified the land, and eventually purposely planted. Early agriculturalists, through direct observation and handed-down wisdom, understood and applied natural processes, i.e., those found in and sanctioned by nature.

Over time, some have moved away from these natural roots. The agriculture-nature interface has become something that requires less natural understanding and a less nature-friendly accommodation. Despite this development, how nature does things is still a good model for agriculture. It is this view that drives science-based agroecology.

AGROECOLOGY

Without putting forth a formal definition,[1] the difference between what may be termed conventional or mainstream agriculture and agroecology rests on a number of foundations, all stemming from various perspectives.[2] Basically, agroecology is a step back from the conventional view, looking instead at broader academic and societal inclusions along with productive and environmental objectives. These differences manifest[3]

- where each crop is an element or component in a larger agroecosystem;
- where each plot is part of a larger, often natural, farm, local, or regional ecosystem that embraces interplot ecological solutions to plot-internal productive obstacles;

Introduction to Agroecology: Principles and Practices

doi:10.1300/5553_01

- through a longer time frame in planning and implementation;
- in a greater sensitivity toward and greater interest in diverse agricultural practices, one born of an understanding of methods, applications, and the cultural role of agriculture;
- in the sanctity of environment and the do-no-harm admonition that accompanies the application of agroecological principles and practices;
- with an accentuation for improving the quality of life for people and the natural fauna that inhabit the land;
- from an academic perspective, the elimination of those partitions that separate agriculture from forestry, recognizing that trees are as much an ecological force in the farm landscape as are crops;[4] and
- from the scientific perspective, in the wealth of theories, concepts, and principles from which to proceed when direct experience or research data are lacking.

In accomplishing agroecological objectives, a wider range of tools are employed; some are outside of conventional agricultural thought. All of these stem from the application and transformation of ecological principles to agricultural practice; all can be viewed by taking this step back into the realm of agroecology.

Some of the differences can be illustrated through a common affliction, an onslaught by crop-eating insects. In mainstream agriculture, success is often measured by how immediate and severe, as in overkill, a solution is.

In contrast, agroecology plans and places the plots, the surrounding landscape, and the temporal sequences such that an insect outbreak is less likely. If one does occur, an environmentally friendly solution should be close at hand. This view, that of a self-perpetuating, long-term, and environmentally friendly plot and landscape mix, applies not only to harmful insects but to all aspects of crop production.

AGROECOLOGICAL GOALS

Another perspective on agroecology stems from stated goals. Yields and positive economic outcomes are important, but more needs to be considered. In its broadest form, agroecology is charged with

- providing the basics of human existence,
- enhancing the quality of life and improving the human experience,
- ensuring that natural flora and fauna can survive and thrive,
- maintaining the totality of the land and climate, and
- having a clear and ecologically sound vision that persists into the distant future.

On the surface, these goals may seem lofty and beyond the reach of agroecology.[5] To comply, agroecology becomes more than just the raising of potatoes, wheat, grapes, lettuce, tomatoes, and so on.

Improving the human experience should be a direct side benefit of the agroecosystems chosen, that is, there is more to be gained than simply acquiring the sought-after crops. In addressing these agroecological precepts, one starts by answering basic human food needs and continues from there. While doing this, some topics that are inherent within agroecology rise to the fore in the quest for a more inclusive understanding.

There are goals at the macro or regional level. The lack of water, a lack of clean water, or water depletion are issues than can be addressed through agroecology. These involve raising crops with less, increasing retention through better land management, and the cleansing of sources. This extends to ocean dead zones, areas of sea that have been sapped of life by farm runoff. These are at the end of a flow that leads back to agricultural fields, a flow cleansed by halting fertilizer inputs and/or runoff. Another escalating problem, global warming, cannot be fixed through agroecology, but individual agrosystems can be made climatically pilable, bending but not breaking as climatic duresses unfold.[6]

Necessities

The pursuit of the basic necessities—food, fuel, and fiber— underlies land-use practice. Farmers cannot survive without core outputs, and this is the first priority of agriculture and agroecology. Doing this within cost and labor constraints ranks high on most lists of underlying cropping requirements.

Civilizations and cultures have risen and fallen because agricultural practices did not keep up with population growth and the demands upon the land. Periodic reports of starvation or severe hunger reinforce the view that events, short or long term, can overwhelm a perilous situation.

This need not be the case, even where population and/or climate are near the margin. History provides numerous examples of food-challenged regions that once produced in abundance. In Roman times, now-forlorn areas of North Africa were, through micromanagement and large-scale land modification schemes, made productive.[7]

Some hostile events cannot be circumvented through good agricultural practices. Man-made calamities (war, social upheavals, etc.) and fully engulfing and destructive natural calamities (severe floods, prolonged droughts, etc.) head this list. Although an omnipresent risk, most climatic stresses and productive threats fall within the corrective scope of agroecology. Often alluded to as a group, a breakdown of combatable threats include

- climate (i.e., drought, flooding, toppling winds, temperature extremes, and fire);
- weeds;
- animals that eat or trample crops (ranging from rodents to elephants);
- herbivore insects; and
- plant diseases.

Surmounting these problems in a sustainable manner, singly or as a group, is a prominent challenge of agroecology.

Profit Motive

The need to profit from agriculture is a driving motive.[8] Even when most of the outputs are for subsistence, products from the land are sold to purchase external goods. Again, the human experience does not take all down the same road.

Some eke a small profit from each plot through low yields and severe cost-cutting measures. The hope being that, across a wide area, small gains multiply so that, in the end, there is a large favorable aggregate between sales returns and the costs of production. Others have the same idea but seek larger yields by increasing inputs. The expectation is that the returns from crop sales will more than cover the added expenses. Still others seek multiple outputs and higher economic gains from a few intensely managed plots.

These are but a few of the possible economic avenues. At the very least, agroecology must be a profitable enterprise, but in trying to achieve this other elements should not be lost.

Quality of Life

Through farming, much can be done to improve the quality of life of land users. Quantity of outputs is fine, especially when shortfalls threaten; however, once the basics are in place, other needs can be looked at.

There is no question that, among the many uses, plants are nutritious, cure various ailments, and are decorative. These and other attributes can be used directly to great effectiveness.

Moreover, a well-formulated rural landscape, one well stocked with biodiversity, can afford clean water, more agreeable temperatures, and pleasing aesthetics. These gains, and more, are part of agroecology and can be promoted at a comparatively low cost. Therefore, a few changes, made not solely on financial grounds but to promote a better lifestyle, can help improve the rural situation.

The Natural Ecosystem

Within the context of agriculture, native plants and animals can survive and even thrive. Not always a competitor to the productive process, accommodation requires some rethinking on the part of land users and the community. These need not be giant steps but merely a reordering of priorities.

The daily melodies of songbirds are enjoyed if added to the list of priorities. Other forms of natural biodiversity, natural ecosystems included, can be similarly appreciated and promoted. The ideal is when the resulting dynamics can be harnessed for agroecological purpose.

MAINSTREAM AGRICULTURE

In some cultures, vast fields of ripening grain epitomize productivity, the good life, and mainstream agriculture. Ingrained in this is the perception that all is well and that humankind has successfully conquered nature.

This is far from the case. Vast fields of ripening grain often invite environmental problems, reinforcing the view that nature is difficult to conquer. Monocultures, short on meaningful and recuperative ecological interactions, are excellent examples of this problem.

The mainstream operating principle is that as problems become apparent corrective inputs provide an immediate fix. This strategy is attractive, there being a certain satisfaction in quick solutions. Also, in confining agriculture to narrow operating principles, users rapidly become familiar with and confident in the methods and outcomes.

The disadvantage is that a vast store of ecological knowledge goes untapped. A few well-chosen alternatives can give superior results in both profit and productivity. These options and alternatives put a practical face on agroecology.

AGROECOLOGICAL TOOLS

Biodiversity is the basis for much of agroecology. Nature does just fine mixing species and so can the farm community. The list of options, and the productive potential of the land, is greatly expanded by intermingling species, an underused tool; this is where much of the progress is made. Outside of this, other, less recognized, options can further enhance agriculture, bringing to the fore a host of supplemental benefits.

Large areas of a single plant species are often chided as a failing of agroecology. Nonetheless, the monoculture is well entrenched in agriculture and there is no reason that it should be entirely abandoned.

Despite the shortcomings, these can be redesigned and brought more in line with ecological principles.

A willingness to examine and utilize wide-ranging solutions adds to the list of tools. Food, fuel, and fiber can come from agricultural or forestry plants, and the separation of agriculture from forestry is enshrined only for academic convenience. This restrictive view does not survive examination from ecological, classification, or human need perspectives.

A case in point, tomatoes and cotton come from annual species, but certain trees can provide a like substitute.[9] Perennially harvesting without the cost of seasonal replanting is enticing; however, consumer preference and the changeover costs may not justify abandoning current production methods.

This changeover may not pass muster, but, in looking further afield, unlikely and uncharted alternatives, trees included, exist. In restricted, or even in mainstream, situations, some may prove economically and ecologically superior. Addressing lofty goals means utilizing lesser-known choices to benefit land users and their working environment.

ALTERNATIVE AGRICULTURE

The cultural branches of the human tree have much in common, but equally interesting are the differences. Just as cultural diversity is intimate with the human experience and agriculture, how individuals view the nature-people interface is also ripe with variation.

As a connect between beliefs and practice, agricultural alternatives offer a peek at the philosophical and practical foundations of agriculture. Often less intrusive on nature, people, through viewpoint differences, continue to utilize a range of ecological tools and, when change is required, build upon natural experiences. If any one ideal exists, it is where individuals and societies fully exploit natural dynamics in the pursuit of environmentally friendly agriculture.

Organic, low-input, regenerative, permiculture, and mimicry head what is a long list of alternatives. Before the underlying foundations of what may be broadly termed organic or alternative agriculture can be explored, the agroecological concepts must be in place. The following chapters set out along this path.

Chapter 2

Spatial Principles

This chapter presents some of the broader concepts and principles that underwrite successful agroecology. In specific, the discussion centers on essential resources, their utilization in a single plant, how resources influence assemblages of unlike species, and how the underlying dynamics are put to task. This introduces, through the development of fundamental spatial principles, the basics of agroecosystem spatial design.

As with most aspects of the natural world, concepts and principles are not laws and, as such, are fraught with exceptions.[1] At times, agroecology may seem to defy logic; some of this lies with the exceptions, much of this with overwhelming and outwardly intractable complexity.[2]

Even with these complications, the theory presented should be deep enough so that a picture emerges. The reader should be able to progress along an admittedly bumpy road from simple agriculture into the realm of complex agroecology.

ESSENTIAL RESOURCES

The vast majority of plants depend on the essential resources, such as light, carbon dioxide (CO_2), water, and elemental nutrients. The latter, the elemental nutrients, include nitrogen (N), phosphorus (P), and potassium (K), along with a host of trace elements.[3]

As with any general biological rule, there are exceptions. Fungi crops, such as truffles and mushrooms, are not light dependent. Other plants may forgo other essential resources. These are unusual cases but, as with mushrooms, some are encountered in agriculture.

Introduction to Agroecology: Principles and Practices

doi:10.1300/5553_02

RESOURCE CONCEPTS

How the different essential resources interact within the physiology of specific plants is a valid and, at times, overriding question. Some theories and general guidelines help us understand their role in growth and yield.

Capture

Essential resource use, that is, those resources converted into plant material, is a function of the amount both available and captured. Seldom is the full amount captured; more often it remains untapped. This is because these are

1. not needed at a given moment (see the limiting resource, this chapter);
2. beyond reach or not found (e.g., outside a plant's immediate root or canopy zone); or
3. bound and not procurable (locked away and nonusable, in chemical compounds, physically fastened within rocks, or part of the active biomass).

Resource capture is a function of the number and diversity of plants in a given area. Although a leap into the unknown as with most essential resources, light collection is a function of the density of the canopies, that is, the leaf area index (LAI)[4] (the canopy density) for each plant is multiplied by the number of plants per area. This produces a nonlinear, asymptotic relationship with regard to the amount of light captured, that is, a few plants per area capture a low percentage of available light, but each has ample light. In contrast, more plants capture a greater percentage of the light, but with less light available for each plant.[5] For other essential resources, capture, rate, and allocation are more of a mystery.

Because the rate of capture can be low, there is often plenty of room for improvement. For chemical fertilizers, the capture rate is in the range of 40 to 50 percent; for cut green biomass, the rate is about 20 percent. It can be safely assumed that more plants and/or more species equate with greater resource uptake.[6]

Production (Growth and Yield)

Reacting to the amount of essential resources available and captured, production functions underlie growth and yield for each plant species and each of the essential resources. Sample functions are shown in Figure 2.1. The exact shape of the resource functions are species and variety dependent. These are universally unknown.[7]

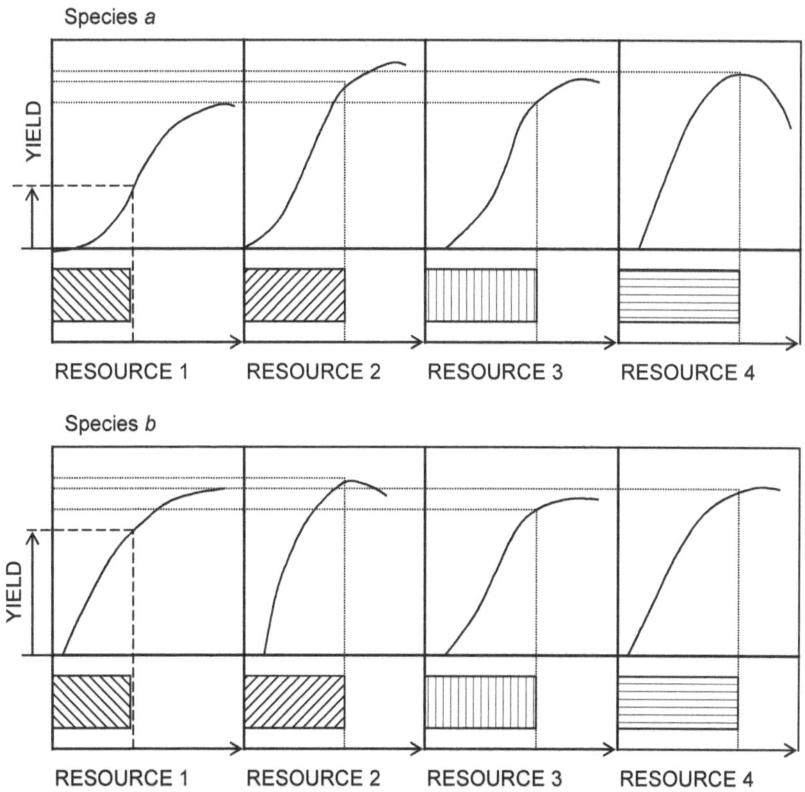

FIGURE 2.1. The essential resource functions for two species (*a* and *b*). The yield from species *a* is limited by resource 1 *(upper left)*. A second species, having the same available resources (along the lower axis) but different use functions, produces higher yields primarily because the limiting resource (resource 1) is less limiting.

Being a mathematical uncertainty, these lack precise application; however, they go far in explaining broad agroecological concepts. This includes when a species is

1. resource demanding (requiring relatively large amounts of a resource to achieve an acceptable growth and yield) or
2. shortfall tolerant of an essential resource (relatively small amounts of one resource can elicit a relatively large growth and yield response).

Commonly, species are thought of as being drought or shade tolerant; other tolerances tend to be less noticed.[8]

Overabundance can also be of interest. Some plants cannot bear more than an optimal amount of one or more resources and suffer losses in the face of large surpluses. Maize can be susceptible to water-saturated soils and, when this occurs, yields increase by taking away some of this resource.[9]

Resource Profiles

Each site has an essential resource profile. Ideally, this matches the need profile for the species and variety planted. Figure 2.1 shows the resources available (in bar form) on a site matched against those needed by the crop (in functional form). The better the fit across the profile, the better a site is utilized by crops (one or more species) that exploit different segments of the resource-use profile. Most often, the productivity constraint comes in the form of a limiting resource.

Limiting Resources

The limiting resource is the single resource that prevents a plant from attaining higher growth and yield. By definition, this resource is in greatest demand.

In Figure 2.1 species *a*, essential resources are relatively more abundant. It is the first resource (resource 1) that sets the upper limit on growth and yield; that is, this is the demanded limiting resource.

In contrast, Figure 2.1 species *b* offers a better site fit. The resource most demanded (resource 1) is relatively abundant and is less yield constraining.

This analysis is based upon the simplest of theorems, that of a single, discrete, limiting resource that directly and unequivocally influences growth and yield. Others see a more complicated picture in which the limiting resource is not absolute but, given an abundance of other essential resource, an otherwise binding lower limit may be slightly superseded or overstepped.[10]

Outside the immediate scope of this discussion (to be looked at in the next chapter as a yield-gaining facilitative effect) are the recognized indirect resource interconnections. A familiar case has light heating leaves, forcing increased transpiration, and increasing water use with a corresponding loss in water-use efficiency.

In addition to the tangle afforded by indirect links, other practical factors intervene to make this line of analysis troublesome. The limiting resource can be time and weather dependent where conditions can change

- hourly, for example, as with a morning dew in a semiarid region;
- daily, for example, where thick, heavy, light-blocking clouds accompany a prolonged rain; and/or
- weekly, for example, where the resource functions, that is, the physiology of a species, change as the plant matures.[11]

After all is said and done, the differences in opinion on functional form along with the practice complexities do not invalidate the concept nor do they nullify the application possibilities. The limiting resource, in league with the concepts of shortfall tolerance or resource-demanding plants, is still useful in explaining certain happenings.[12]

Balance

Adding or taking individual resources from a plant can increase yields. For best results, the growth and yield function for any one species should match the essential resources available for the site in question. The balance can be natural, as when a species or variety is site compatible (more so with species *b* than species *a,* Figure 2.1), or when resources are added.

There are economic advantages in precisely approximating or knowing, but not exceeding, the needed amounts of costly nutrients or water inputs. Dai et al. (1993) showed a small, but commutative, monetary gain in being mindful of which essential resources contribute to increased yields and adjusting fertilizer inputs accordingly. For maize, profit (revenue – costs) increased between 5 and 15 percent as a result of applying the correct balance of essential resources. In this case, nutrients, not rainfall, had be to limiting and therefore contributing. Figure 2.2 shows such a situation. The added resource (right) does not contribute, only imposing an added cost (unnecessary nutrients) without any corresponding yield gain.

BIODIVERSITY

To this point, the focus is more on a single species occupying one site. More control can be exercised, and soil and plant resource pro-

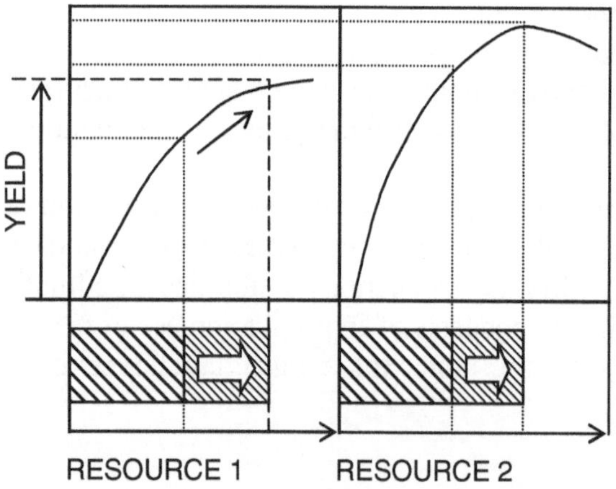

FIGURE 2.2. Two resource functions for a single species. Adding a resource (resource 1) is advantageous whereas adding the second (resource 2) yields nothing additional. Any associated cost for resource 2 (as with a commercial fertilizer) is wasted.

files better matched, by dropping the single species assumption and looking at systems of two or more productive species. The express goal is to increase overall yields. In this case, the role of limiting resources can intensify or lessen, depending upon the niche relationship between unlike species.

Niches

The niche concept underlies most of ecology and applies to agroecology. Sidestepping views on niches in natural ecology,[13] the agroecological picture, especially for basic one-on-one, plant-plant relationships, is better focused; that is, because the specific mechanisms can be individually enumerated and applied, the concept can be considerably narrowed.[14]

Although only a small segment of the full niche concept applies to the agroecosystem spatial picture, the full ramifications of the niche do have broad significance. This occurs when native flora and fauna interact within and, ideally, positively contribute in an agricultural setting. In-use examples of the full niche concept are given in subsequent chapters. To preview a few, there are niches which differentiate co-residing companion species (Chapter 4) and those of predators as they prey on harmful insects (Chapter 5).

Complementarity

The immediate alternative to the full niche is the concept of plant-plant complementarity. This has unlike species interacting positively, and yields increasing, through

1. *system or absolute gains,* in which two or more species utilize site resources better than a single species;
2. *facilitative gains,* in which, in the simplest condition, one species augments the productive (harvestable yield) capacity of a second;
3. *risk reduction,* in which the presence of other species (one or more) cause the productive plants to perform better in the face of the various climatic stresses and productive threats; and

4. *economic gains,* in which an agroecosystem offers greater profit potential through revenue increases and/or cost reductions.

These involve not only direct mechanistic relationships but encompass most of what can happen. This begins the process of identifying those species pairings that have complementarity, the limits to this complementarity, and the field (spatial) layout that can extract the most from any given pairing. Well-defined complementarity helps land users increase the relative yields by varying the planting density, planting times, and other management factors or resource inputs.

A common interspecies intercrop is that of maize grown together with beans. Although generally a robust combination with regard to site and capable of risk reduction and absolute yield gains, there are limits to be considered. Maize and beans become competitive when seasonal rainfall falls below 400 mm.[15]

There are also ceilings on the relative populations of each species, i.e., where too many of one species may overwhelm the second and harm an otherwise harmonious relationship. With any intercrop combination, land users change planting densities to accommodate site and output needs.

MEASURING GROWTH AND YIELD

Without some means to cross-compare diverse crops, an analysis is difficult. This is the apples-oranges dilemma, the need to reconcile the yields of unlike outputs to obtain a good picture of the limits of complementarity and to direct the underlying dynamics to their fullest benefit.

One approach to measuring growth and yield is to translate different crop yields (e.g., apples and oranges) into monetary values and conduct classic financial analysis (e.g., revenue minus costs). This is commonly done, but assigning market values or selling prices has little to do with the underlying dynamics and may even cloud the issue.

The alternative to a financial calculation is the land equivalent ratio (LER).[16] With two crops, the equation for calculating LER is

$$\text{LER} = (Y_{ab}/Y_a) + (Y_{ba}/Y_b) \tag{2.1}$$

where Y_a and Y_b are the monocultural yields of species *a* and *b*. Y_{ab} is the yield of species *a* grown in association with species *b*, and Y_{ba} is the yield of species *b* in association with species *a*.[17]

A hypothetical case demonstrates use. If species *a* has potential monocultural yields of 5,000 kg·ha^{-1} and, on the same site, species *b*, as a monoculture, has an output of 8,000 kg·ha^{-1}, these are the denominators. If intercropped the two species, *a* and *b*, respectively, offer harvests of 3,000 and 4,800 kg, the LER calculation will be

$$\text{LER} = (3{,}000/5{,}000) + (4{,}800/8{,}000) = 1.2 \tag{2.2}$$

By definition, a monoculture has a value of 1 (i.e., Y_a / Y_a). If the ratio value (LER) for the species mix is greater than 1 (as in Equation 2.2), then growing the two species (*a* and *b*) together is the better alternative. If the LER is less than 1, monocultures of either species are the better option.[18]

This equation represents a line of development that stretches across most agroecological economics. Figure 2.3 illustrates the underpinnings of the LER calculation. The balance is between the number of plants per area and average yield of each. Somewhere along this, an optimal LER occurs (lower function, Figure 2.3).

As plant-plant complementarity increases, the number of plants per area can also increase. If this comes about without harming average yields, this produces an upward shift in the LER. This occurs in Figure 2.3 with the dotted curves. Generally, LER values above 1.5 qualify as being high. The uppermost values seem to be about 3.5.[19]

PLANTING DENSITY

How many plants to put in a given area has confounded farmers since the dawn of agriculture. Too few and essential resources may go unused, too many and crowding will lower yields or produce an inefficient harvest ratio (i.e., as competition for light can force taller stems, less fruit). The ratio of biomass to useful output (e.g., fruit) is the harvest index.

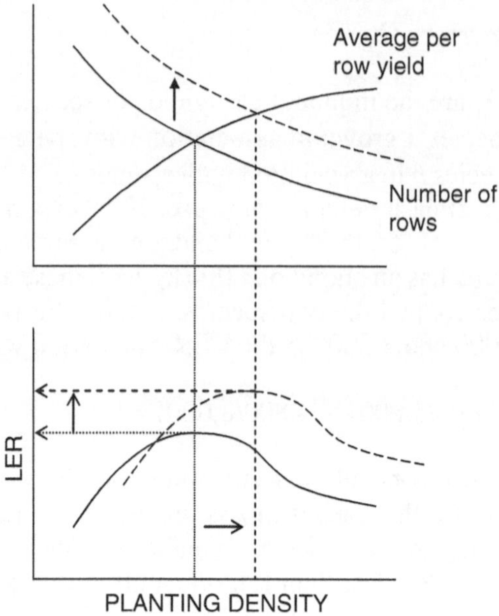

FIGURE 2.3. Derivation of LER-planting density curve. The upper graph shows how the number of plants per area and average yield per plant produce the LER-density curve. The change *(dotted lines)* shows what happens when populations are increased because of better species-site compatibility and/or greater plant-plant complementarity.

In nature, sparse plant populations define arid zones. Luxurious growth occurs in rain forests where water is abundant and where many of the species have a high degree of shade tolerance. These extremes guide planting density in agriculture: wider spacing when there is less of the limiting resource, closer spacing when the limiting resource is relatively abundant.

Some species combinations are better densely planted than others. Extrapolating from Figure 2.3, different plant pairings, depending upon the degree of complementarity, produce different optimal densities.

The fundamental rule of complementarity is that, when unlike species pair well vis-à-vis the limiting resource(s), these should be planted close together (i.e., with a close interspecies interface distance; see, e.g., Plate 2.1). Where realizable complementarity is less,

PLATE 2.1. Two plant-plant interfaces: *(a)* onions with lettuce and *(b)* poplar trees with maize. The distance between like and unlike species regulates the interactions and productive outcome.

the best results (highest LER value) will occur when the interspecies interface distance is decreased. These guidelines translate to practice through planting ratios.

PLANTING RATIOS

A comparative number of each species in a given area is the planting ratio.[20] Planting ratios are not constant but can be expressed as percentage of optimal monocultural density. The optimal monocultural density varies per species but is manifested as the 100 percent density. For two species, a 100-100 percent planting density is where both are intercropping at the full monocultural planting density. In a competitive intercrop situation, values less than 100 percent are often the better option, and this means sliding down ratio lines.

Using a hypothetical planting, the various ratios, starting at the 100 percent level, are given here:

Primary species (*a*) Plants per area (percent)		Secondary species (*b*) Plants per area (percent)
10,000 (100)	<——————>	20,000 (100)
7,500 (75)		15,000 (75)
5,000 (50)		10,000 (50)
2,500 (25)		5,000 (25)
2,000 (20)		4,000 (20)
1,000 (10)		2,000 (10)

The 100 percent values, as the optimal monocultural planting densities, are at the end of, and anchor, these values. Intermediate points are, for example, 5,000-10,000 (50-50 percent), and 2,500-5,000 (25-25 percent).[21]

Similarly, the 100-50 percent, 100-25 percent, and 25-100 percent ratios head or anchor other ratio lines. Using these density figures, representative density pairings (points along the ratio line) for the anchor ratios are, for species *a* and *b*, respectively,

100-50 percent = (10,000-10,000) or (5,000-5,000) or (2,000-2,000)
100-25 percent = (10,000-5,000) or (5,000-2,500) or (1,000-500)
25-100 percent = (2,500-20,000) or (1,250-10,000) or (625-5,000)

Each of these lines represents an LER-planting density curve as shown in Figure 2.4. Although optimal monocultural density (the 100 percent value) is presumed to represent the highest yield, this is not true along the ratio lines. Generally, because of increased interplant competition, the LER is maximized when one or both species are below 100 percent density.[22]

Each curve (planting ratio series) has a different LER maximum. These curves combine to form a surface, and somewhere along this is a curve and density value that is the global optimal (best possible LER). This is shown in Figure 2.4, topmost (solid) curve.[23]

In practice, there seems to be much flexibility and, at times, little difference between fewer higher-yielding or more lesser-yielding plants. The inefficiencies (yield losses) associated with incorrect planting densities are in the 5 to 15 percent range. This assumes that the planting densities are reasonable (only slightly off the mark) with regard to local planting conditions, for example, nutrient- and moisture-holding capacity of the soil.[24]

Given normal climatic variation and other productive threats, this may be relatively insignificant and a small price to pay for an inaccu-

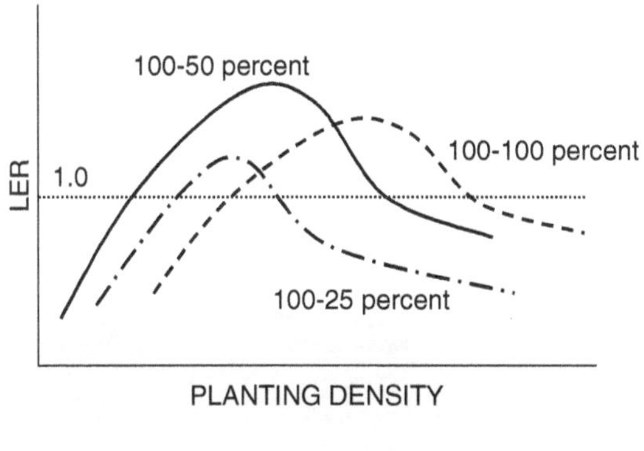

FIGURE 2.4. LER-density curves for three bicultural planting densities. The stronger of the three *(uppermost)* allows for more in-use flexibility because, for more of the density range, the LER is above 1.0. These analyses also apply to different spatial patterns and pattern arrangements.

(a)

(b)

PLATE 2.2. Two spatial patterns: *(a)* eggplant, hot pepper, and celery in a close interface checkerboard pattern and *(b)* multiple vegetable species in a row pattern.

rate assessment. Farmers, many being optimists, do not want to miss out on the good years and generally accept higher-density, more-resource-demanding recommendations.

SPATIAL PATTERNS

Mixing two species with inherently high LER potential can, if done well, produce a very high yielding system. Even with a strong pairing, this is far from automatic as other factors intercede. Most often, once a favorable interspecies pairing has been identified, the starting point to implement complementarity is the spatial pattern.

Following the fundamental theorem of complementarity, that is, if two species can coexist well on a site, the relationship is best exploited by planting these species in close proximity. Close proximity means a spatial pattern with a high degree of interspecies interaction. Accordingly, stronger complementarity is best accommodated through higher proportions of interspecies interactions.

The closest physical associations between unlike species occur when (1) one species is entirely surrounded by a second species, for example, as in a checkerboard pattern and when (2) planting density for each species within an intercrop is near optimal for that species (i.e., the same as that recommended when each is planted as a monoculture).

Even if complementarity is high, patterns that maximize the amount of interplant interface (e.g., checkerboard pattern; see Plate 2.2) may not be the best option. Other factors intervene, for example, a large interspecies size differential and practicable sowing and harvesting issues that arise when two unlike species are grown in close proximity. Still, the spatial patterns are a first step in reconciling these and other on-farm, utilitarian concerns.

Basic Patterns

Underlying patterns start this process. Those illustrated in Figure 2.5 are

1. individual,
2. border,

3. row,
4. groups or clusters,
5. border strips, and
6. row strips.[25]

All are presented here in bicultural form, that is, systems or plots containing two species.[26]

The first three (Figure 2.5, top) are fine patterned where each plant has a unlike species as a neighbor. Complementarity is brought to the fore through close plant-plant associations and fine patterns.

Where complementarity is not assured or is relatively weak, a coarse pattern may be best. The second three (Figure 2.5, bottom) are coarse patterned, having like plants grouped such that some are bordered on all sides by a like species.

As seen, patterns are a fluid concept. There are practical reasons for selecting a basic pattern and equally valid reasons for deviations;

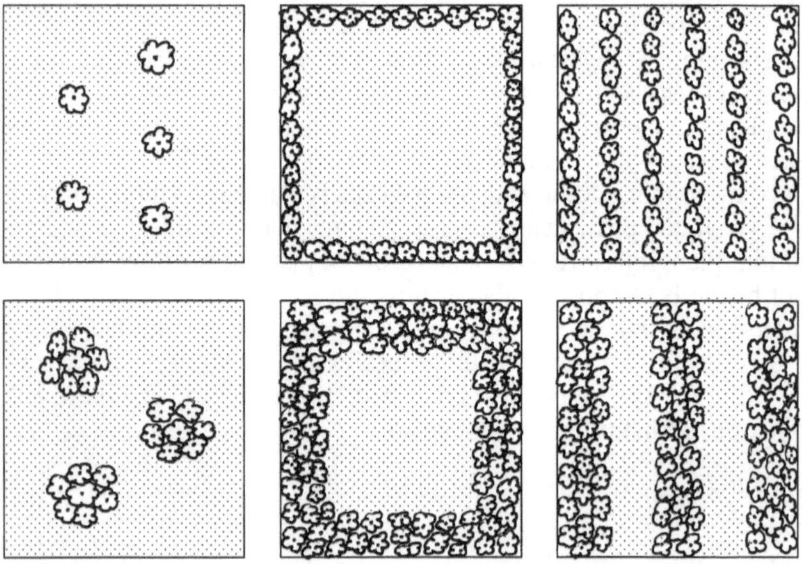

FIGURE 2.5. Cross sections for six possible bicultural spatial patterns. The upper three diagrams are examples of fine, high interspecies interface patterns (*from the left:* individual, boundary, and row). The lower three (*from the left:* cluster, strip boundary, row strip) are some corresponding coarse patterns.

much of this starts with the ecological underpinning, for example, whether to attempt a high LER through the ecological gains from close plant-plant associations. A pattern can also be chosen to fit land contours or better accommodate farm needs. In the latter case, mechanized agriculture may require wide strips for efficient plowing, sowing, and harvesting.

Pattern Arrangements

The patterns underlie and can extract much in terms of positive plant-plant relationships and in addressing site and farm needs. More can be done working with patterns in which each spatial pattern has any number of arrangements. To emphasize this point, Figure 2.6 shows some of the bicultural arrangements for the row pattern.[27]

Height inversion adds to the array of options. In Figure 2.5, a boundary fence can surround a pasture or a taller orchard. More profound, a gap pattern is the height inversion of a cluster pattern.[28] There are also mixed patterns; strips alternating with rows is a common example.

The reasons for selecting a pattern arrangement are many and varied; most relate to the implementation of complementarity. Keeping two species in close proximity, through a narrow interface distance, extracts the most where two species are naturally compatible. Where less so, this distance may be increased. The more complementary the two species, the closer will be the optimal interface and the higher the potential LER (see Figure 2.3).

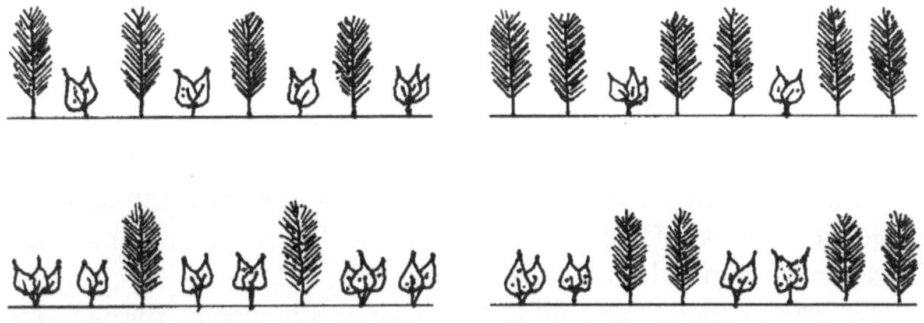

FIGURE 2.6. Four possible bicultural arrangements for a row-strip pattern. These are *bababa (upper left), bbabb (upper right), aabaa (lower left),* and *aabbaa (lower right).*

The same may occur with uneven plantings (as in Figure 2.6). Rather than having species *a* and species *b* alternating, as in *ababa,* the better option may be *aabaabaa* or *abbabba*. There are reasons for switching the planting ratio: the better distribution of light; a nutrient imbalance that favors more of one species, less of another; or from an economic perspective, one species having a greater market value than the second.[29]

As one proceeds in incorporating population imbalances (more of one species, less of another), a point may be reached at which a monoculture seems the best. However, before going down this road, one must keep in mind that, in having two species with a multitude of cross effects, ecological advantages may exist that are not immediately apparent. The ability to counter climatic threats and other risk factors are instances in which a well-chosen intercrop almost always trumps a monoculture.

The patterns and arrangements do enhance the analytical complexity. In addition to representing density ratios, there are separate curves (see Figure 2.4) for each spatial pattern and arrangement.

Pattern Options

Not being content with patterns and arrangements, there are other facets to consider. These expand the the range of options and provide more alternatives in addressing environmental and other issues. Those looked at here are spatial disarray, upper-level canopy patterns, and buffer species.

Canopy Orderings

As the number of species in an agroecosystem increases beyond two (a three-plus polyculture), another pattern is introduced. This is the canopy ordering with regard to relative height. Two fundamental sequences are possible: (1) midpoint and (2) minimum interface. These are shown in Figure 2.7.

The first of these, the midpoint, places each species midway between two taller plants. This continues until the full complement is in place. This pattern evenly distributes light as long as the canopies of the uppermost species do not extend over neighboring plants.

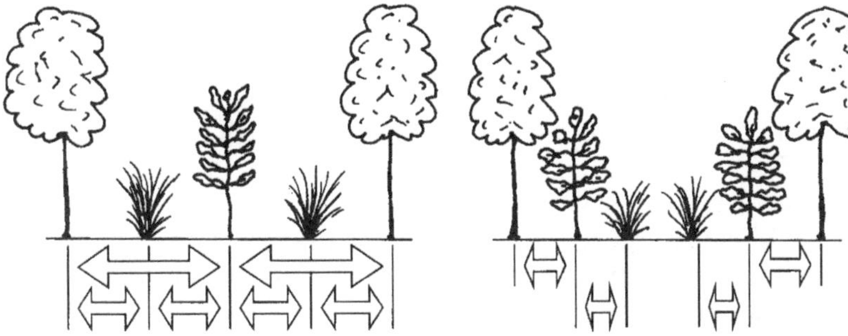

FIGURE 2.7. The two fundamental canopy patterns for systems of three or more species. Left shows a midpoint design where the next shorter species is located between two taller plants. On the right is a minimum interface design where the next shortest species is immediately adjacent to a taller plant.

In contrast, the minimum interface has each species immediately adjacent to a tall plant such that, in cross section, each clump or cross-sectional row series is an outside-in progression from short to tall. This ordering may be best where the taller species (one or more) can extend over others but is prevented from doing so by immediate competition for light and belowground resources. In use, this placement option allocates more resources to lower species while helping to contain canopy spread in the taller neighbors.[30]

Buffer Species

For economic reasons, the mutual presence of two competing species may be desirable. It may not be possible to reconcile a competitive overlap through pattern and pattern arrangement. Instead, a buffer species can be inserted. This is more common with coarse patterns and/or larger perennial plants, for example, one strip buffer protected from a neighboring strip.

Buffer species serve these functions:

1. to keep the canopies of a tall species from overhanging those of shorter plants;

2. to prevent roots from one species spreading above, below, or into the root strata of a second species;[31] and/or
3. to block horizontal light from reaching ground level beneath a tall canopy, triggering unwanted weed growth (part of an edge effect).

The ultimate purpose is to allow competing plants, through the insertion of a buffer species, to be planted closer together than otherwise economically feasible, thereby increasing the LER. Figure 2.8 shows the before (left) and after (right) of buffer placement and the desired effect. As use is heavily weighed toward perennial systems, most buffer species are trees or shrubs.

Buffer trees can have another task, that of a guide species. Guide species, planted around or next to a tree species, hem or hug the trunk of the tree. Irrespective of cropping benefits, this forces height growth while eliminating side branching (Figure 2.8, upper drawing).

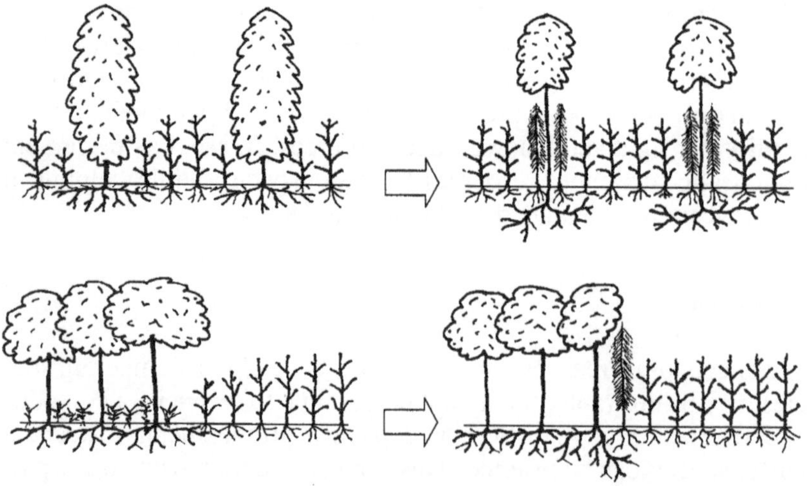

FIGURE 2.8. The before and after use of buffer plants. On the left *(top and bottom)* are unbuffered agroecosystems. Adding buffer plants (right) improves the understory productivity, tree (stem) form, and other system properties. This mechanism, as shown, constricts the spread of both the overstory canopies and belowground roots.

In forestry situations, guide species result in quality logs of higher market value. In agroforestry, an elevated overstory canopy without lower side branches interferes less with manual or machine chores, for example, crop planting and harvest. With scattered trees, this also brings about a more even distribution of understory light, provided that the guide species does not spread horizontally.

Disarray

Many find spatial disarray to their liking. Broadcast grains result in spatial disarray but, because of cheaper planting costs, their use is more of an economic than ecological incentive. Some find advantage, other than the cost of planting, in nonordered arrangements.

Although observable in many parts of the world, little applied work has been done on the ecological underpinnings and outcome. That presented here relies upon theory.

The idea is that the best yields from an ordered system occur when planting densities closely match the resources levels. This is shown in Figure 2.9 as the inner taller curve, where, at less than optimal resource levels, the LER is less than optimal.

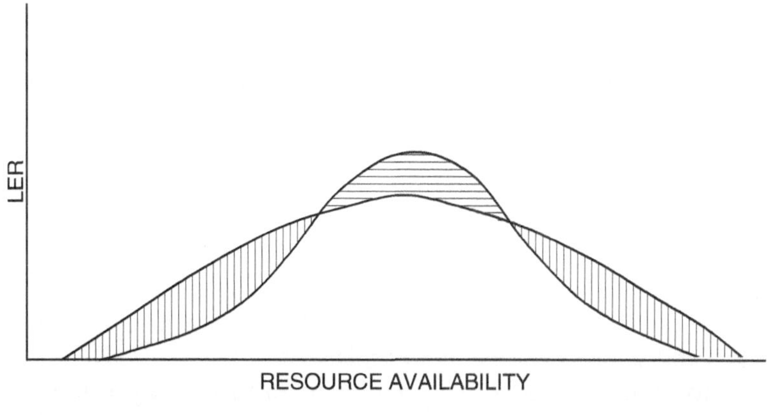

FIGURE 2.9. The outcome of disarray where the LER is better *(horizontal striped area)* for an ordered system *(inner curve)* when matched with optimal resources. When resources are more variable, LER gains accrue *(vertical striped area)* with disarray *(outer curve).*

With disarray (the shorter, sprawling, outer curve, Figure 2.9), the potential LER will not be as high as with an ordered system with optimal inputs (the horizontal striped area, Figure 2.9) but will be higher when resources are suboptimal (vertical striped areas, Figure 2.9). In other words, disarray better captures the good and bad moments, but is not as good when the resources are optimally abundant.

As a consequence, disarray may be the best option on highly biodiverse plots when

1. the site is highly resource variable;
2. the optimal planting density for the component species (two or more) is not known; and/or
3. the interactions between closely interplanted species are a mystery.

In this combined setting, the only alternative may be to start with disarray. Once an ecosystem is established, it is quite possible that varying degrees of plant-plant complementarity will be observed. What commonly happens is that these favorable plant-plant relationships are subsequently exploited through companion plantings within a disarrayed setting. This strategy tends to increase productivity while maintaining the risk-countering advantages of disarray.

Canopy orderings are often the first step along the favorable pairings road. Figures 10.5 and 10.6 show midpoint and minimum interface canopy layouts with disarray. As seen in these drawings, the underlying patterns are masked through disarray and can be hard to spot.[32]

Topographical Adaptations

Much of agriculture is topographically dependent. This can have a major or minor influence on the layout of the system. The orientation of rows and strips is a large part of this.

Erosion-prone hillsides must be planted in such a way as to retain the soil. Some perennial systems are strong enough to resist these forces. Others, especially those with a dangerous bareground (post-plowing, preplanting) phase, must be topographically accommodated.

Hedgerows and strip cropping specifically address the erosion issue and, depending on the severity of the situation, topography can dictate densities and spacing more than the plant-plant interactions. It is best if both (complementary and topographical needs) work in concert.

Topographical placements offer multiple options. Figure 2.10 (left to right) shows four plot layouts, one contour conforming, one semicontour, and two cross-contour designs (see also Plate 2.3). The most severe erosion-prone situations require contour-conforming rows; less menaced sites may be approached with semicontour layouts. In locations where erosion is less of a concern, for example, where the land relief is not severe, cross-contour placements serve well.

Allowing sufficient light to reach productive species, along with erosion control, is also a concern. This raises the question of row and strip orientation. When the relative height between the component species (in the row and alley or in neighboring strips) is small, this is a lesser or an unimportant factor. Greater large height differences may require north-south or east-west facing rows so that the shorter species receives light in sufficient amount and at the right time.[33]

SPECIFIC INTERACTION ZONES

A small body of theory revolves around specific interaction zones (SIZs).[34] For example, insect-repelling marigolds can be interplanted

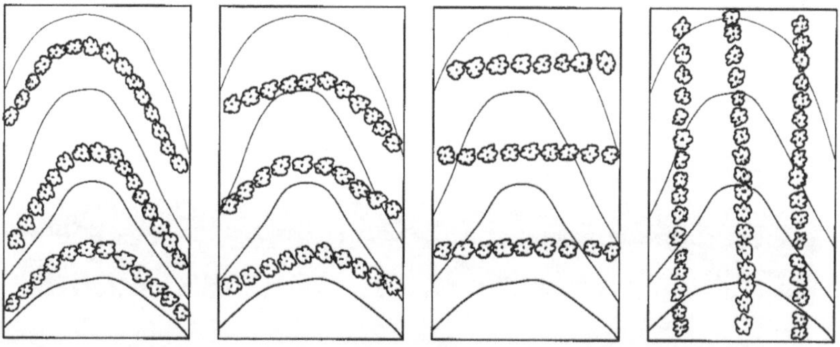

FIGURE 2.10. Four possible spatial patterns for accommodating topography. The patterns *(from left to right)* are contour, semicontour, and cross-contour. The highest elevation is at the top of each drawing.

(a)

(b)

PLATE 2.3. Two contour patterns: *(a)* strictly contour and *(b)* cross contour. In the latter, the two features, the line of vegetation and the perpendicular ditches, do not precisely follow the land relief.

to great benefit where the density is based on a high degree of plant-plant (crop-marigold) complementarity. Having a large population of marigolds can be overkill. It may be better to allocate more area to productive species, less to marigolds. Specifically, this involves having a per area marigold density such that the insect-repelling SIZ produced by each plant touches or overlaps.

This is not the only type of SIZ. Windbreaks offer an affected, wind-protected zone, and the downward heat reflection off an overhead canopy (a frost countermeasure) produces a slightly heated zone. SIZs exist for most interplant influences and frequently determine the spacing and density for an added facilitative species.

Ecosystem and landscape are cross-hatched or overlaid by the different SIZs. A plot design strategy is to maximize SIZ advantages or SIZ gains while living with the precepts of complementarity. Figure 2.11 shows the SIZ for an insect-repelling species that has been overlaid within a biculture. To maintain a high LER, the species most complementary with the repelling plant is located next to this plant.

The SIZ concept permeates much of agroecology in different forms and proportions, and in multiple variations. A case in point, there is also a temporal version. Again citing marigolds, these repel in-soil nematodes for several years after removal. This temporal SIZ has use in rotational planning, providing protection without spatially mingling this plant with a current crop.

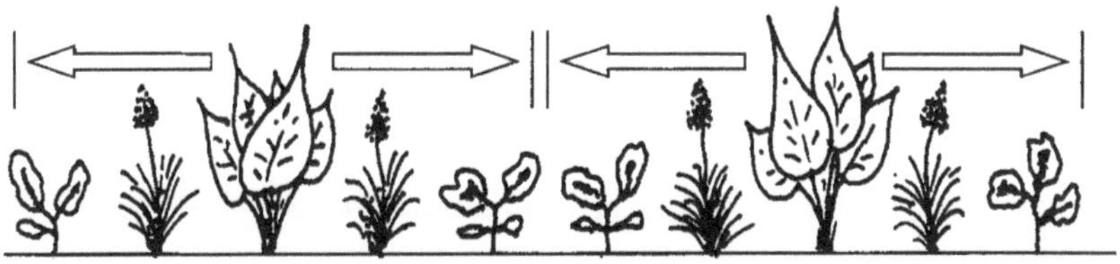

FIGURE 2.11. Specific interaction zones (SIZs) where, in this case, the repellant effect, as marked with horizontal arrows, determines the planting density for a repellant species. SIZs often decide the planting density (i.e., intraplant spacing) for many facilitative (i.e., nonproductive) species.

SUPPLEMENTARY ADDITIONS

A few crops can be inserted without interfering with established ecological dynamics. These supplementary additions are more of an afterthought but, because they add value, often without much effort, are fairly common. Within this category are rattan vines, truffles, and mushrooms, all of which can be included without interfering with or modifying the layout and timing of the intended ecosystem. As with high-value truffles, some addition can be economically significant. The only requirement is for an accommodating ecosystem. Where appropriate, mention is made within this text when such intrusions offer a clear economic gain.

Chapter 3

Agrobionomic Concepts

The concepts presented in the previous chapter focused on spatial agroecology. This chapter goes to the heart of productive agroecology, enumerating the mechanisms of complementarity.

Growing multiple species in close proximity, as an offshoot of what nature intends, undoubtedly goes back to the earliest days of agriculture. Farmers then may have lacked a formal ecological understanding, but they implemented what they observed. Those who documented these efforts, for example, Evelyn (1664), were not privy to current terminology, and instead mentioned good and bad neighbor plants (i.e., competitive production) or when one species needs the comfort of another (i.e., facilitation).[1]

Irrespective of the vocabulary utilized, these ecological concepts apply whenever two of more plants of different species or niche-divergent varieties of the same species interact. Plots of niche-identical or nearly identical plants (as with a monoculture of genetically identical clones) do not convey biodiversity gains and generally fall outside the agrobionomic or agrobiodynamic principles discussed here.

Working under the premise that categorizing nature is not always a precise endeavor, the mechanisms of gain and loss, that is, governing or agrobiomonic principles, can be itemized.[2] Starting with the forms of governance, these are subdivided as follows:

- Species governance
  - —Competitive production
  - —Facilitation
  - —Exclusion
- Ecosystem governance
  - —Top down
  - —Bottom up

Introduction to Agroecology: Principles and Practices

doi:10.1300/5553_03

SPECIES GOVERNANCE

Governance by a few occurs when a limited number of component species or individual plants set most of the ecological agenda. The essence of species governance is a form in which most of the ecological strength (productive capacity) of an agroecosystem derives from and is traceable to a few key interspecies interactions. A common case is where productive outcome depends upon the favorable dynamics with regard to a single limiting resource.

Species governance can be, in turn, subdivided to three main categories with subdivisions and specific mechanisms. These are

1. competitive production: (a) acquisition and (b) partitioning and conversion,
2. facilitation, and
3. exclusion.

With two unlike species, competitive production as well as facilitation often coexist. If resources (one or more) are utilized solely by the capturing organism, then it is competitive acquisition; if made available to others, then it is facilitation. These obscure categorization but do not negate the underlying principles.

As a clarifying point, the gains from competitive production can be positive or negative. The mechanisms of facilitation, by definition, are only positive.

Competitive Production (Acquisition)

When competing species do not take fully from the general resource pool, but have their own source of resources, this constitutes an acquisition advantage. This occurs at a number of levels:[3]

1. One or more essential resources are utilized in a different time or season.
2. The resources are utilized at the same time or season, but
   a. are collected in different zones, for example, plant root profiles are in different zones (for nutrients or water) or canopies collect different forms of light (horizontal or vertical), and/or

b. resources are derived from different sources, for example, from air fixation of N_2 by legumes or in-soil acquisition of NO_2 by nonlegumes.

These are discussed in the context of timing and above- and below-ground effects.

Timing

A temporal resource division can occur when two species seek the same resources but do so with enough temporal separation so that growth proceeds unabated.

In nature, some plants start growth early to avoid the more intense competition during the peak growing season. In agriculture, this happens when two species are simultaneously planted: one takes most of its needs early whereas the other requires a heavier dose of resources later in the growing season. Plate 3.1 shows an early-season resource demander (lettuce) with a late-season crop (tomato).

PLATE 3.1. Resource allocation through timing. This photo has short-duration lettuce beneath longer-term tomato.

More biocomplex versions are possible as with a maize, millet, and sweet potato intercrop. After harvesting maize and millet, the sweet potato has full access to the remaining essential resources.[4] Management options can return similar results, for example, by staggering planting times or by pairing a fast-sprouting seed with one that germinates slowly.

Root Strata

Other gains in partitioning occur when plants hunt for a distinct derivation of resources. This happens when roots are in separate zones or strata.[5]

This mechanism is found in nature. In the drier regions of Hawaii, the shallow-rooted trees *Diospyros sandwicensis* and *Nestegis sandwicensis,* both of which collect surface water, coexist with the deep-rooted *Metrosideros polymorpha* and *Reynoldsia sandwicensis,* which collect deep moisture. This relationship can be exploited with crops (Figure 3.1, top). A tree species from China, paulownia, has few surface roots but many deep spreading roots than grow deeper than those of most crops.[6]

Vertical root specialization can occur with monocultures or polycultures. In grain fields, competition for belowground space can drive all roots into vertical zones with less overlap than expected (Figure 3.1, bottom).

It is possible to force roots into respective strata and bring about a favorable separation through spatial pattern and/or the timing of the planting. Timing or an initially large interspecies distance allows deeper roots to penetrate without initial competition.[7] For this reason, clover is late planted with beets or onions; the sowing of barley trails that of onions; and the planting of rye is delayed when between onions, peas, or cabbage. Among these intercrops, clover, barley, and rye are the shallow-rooted additions.[8]

Vertical root specialization is a characteristic of root buffer species. Most will have a curtain of dense roots that discourage other roots from breaking through (as in Figure 2.8, top and bottom).[9]

Light Sources

Since aboveground effects are observable, light management can be important for deriving the most from any intercrop. In some cases,

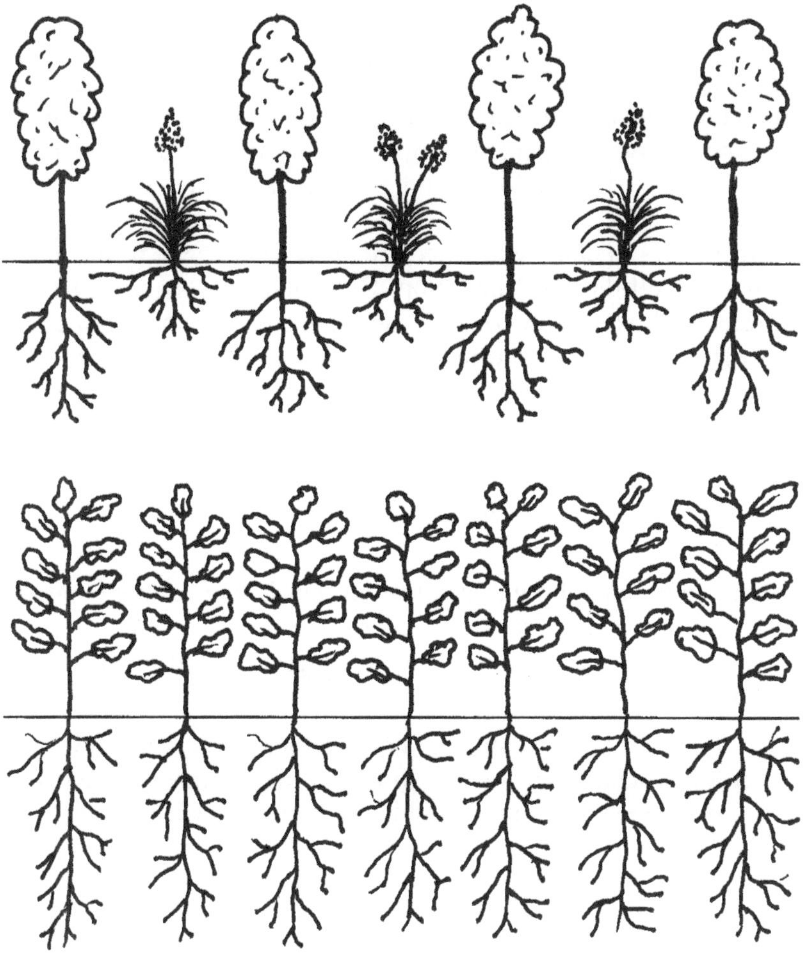

FIGURE 3.1. Forms of root separation. The top drawing shows horizontal stratification, the lower shows vertical root partitioning.

for example, when soils are wet and well endowed with nutrients, light can be the overriding and governing influence.

As with the belowground, there are different sources. Light can be categorized into (1) horizontal (encountered in the early morning or late afternoon) intercepted by plants with vertically formed canopies and (2) vertical (that is found at midday) gathered by plants with wide, thin, horizontal canopies.[10]

Light preference can also be seen in the leaf angle. Horizontal leaves are more exposed to vertical light. Leaves that hang vertically are more exposed to horizontal light. These differences can be manifested within a species, for example, a trait associated with varieties of wheat.[11]

Differences in canopy architecture or leaf angle can be exploited through a well-designed intercrop. One such light-maximizing design has north-south rows of a tall, horizontal light gathering species. Between the rows is a shorter species that collects vertical midday light. The north-south orientation ensures efficient light partitioning based on what the plants themselves seek.

Competitive Production (Partition and Conversion)

There are gains through acquisition, but if two or more species share a common resource pool, there can be effective results, such as better yields, as measured by LER, or a higher valued product. This happens when[12]

1. a single resource is utilized more efficiently by two species,
2. a combination of resources, through different resource-use profiles, are exploited more efficiently by two species, and/or
3. a reduction, caused by one plant, in the amount of one or more resource going to a second species can, for the second species, increase the
   a. overall productivity,
   b. the harvest index, and/or
   c. the quality of the output.

Most of the mechanisms of competitive production are based on the concept of marginal gains. As an extension of functional form for the limiting resource (see Chapter 2), it holds that, as essential resources are added, each additional unit is not as potent as the last and there is a decline in yields per unit of input. Figures 2.1 and 2.2 incorporate these relationships, showing the declining effectiveness when essential resource levels are high.

Gains can also come through resource removal. This is shown in Figure 3.2 in which taking an overabundance from species (right) leads to increased yields from this species. As mentioned, maize is one plant that suffers greatly from waterlogging and benefits when the soil moisture is reduced to more optimal levels.

Although the exact functions may be in question, the concepts are not. As a general rule, a wealth of essential resources can be best exploited by multiple species. Interspecies complementarity is a better approach, but this general rule holds even if the pairing is slightly competitive.

Marginal Gains

The efficient conversion of essential resources into usable output can be due, in large part, to increased marginal gains. This is facilitated by taking abundant or surplus resources from one species and giving these to a second; as a result the LER increases. This follows from the limiting resource concept discussed in the previous chapter.

In the simplest form, this can be when the two species share a limiting resource. Complexity increases if the two species have different resource use profiles, for example, if water is limiting in one, nitrogen

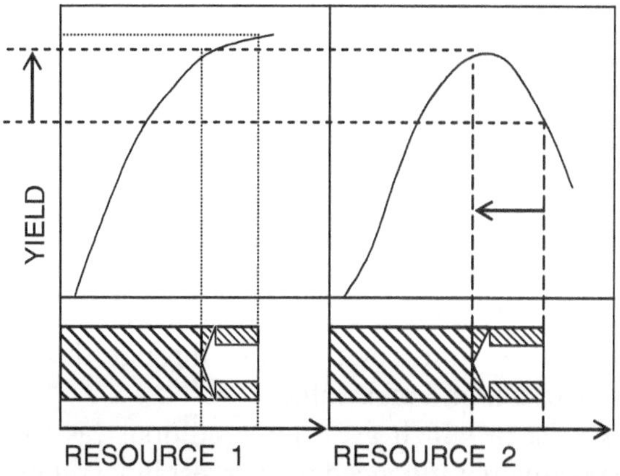

FIGURE 3.2. Resource removal can, in the right situations, provide a yield boost. In this illustration, the resources taken by a second species (not shown) reduced an adverse oversupply in Resource 2, increasing yields.

in the other. Marginal gains can be improved by allowing plants that are water limiting to capture water, while the species that is nitrogen constrained can seek the surplus nitrogen.

Margin gains, in the simplest form, has a portion (in Figure 3.3, this is one-half of each resource) of each essential resource expropriated from species a by species b. Instead of a monocultural LER of 1.0, the two species combine for an LER of 1.20 (0.7 + 0.5).

There is evidence for these efficiency gains. In annual intercrops, Morris and Garrity (1993) found that intercropping was not capture efficient, as compared with sole cropping, was not high (–6 percent to +7 percent better) whereas intercropping exhibited high use (marginal) efficiency (+18 percent to +99 percent better).

The partitioning of resources does not have to be haphazard. Increased LER values are obtainable by taking advantage of divergent resource requirements. That is, it is best if, in a coplanting of two species, each has a different limiting resource, for example, for one key resource, one species demands this resource, whereas the other tolerates a shortfall.

Figure 3.4 illustrates how two species with complementary demand levels and shortfall tolerances can successfully coexist if resources are allocated such that each species benefits. A favorable allocation can be further enhanced through spatial and other ecosystem design variables.[13]

Harvest Index

For most agriculture crops, farmers prefer to have shorter stems, smaller roots, branches, and leaves with more grain or fruit. This is the harvest index: the ratio of biomass to usable output. This can be adjusted through growing conditions or through the selection of a species that gives a high index value.

Examples exist in which growing conditions (as found in intercrops) have positively influenced the harvest index. It has been demonstrated that millet with groundnuts gave an biomass LER of 1.2, but for the grain, the LER was 1.33.[14] This is a clear improvement in the harvest index.

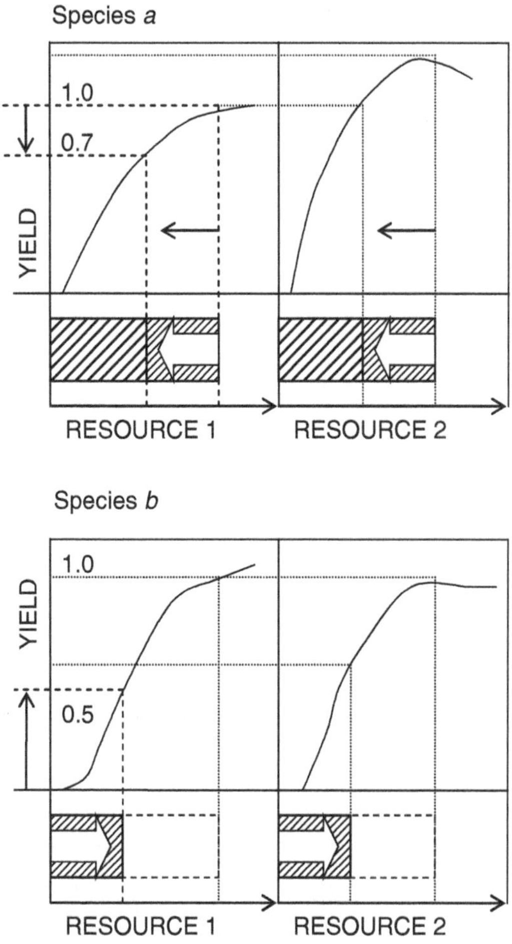

FIGURE 3.3. Marginal gains occur when two unlike species divide more or less equally the same essential resources. Here half the essential resources are taken from species *a* (top) and allocated to species *b* (bottom). The resulting LER is greater than 1.0 (in this case, the LER = 1.2; i.e., 0.7 + 0.5 as shown).

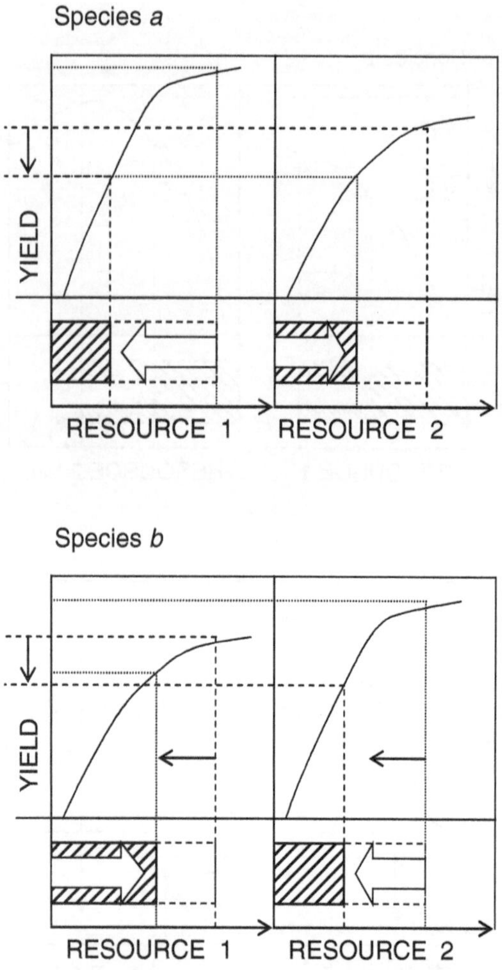

FIGURE 3.4. More sophisticated resource allocation happens when, through design (pairings, patterns, management, etc.), resources are allocated and yield reductions from the monoculture are comparatively minor. Here, Resource 1 is taken from species *a* and directed to species *b*; for Resource 2, the reverse occurs.

For leafy crops, light shade is known to improve leaf size and number.[15] This quality, more leaf area to gather more light, can be beneficial in leaf crops such as cabbage, tea, and clover. This does not apply to fruits; here the response to a light shortfall is increased fruit tartness.[16]

Quality Gains

Overriding many aspects of agroecology is the profit motive. The agrobionomic principles are no exception. Yield gains directly translate into increases in revenue, and these are normally expressed through the LER.

As one gets into the practical aspects, quality becomes a revenue issue. Depending upon consumer preference, fruits can be sweet or tart, and there is always a need for more nutritious grains and vegetables. Farmers are generally paid more for delivering what consumers want.

Quality can be improved through varietal selection or when farmers have the opportunity to adjust quality through growing conditions. Wind protection can result in sugarbeets with a higher sugar content, and shade (in not having to face taxing midday sunlight) can raise the nutrition content of pasture grasses and oats or, with coffee, produce larger beans and a better tasting product.[17]

Although not reflected in the LER, quality can be quantified through increased market value. Adding revenue values (selling price) to the LER goes some distance in accommodating quality differences (assuming these values do reflect improved quality). The relative value total (RVT)[18] adds selling price such that

$$\text{RVT} = (p_a Y_{ab} + p_b Y_{ba})/p_a Y_a \tag{3.1}$$

Here p_a and p_b are the output or selling prices for species a and b, respectively. The primary species (here species *a*) or that offering the greater monocultural return for the site being considered (i.e., where $p_a Y_a > p_b Y_b$) becomes the denominator.

Using the numbers from the sample LER calculation (Chapter 2)

$$\text{RVT} = [1.50\,(3{,}000) + 1.25(4{,}800)]/[1.50(5{,}000)] = 1.4 \tag{3.2}$$

The added values, 1.50 and 1.25, are the selling prices for outputs *a* and *b,* respectively. As with the LER, a result greater than 1.0 indicates favorable outcome.

Facilitation

There are situations in which resources are partitioned, but only after the essential resources flow from one species to another. There are also situations where the mere presence of one plant aids the growth of another by indirectly solidifying the resource situation (e.g., preventing or slowing the loss of resources from a second species). These two mechanisms, transfer and protection, often overlap or combine as a facilitative influence.

In the previous chapter, the indirect links between essential resources are briefly mentioned. A large part of the facilitation is in neutralizing negative indirect influences and advancing those that are positive. As previously mentioned, sunlight that increases transpiration and lowers water use efficiency (an indirect relationship) can be neutralized through a more humid microclimate, through shade, and/or some mechanism of increased water capture and retention.[19]

Whereas partitioning gains are more prominent in a resource-rich situation, facilitation finds its footing when resources are less abundant. Most often this involves the limiting resource (one or more).

The mechanisms of facilitation are as follows:

1. Nutrient capture-transfer
   a. Nutrient pump
   b. Airborne nutrients
   c. Fixation of nitrogen
   d. Chemical conversion
   e. Timing of release
2. Water capture-transfer
   a. Hydraulic lift
   b. Microclimatic change
3. Retention
4. Accumulation-concentration
5. Site improvement
6. Parasitic

Basically, these are mostly a restatement, in greater detail, of the mechanisms of competitive partitioning. As plant species are capable of gathering resources not available to another, instead of utilizing these directly (as with competitive partitioning), facilitation gives them, through preplanned interactions or active intervention, to another species.

Often observed is the direct handing over as when decaying biomass from one species goes to a second. More bewildering are the indirect mechanisms in which one change produces another, which in turns aids a different species. These are found occasionally and require some researching to diagnose. The proliferation of unseen, but beneficial, microorganisms (microbes, earthworms, etc.) due to better soil conditions fall under this indirect facilitative heading.[20]

Nutrient Capture-Transfer

There are a number of mechanisms and routes that allow nutrients to move from one plant species to another. The main routes are though aboveground biomass. When one species sheds leaves and branches, this can be a natural loss or the result of active pruning. The latter can be a more potent nutrient source. Active biomass can, at the right time, be flush with nutrients. In contrast, the parts of the plant that are naturally shed are those less active and not always replete with a full complement of elemental nutrients.

Another source is belowground. This source can escape notice, but pruning of roots, through normal plowing or as planned activity, can provide nutrient-rich organic material well placed for decay and capture.

These mechanisms, good as they are, are still weak. Palm (1995) found that only 20 percent of the nutrients released by one species is captured by a companion plant.[21] Still, more nutrients in an ecosystem translate into greater yields. The following mechanisms can result in yield gains.

Nutrient pump: Clearly, plants with deep roots can capture elemental nutrients in deep soil strata and carry these to the surface. Subsequent death and decay of the transporting plant can make these available to the plant community. Although this mechanism has not been scientifically validated, it is generally accepted as true. If active, this would not be a major nutrient source.[22]

Airborne nutrients: Nutrients are found in the air, sufficient for some species that exist solely on this source, for example, bromeliads.[23] Although the mechanisms (ecosystem design features) that enhance the capture of airborne nutrients have not been patently identified, it is thought that a few taller plants can slow the wind speed and therefore increase deposits. Although the overall effect may be uncertain, it does contribute.

Fixation of nitrogen: Plants that gather in-air nitrogen and, through leaf or root decay, transfer this to companion species fall under this heading. This can be a significant source of this element.

Chemical conversion: Some elemental nutrients are bound in chemical associations or are physically trapped in rocks. Some species are better than others in unlocking these elements. However, their overall importance in agriculture is less known.

Timing of release: Within ecosystems, nutrient cycling is controlled, in part, through ground-level microclimate. Once plant material is shed, it can lie on the ground for an extended period or be put to immediate use.

A second species can be added, both to provide a abundance of such material and to encourage rapid decay and speedy recycling. Part of the mechanism can be carbon ratio coupled with microbial decomposition. Certain species can promote a timed, that is, more rapid, release of nutrients when environmental conditions warrant.[24]

Water Capture-Transfer

In addition to elemental nutrients, there are facilitative mechanisms associated with water. As listed here, these too involve capture and transfer.

Hydraulic lift: Natural ecosystems can be a mix of shallow- and deep-rooted plants. The shallow-rooted species survive on light rains, and those with deeper roots seek deep sources. In the process of dissolving surface nutrients, deep-rooted plants need to deposit water within the heavily root-populated surface soil strata. This free moisture can be usurped by other plants. Although the relative importance of this mechanism is in question, its validity is not.[25]

Wind reduction: By altering the microclimate around plants, water use and water requirements can be changed. The actual mechanism can be direct protection (a reduction in sandblasting where windborne soil particles physically damage plants or a reduction in soil wind erosion) or indirect protection (nearby foliage blocks winds, decreasing the evapotranspiration rate and increasing water use efficiency). The level of protection can be significant.

As a specific topic, the gains are discussed in detail under windbreaks (Chapter 12). It should be mentioned that windbreaks are not the only source of wind protection; this is often a result of close spacing between diverse species.

Retention

The loss of essential resources from a farm plot is an agricultural problem with facilitative solutions. Part of this solution is the nutrient net in which minerals that stray below the root zone of crops can be captured and returned to the surface (through biomass incorporation and subsequent decay). Some species make good facilitative additions because of their lower strata roots. As previously discussed, these roots need not compete directly with crops and can provide last stop nutrient capture.[26]

There are a host of other mechanisms that contribute to retention. Having greater biomass and biodiversity is foremost where nutrients are temporarily locked safely away in organic materials. Babbar and Zak (1994) describe a high biomass situation in which, despite high levels of easily leached nitrogen, losses were less than in plots with less biodiversity.

Accumulation-Concentration

Whereas retention keeps nutrients on site, accumulation allows these to be collected, at times from afar. Fallows, where land lies idle for some period, allows nutrients to accumulate across time periods, enough so that the increased yields allow for profitable agriculture.

Among the species utilized are nitrogen fixating species that grow rapidly, concentrating and storing large volumes of this nutrient. Nitrogen is not the only nutrient gathered. The African tree, *Tithonia diversifolia,* is able to accumulate higher than normal volumes of phosphorus in its leaves.[27]

Site Improvement

Poor or imbalanced soils are hostile to many agricultural species. When this occurs, a site can be bioremediated, that is, a plant species is chosen that can, over time, rectify a hostile soil situation.

Among the possibilities are the buffering of acid or alkaline soils and of high concentrations of other damaging elements or compounds.[28] The mechanism uses a tolerant species which through accumulation of neutral or chemically buffering biomass fosters a more agriculturally friendly site.

Parasitic

Some species take advantage of others through a parasitic association. This has one species serving as an unwilling host, supplying essential minerals to another to the detriment of its own growth. This also qualifies as a facilitative relationship.

Pure and positive parasitic tie-ins are rather rare in agriculture. One such case is the decorative plant mistletoe, found in southern North America. This is parasitic on oak trees.

Exclusion

As a form of facilitation, the removal of resources from one species can benefit another. With facilitation, a species, purposely planted, serves a supply role. In contrast, exclusion forestalls the flow of nutrients in an unintended direction, from a primary to a secondary species.

Full exclusion is primarily associated with weeds. Costly if allowed to remain, costly to remove, weeds may be the bane of agriculture. Full exclusion is often the sought-after solution. Due to their importance, weed-countering options are discussed as a separate topic (Chapter 4). This is not the only application of resource denial.

Others have refined the exclusion concept to encompass the situation where secondary plants remain, but with reduced competitiveness. This is partial exclusion. A vegetative cover, smothered below the primary crop during the growing season, subsequently released when the primary crop is harvested, is the common case. Similarly, many cover crops are expected to survive across growing seasons by reappearing once the primary crop has been removed.

ECOSYSTEM GOVERNANCE

Species governance may dominate much of biodiverse agriculture but is not the only alternative. The other, lesser acknowledged, flank of the agrobionomic principles is ecosystem governance.

Rather than having the individual species set the ecological agenda, ecosystem dynamics takes hold through multiple, intended plant species and a host of inadvertent and often unrecognized benefactors (weeds, fungi, bacteria, worms, ants, birds, etc.). In essence, this is when there is enough immediate biodiversity, flora coupled with fauna, so that the ecosystem takes on a life of its own. The multiplicity of interspecies interactions do not define ecosystem governance; this comes about only when the many one-on-one interactions are transcended and exceeded. All effects contribute, such that

1. it is not possible to meaningfully separate the specific (species) mechanisms;
2. the system becomes more than the sum of the parts or individual species; and
3. external management, lacking full control, must accept the ecosystem as is or risk losing or weakening the unique properties and the self-generating ecological mechanisms.

This brings about discussion as to what constitutes a functioning ecosystem. A strict interpretation is where the species have coevolved and any one does not ecologically dominate another. Most natural ecosystems, without introduced exotics, fall into this category. In agriculture, mixed species grasslands (pastures) are often natural ecosystems commandeered for agricultural purposes (see, e.g., Plate 3.2).

In most of agriculture, the ecological supremacy of one or more species is the desired outcome. Less found are those with considerable biodiversity. Although not coevolved, the intended species can be made to interrelate, and an ecological balance can be maintained,

PLATE 3.2. The classic example of ecosystem governance in agriculture: A mixed species perennial pasture.

through management intervention. In agriculture, agroforests are a collection of exotic species with undisputed ecosystem governance. Another example, household gardens, those containing a copious species mix, also tend to be ecosystem governed.

Parameters of Use

There are questions on how best to capture the in-use benefits of ecosystem governance. Without a strong native component, the stage is set for ecosystem governance, and limits established, through three spatial parameters: (1) biodiversity, (2) biodensity, and (3) biodisarray. On the temporal plane, there is a fourth parameter, duration. Duration (i.e., bioduration) allows time for contributory organisms to be introduced, to reproduce, and to become an ecological factor.[29]

Biodiversity is simply the number of niche-diverse species present. This is a case where more is better. There is a lower threshold but no upper limit.

Biodensity has the contained species growing in close proximity. Outside the parameters are those wide spacings that direct ample essential resources to all plants. On this point, farmers must accept crowding as nature intends.

Biodisarray ensures that all species are intermingled and the interactions scrambled. Full randomness is a possibility but not a requirement. The pattern should be such that interspecies interactions occur on a wide and broad scale without any dominating one-on-one relationships.

Density, diversity, disarray, and duration may underlie ecosystem governance, but how the inherent dynamics take hold is an open question. In-place natural ecosystems come fully equipped for the tasks at hand; collections of exotics may not. For the latter, larger plants may be sufficient to accomplish what is ecologically desired (a top-down approach), but the smaller, often unobserved, microorganisms may be lacking. Microorganisms may be important in triggering ecosystem governance (the bottom-up view). In presenting these arguments, simultaneous occurrence (ecosystem development in both directions: top and bottom) may trump both views.

Top-Down

The top-down approach assumes that the major (larger) species are the driving force for a fully fledged ecosystem. The key elements, and the visual clues that all is well, are biodensity, biodiversity, and biodisarray. Since much is visual, some guidelines are possible.

For biodiversity, in addition to having many different species, this includes like plants of differing ages. The prevailing view is that seven to ten niche-diverse species are needed to reach the ecosystem threshold.[30] Although not thought of as positive, weeds, in modest amounts, can be a proclaimed component in ecosystem governance.

As what may be the least important of the threshold criteria, disarray ensures that enough randomness exists to surmount repeating and ecosystem defeating plant-plant interactions. That is, if the same competitive production and facilitation mechanisms are continually repeated throughout an ecosystem, the full scope of interaction gains will not be realized.

Other inefficiencies should be eliminated. Light striking bareground is a visual indication of a lost resource. When this occurs, an ecosystem may be better served by planting or letting something grow. The increased density goes not to the point that crowding and poor yields result, but to the undefined point where an ecosystem is most efficient. Natural ecosystems in a local area reach an optimal, not overcrowded, density plateau.

Adding to the ecosystem are herbivore insects. Mostly a negative, these draw predator insects that further balance and shape what is in place. Ever smaller flora and fauna continue to refine what is wanted and what is attained.

With the top-down view, the principal interactions, as defined by larger plants and animals, contribute to the bulk of the interactions. The key and measurable influences are amplified by seemingly insignificant interactions from ever-smaller sources. The subsequent gains from ever-smaller plants and animals down to the microbial level, although significant, are subordinate to the ecological interactions obtained from the principal components (the dense, diverse, and disarrayed larger ecosystem).

Bottom-Up

The bottom-up view of ecosystem governance puts more emphasis at the microbial level. Plants provide an inviting and favorable base and, from this, conditions exist for a flourishing, beyond-expectation agroecosystem. Although threshold values for density, diversity, and disarray set the stage, these do not dominate the enfolding events. With a bottom-up ecosystem, duration may be more important, especially if the microcommunity is not in place and needs time to achieve full potency.

It is known that microbes can improve soil moisture holding capacity, decompose different organic materials (e.g., the compounds found in stems, leaves, etc.), and make available a range of minerals. Microbes also counter plant diseases, reduce the impact of temperature extremes, and undertake other facilitative chores.[31]

Questions exist on how to best capture these effects. Some facilitative microorganisms are associated with specific plant species; others, more public, exist in ecosystem types with little regard for the plant species contained.

Species-associated mycorrhizae are common with some agricultural species. Other associations may be less noticed. Garbaye et al. (1988) found that exotic trees grew better if the surrounding soil is inoculated with microorganisms imported from their ecosystem of origin.

Others may originate from, and may prefer, certain types of ecosystems. Cleveland et al. (2003) observed a large difference in microbial activity between neighboring forests and pastures. Although the full picture is not overly clear, it may be the background ecosystem that brings much of this to the fore.

The picture does not come together easily. Despite their in-ecosystem potency, microbes alone may not be capable of undertaking all the tasks required. The bottom-up view has microbial strengths augmented by the presence of larger organisms.

Many species of ants and worms are clearly facilitative.[32] The ecological gains from earthworms include soil enrichment, recycling of organic materials, and improved water absorption. Similarly, ants enrich the soil by breaking down plant residue, by improving the soil structure, and by adding water infiltration and holding capacity.

The bottom-up view has ants, worms, and the like as part of and intertwined with microbial life. Under this image of ecosystem governance, the even larger organisms are further interwoven, building upon and solidifying the already firm base. Among the questions posed are how far up the ecosystem (starting with single-celled organisms) one must proceed before a fully functioning facilitative threshold falls naturally into place.

Implications

As one enters the world of biocomplexity and ecosystem governance, one finds a highly productive, relatively risk-free setting in which nature, at little cost, offers critical inputs. The tradeoff has the farmer as more of an active bystander, rather than the reigning force. Many find this appealing for the economic and ecological benefits these systems bring to a rural landscape.

The form of ecosystem governance, bottom-up or top-down, may seem academic but can have profound undertones. A top-down view is easier to implement; this is just the close intermingling of many niche-diverse plant species. A bottom-up ecosystem is more problematic but may ultimately prove the better agroecological choice.

The problem with bottom-up governance is that microflora, nonweedy larger flora, and micro- and small fauna do not move easily from one spatially isolated ecosystem to another.[33] Building a functioning, self-governing ecosystem from the bottom up requires manually appending extraneous, nonweedy plants; microbe-containing soils; and a variety of small creatures. The other option has connecting corridors so that bottom-up flora and fauna can move across the landscape, colonizing where absent.

Chapter 4

Productivity Threats

Land-use practices often reflect a desire in not going without. The land must yield or the farm family, farmer, and others face hard times. Productive certainty and the mitigation of risk are strong selling points for agroecology, more so when the countermeasures to threats are productive and environmentally friendly.

Frequently, overcoming threats and ensuring yields entails a compromise between risk mitigation and productive outcome. If the failure penalty is high (e.g., starvation), farmers will gladly sacrifice potential yields to ensure a basic output level. Commercial farmers, even those with crop failure insurance, face financial difficulties if the land does not yield and will embrace countermeasures if not too intrusive.

Keeping agroecosystems secure over time and in the face of continuing threats is one aspect of sustainability. Accordingly, risk, immediate yields, and long-term sustainability have an undefined relationship. In the drive for productive security, various compromises, long and short term, are involved.

Agroecology offers many on-site techniques to address the common climatic stresses and productive threats. These vary in short-term effectiveness and in their long-haul implications. Of those listed in Chapter 1, climate, weeds, and animal assault are discussed in this chapter. Given the multitude of options, countering damaging insects merits a separate chapter (Chapter 5).

CERTAINTY-SUSTAINABILITY

All agricultural producers want to operate in a risk-free environment. This being impossible, most spend considerable effort to en-

Introduction to Agroecology: Principles and Practices

doi:10.1300/5553_04

sure minimal damage from the various threats to yields and the environment. This requires continual inputs, or, if farmers do not want to go down this road, ecosystems or farm designs that, without active intervention and inputs, can self-counter the threats. This links immediate productive certainty with sustainability.

Sustainability in the longer term can be viewed and defined in a number of ways:

1. The simplest is the ability of the land to produce over time. If an agroecosystem is not totally secure and soil is continually lost, sustainability is far from assured. Flowing water or strong winds could be responsible for the soil loss.[1] The removal of mineral resources through harvests, when these exceed replacement, also precipitates negative sustainability.

   The best practices do not diminish the productive capacity of the land. Better yet are those practices that actually improve the land. This is an offshoot of the type of agroecosystem selected, the rotational strategies employed, and implementation of an ecologically positive management regime.
2. Another factor is a do-no-harm admonition.[2] Such a forewarning is provided in order to retain or improve in-soil resources. Plant-site negativity is illustrated in semiarid China with the introduced tree species *Salix matsudana.* This tree freely soaks up scarce groundwater whereas native species absorb less. When water is limited, soaking up this essential resource is not a good plant characteristic and not good for nearby agriculture.[3]

   Additional requirements lie in refraining from doing anything that would hurt neighboring ecosystems (such as drifting chemical sprays) or endanger the health of farmers, workers, or persons living nearby, while still maintaining the purity of water originating from farm fields. Along these lines, the foremost danger is in applying environmentally harsh chemicals.

   Part of a good land and ecosystem policy is to encourage those natural cycles that promote positive ecodynamics. This can proceed on a number of fronts:
   a. nutrient cycling to replace overapplied chemical fertilizers;
   b. substituting insecticides with predator-prey or other natural means to control plant-eating insects;
   c. using biodiversity to downgrade a weed threat.

3. A stronger pronature overture and more inclusive form of sustainability is undertaken with a deference for natural flora and fauna. This most stringent of the sustainability definitions is also cumulative with a positive on-site, long-term performance and the do-no-harm admonition. Under strong sustainability, soil and health issues are encompassed and native flora and fauna (small and large) are protected. The restriction, and the challenge, is the need for concurrent and viable harvests.

GENERAL COUNTERMEASURES

In agriculture, different views exist on risk containment. For most, productive uncertainty is managed by countering immediate threats and ensuring adequate and controlled essential resources. This can mean adding water in times of drought, draining when flooding occurs, spraying when insects threaten, or culling the diseased plants when affliction strikes. Over time, a reliance is placed on one or two threat countermeasures, and this might bring on other problems and raise other issues.

A case in point is irrigation water brought from afar or from deep sources. This alleviates drought danger by permitting plentiful harvests where few have occurred. Irrigation systems can introduce other types of risk. Salt contamination is one peril that is difficult to remedy.[4]

Rather than one or two countermeasures, agroecology suggests the implementation of a range of threat counter-options. Those described in this and subsequent chapters are often a defensive series with less reliance on a single and immediate yield-ensuring alternative. This may require some rethinking on conventional agriculture.

In agroecology, insects can eat crops as long as they do not eat too much, and drought can reduce yields as long as this is within acceptable parameters. Minor losses are tolerated or may even be part of a defensive strategy.

There is the general assumption, subscribed to in this text, that some agricultural practices are better than others at containing risk. Properly instituted organic agriculture falls under this heading.[5]

A general landscape countermeasure that some undertake is overplanting and scattering, sowing crops in widely located, diverse settings. This insulates fields from one other; if some fail, there is the hope that a few discrete plots would survive and yield.

Even the best-planned farm landscapes can go awry. Swarms of locust, those of biblical proportions, may have far-reaching and severe consequences, completely overwhelming defenses. When the magnitude of the threat is coupled with infrequency, few plan ahead and do not have prepositioned or even the interjected defenses at the ready.

To counter exceptional threats, many opt for secondary outputs that survive or thrive when staple crops fail. Alternative crops, those that persist when others fail, help in this count. Cassava, common in humid tropical regions, is tolerant of climatic variation. The roots lie protected below the ground and can be harvested over long periods. This secondary and, at times, survival harvest can be intercropped or grown in detached plots.[6]

Trees often bear a standby crop and, although not as tasty as staple foods, do sustain people and animals. Across many centuries in Europe, the lowly acorn sustained life when disaster loomed. In sub-Saharan Africa, the baobab and shea trees come to the rescue when insects or drought decimate everyday crops.[7]

THREAT LEVELS

The measurement of climatic stresses and productive threats can be problematic as not all threats are easy to define. Some are too infrequent to consider and some overwhelm normally sturdy defenses: countermeasures not always being immediate and resolute.

Generally, threats can be expressed in (1) the number of events (incidents), (2) their magnitude, and (3) the area struck. Duration may constitute a fourth circumstance, but not for all threats: fire lacks duration; drought may vary in duration and severity (magnitude); however, in the case of flooding, magnitude and duration have a separate presence.[8]

In establishing in-place and interjected (i.e., temporally brought to bear) defenses, one hopes to reduce the number of incidents, their magnitude, and the area impacted. Corresponding defenses can vary in their resistance strength, for example, with a few potent or a series of weaker but overlapping safeguards. Defenses cover small or wide areas. In the case of shelterbelts, these are generally more shielding than windbreaks but monopolize by taking from production larger areas. The main risk-mitigation mechanisms as discussed in this and subsequent chapters.

SPECIFIC INTERACTION ZONES

Agroecology involves managing multiple biological interactions. These often have a demarcated area, originate from one or more sources, criss-cross individual plots or farm landscapes, and exert a positive or negative influence.

Root spread, insect-repelling plants, and windbreaks all have specific interaction zones (SIZs). Windbreaks have a location and accompanying wind-protected SIZ. The SIZ for insect-repellant plants may not be as large but may be an equally valid influence. The patterns in well thought out agroecosystems utilize, augment, and extend various SIZs.

The gains and losses for SIZs are shown in Figure 4.1. Roots can confer a negative (e.g., drawing water away from other plants) or a positive (e.g., contributing nitrogen) zonal effect. Close to a plant or structure, there can be a high competitive zone. A bit farther away, the positive overtakes the negative; yields respond accordingly.

These zones are commonly thought of as a purely spatial influence. As mentioned in Chapter 2, these also have a temporal presence, as when residue from a nutrient-contributing or insect-killing plant is retained in the soil for some months.

Through agrodesign, SIZs can be expanded. Whether spatial or temporal, SIZ influences can extend across a larger area than normally possible. The typical case is where insect-eating birds find habitat in forest fragments. Trees, planted within and at the edges of fields, may lengthen the feeding range of these birds (an expansion of the SIZ) or may enlarge or amplify the habitat possibilities (an augmentation of the SIZ). This also occurs on a smaller scale with predator insects.

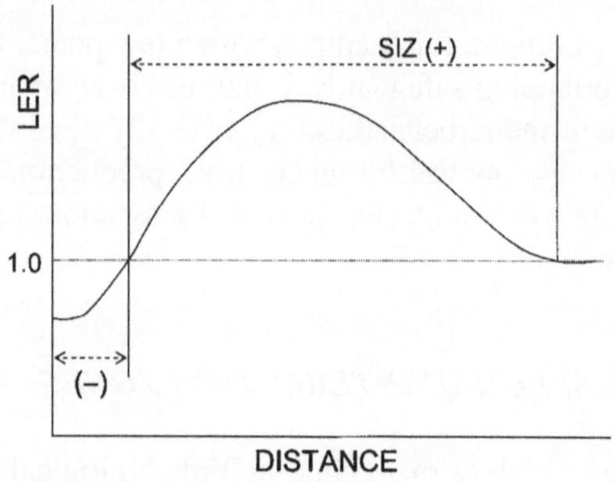

FIGURE 4.1. The SIZ-productivity gains that relate yield gains as a function of distance away from a protective plant or structure. Gains can be expressed directly in absolute yields, LER, and, indirectly, as a risk reduction and potential increase in harvests.

CLIMATIC THREATS

The threat categories, whether immediate or long term, are the basis for ecosystem selection. The climatic threats are as follows:

- Drought
- Heavy rains
  - Erosion
  - Nutrient leaching
- Flooding
  - Waterlogging
  - Scouring
  - Leaching
- Winds
  - Constant breezes (drying)
  - High velocity gusts (toppling)
  - Downward bursts (topping)
- Temperature extremes
- Fire

Each of these is countered in its own way. The advantage is that the countering mechanisms overlap one another and that each may address more than one climatic threat.

Drought/Excess Rainfall

A periodic lack of rainfall has plagued humankind since agriculture began. The danger of crop loss may be so great as to force the expense of constructing and maintaining irrigation.

A lack of water may seem absolute, but subtle differences can be important in devising a counterstrategy and selecting an appropriate agroecosystem. Some may experience regular rain but in lesser amounts. There are also situations where heavy rains occur followed by a severe shortfall. Typically, the history of an area provides insight into form, frequency, and severity. In exceptional situations, as with an oppressive and prolonged drought, there are few on-site counters.

Shortfalls are mitigated by keeping available water on site through (1) lessening runoff, (2) storing water, and (3) reducing use. Irrigation aside, many of the countermeasures for drought also counter heavy and damaging rains. For example, capturing and controlling rainfall on site can lessen a subsequent drought. Capturing and controlling heavy rains also prevents damaging erosion and mineral leaching.

Runoff can be controlled by slowing water movement, encouraging the water to percolate into the ground. Vegetative rows or strips spaced at regular intervals, contour ditches, and bunds accomplish this. Once underground, movement is slowed, the water is stripped of transported soil, and water-carried nutrients are subject to capture. The plot in Plate 4.1a incorporates a barrier strategy.

Barriers may be most effective with brief periods of heavy rains. There is the proviso that barriers should be deep or high enough and at intervals so that these are not overwhelmed. The latter, that is, the interbarrier distance, is the SIZ.

Runoff is also contained by ground-level vegetation. This can be living or dead plant material. The idea is to directly protect the surface from being hammered by raindrops to slow water movement, encouraging absorption into the soil and to keep water on site, shielded from the sun and evaporation and available for capture by plants.

(a)

(b)

PLATE 4.1. The two antierosion strategies: *(a)* a series of contour barriers protecting hillside crops and *(b)* a dense cover species protecting a hillside. The barrier photo is from Kenya, the cover crop photo from Nicaragua. The cover species is *Arachis pintori.*

The cover approach may be better with more frequent, lighter rains. To be effective, vegetation (living or dead) must be at ground level and uniformly cover over 50 percent of the area. The more vegetation, dead or alive, the better. Large, slow-decaying leaves tend to be, with exceptions, superior to smaller, short-lived foliage.[9]

Drought mitigation often involves a reduction in evaporation. By controlling winds and keeping a humid atmosphere, plants absorb less water and prolong what is available. Shelterbelts and windbreaks function this way (see Chapter 12).

Flooding

Excess water can be as much of a problem as a lack of water. With to much rainfall, the most susceptible topography is that of the hillsides; flooding is more a bottomland event. The dangers still lie in erosion, the carrying away of nutrients, and, with some agricultural species, waterlogged roots, which will ultimately kill the plant.

Many of the countermeasures for drought and heavy rainfall apply also to flooding but with changes in design. Ditches and barriers still slow water movement but are employed more to prevent scouring and for the on-site deposit of waterborne soils. Earthen barriers (contour ditches or bunds) are less utilized. Instead, hedges and other vegetative rows are formulated especially as waterbreaks (see Chapter 12).

Infiltration still has a role, mainly where flooding is less severe or when water is receding. On level but undulating sites, where water can pool, deep-rooted trees or shrubs, planted in the depressions, can poke holes in impermeable subsurface soil strata, allowing for more rapid drainage.

Winds

Winds come in various forms and with mostly negative consequences. The negative effects include plant lodging, leaf and branch rubbing, increased evaporation and evapotranspiration, and sandblasting.[10]

There are also a few positive wind effects. Drying may lower the threat from fungal diseases and may inhibit some herbivore insects, provided the winds are not strong. Despite the few gains, winds are mostly viewed as a negative and treated likewise.

As with other threats, it is more effective to design counters around the prevailing problem. As winds come in different forms, an appropriate design best addresses a specific need.

Where wind is a problem, windbreaks come to mind. Although useful in many situations, these do not exhaust the list of countermeasures. Internal plot design plays a large role in wind control and can be the first line of defense.

Constant Breezes

The constant blowing of winds is endemic to some sites and can bring on unsatisfactory consequences. There is an increased water need to counter increased moisture loss; this must come directly from the soil or from plants.

Although seldom thought of, branch rubbing wounds plants, allowing for the entry of enemy organisms, requiring effort on the part of the plant to heal the wounds and counter the invading organisms. Although they appear to be minor, such wounds can be a major productive handicap. Brenner (1996), in summarizing previous work, stated that wind-induced, interleaf rubbing has reduced wheat yields by 11 percent, maize yields by 28 percent, and bean yields by 45 percent. The counter is an internally breeze-protected plot or some explot defense.

High-Velocity Winds

Occasional high winds are not a blessing but a curse. The most serious damage is through toppling or lodging. Plants that collapse produce less or nothing. High winds from storms are also a source of sandblasting (the slamming of sand particles into leaves or stems). The accompanying branch rubbing and sandblasting wound plants and must be healed at a cost. The cost to the farmer is the loss of potential outputs and the danger of introduced harmful organisms.

High winds require sturdy defenses often located next to the endangered area. Again, the windbreak is the bulwark defense.

Vertical Bursts

Fields and plots may be protected from horizontal high winds, but vertical bursts can impact parts of a field. The symptom of microburst is a small flattened area in a large exposed plot. Although the impact is less, pertaining only to the affected area, the negative consequences are the same as with vertical high winds.

Countermeasures include a field laced with wind-intercepting trees or composed of smaller, less susceptible plots. With large plots, ex-plot windbreaks are less effective at controlling such bursts.

Temperature

Frosts can be a common occurrence in mountainous, semiarid regions where temperatures can plummet on clear nights. Temperature can be important in temperate zones when, near the beginning or end of a growing season, a late or early frost can prove costly. Even semitropical regions experience brief periods of cold when winter weather penetrates.

Periods of high temperature also affect plants. These spikes in temperature, especially in drier regions, are negatively manifested by the midday wilting of leaves. Such temperatures are detrimental to the plants and are countered by providing more water or, better yet, reducing ambient temperatures. There is a dual advantage in the high and low temperature safeguards in that the methods to secure from frosts also serve to deter high temperatures.

The periodic frosts of mountain regions are countered by retaining day heat. Rocks and water, exposed to daytime sunlight, can provide a respite from freezing when this heat is released during the night. To ward off frosts, people of the Altiplano of Bolivia plant crops on mounds in flooded fields. In the mountains of Pakistan, heat-absorbing, heat-releasing stone terraces protect wheat from chilly nights.

Tree canopies can reflect radiant heat downward. In China, tea is protected from frost by alternating tea with strips of heat-retaining forests, and in South America, coffee is protected in the cold-susceptible highlands by trees directly overhead.

Even minor elevation differences serve this function. To guard valuable plants from cold air, mounds are employed in the highlands of New Guinea.[11]

In parts of North America where temperature plunges occur, susceptible fruit trees can be guarded by spraying them with water. Although ice-covered trees may seem counterintuitive, this protects by keeping the temperature of the trees and fruit above the critical killing threshold.

Frost and heat counters generally have very small SIZs. As such, there is an intimate in-field association between the plant and the counter.

Fire

Burning in agriculture is both a useful tool and a destructive force. As a useful tool, fire controls plant diseases and insects, and aids in weed control.

As a negative influence, fire is less documented as few agricultural fields burn during periods of growth. It should be noted that grains are susceptible after in-field drying and before harvest. Thus, fire is a climatic threat to agroecosystems.

INTEGRATED WEED MANAGEMENT

Of all the threats to crops, unwanted biodiversity, in the form of weeds, may be the most pervasive and costly. Weeds, if left unchecked, can overwhelm the crop, reducing harvests to near zero.

Blanket control, as with herbicides or hand weeding, may not be the best answer. Cost or environmental concerns can make these options unattractive. Integrated weed management, on the other hand, offers various measures best aimed at key weed species. As with most agroecological initiatives, a perfect outcome is not expected, the penalty being a less-than-optimal harvest.

Two concepts underwrite a composite approach: (1) not all weeds are competitively equal and (2) a series of steps can be taken to ensure that the crop dominates.[12] With regard to the first concept, integrated management targets only the more competitive weeds. It is the competitive characteristics, not the plant, that define a weed. Useful cover crops, because they are not overly competitive with desirable crops, are weeds put to good use. This is demonstrated by the pervasive weed purslane which, because it has a lesser effect on some primary species, can serve as a cover crop.[13]

By ignoring some weed species, the scope of the problem is reduced. If not part of the problem, the unwanted additions are part of the solution. Noncompetitive weeds are a form of biodiversity that can be exploited to, for example, harbor beneficial insects.[14]

As for the second concept, integrated weed management consists of a series of steps starting with prepreparation, continuing on through preparation, postplanting, and crop growth phases. Site prepreparation can involve fire, rotations, an off-season cover species, or other methods to suppress potentially threatening plants. The preparation phase includes tillage and timing, again each designed to disfavor certain unwanted plants. Anderson (2003, 2005) has gone down this road, integrating various ecological mechanisms (i.e., rotations, tillage, and plant residue management) seriously impacting weed populations and the costs of removal.

Integrated weed management is not a fully developed topic. There are many unexplored facets, many involving the niche relationships between one or more primary species and these unwelcome and unwanted quests. Because weed control is highly species and niche dependent, only the key aspects of, and individual elements in, an integrated control method can be discussed.[15]

Rotations

Changing the location of seasonal crops each year is a form of alternative agriculture. As such, this is a potent strategy that counters a number of threats, weeds included. Grodzinsky (1992) discusses a weed-countering rotational strategy where a crop of *Brassica* spp., either winter rape *(B. napus)* or spring colza *(B. campestris),* protects subsequent crops from weeds.

In Africa, the feared, hard-to-remove parasitic weed striga attacks the roots of maize and sorghum. If growth is triggered by a false host or suitable biomass, the danger passes. This may be one of the best examples where the most competitive weed species has been pinpointed and countermeasures devised.[16]

Timing

The question of when to sow may depend on the weeds. One option is to plant well before the weeds establish so that the crop can compete favorably. The second option is to delay planting, allow weeds to establish, and hope that land preparation will decimate this aspiring but unwanted element.[17]

Tillage

Land preparation further disrupts the timing of weed species. Deep plowing buries seeds and sprouting weeds, and some of these do not recover.[18] Although it does not provide a large measure of protection, tillage is part of a larger, comprehensive antiweed, anti-erosion scheme useful to control the initial bloom. Practical aspects of the tillage-cover crop relationship are advanced further in Chapter 8.

Weed-Consuming Organisms

Less used but within the realm of possibilities are those organisms that feed on weeds. Ground beetles have a taste for weeds and, as an added benefit, also eat cutworms, armyworms, and corn rootworms. Field crickets can consume over 200 pigweed seeds in one day. Muscovy ducks and geese are voracious weed eaters and, if deterred from consuming the crop, serve well.

From weed-eating insects to larger grazers (sheep, goats, llamas, etc.), the possibilities in employing weed-consuming organisms abound. Not outside the realm of possibility, people also consume edible weeds. Added to the control possibilities are fungi and other plant diseases that afflict common weeds.[19]

Mulch

Mulch from selected plants can impede the growth of weeds. As part of a cut-and-carry strategy, appropriate mulch can provide nutrients as well as diminish weeds. In one study, biomass mulch from the tree gliricidia reduced weed density and size. In contrast, leaves from the tree species leucaena were not good at this task.

In other examples, rice straw inhibits a range of weeds while straw mulches from barley, oats, and wheat have served to suppress weeds without affecting newly planted conifers.[20] The reason may be allelopathy, the chemical composition of the mulch inhibits the sprouting of weed seeds.

Companion Species

It is possible to densely plant an area with a weed-excluding, facilitative companion species such that weeds have difficulty in establishing and in finding the essential resources needed to thrive. Cover crops, as a facilitative species, fall under this heading. The countermeasure employed is one of niche exclusion where unwanted species are denied one or more critical niches so that they do not pose a threat.

Companion species have been planted on numerous occassions with success. Wild mustard *(Brassica kaber)* planted under apple trees virtually eliminated weeds while having a positive influence on apple yields. From observations in southeast Asia, a dense-canopied, fast-growing tree species, *Peltophorum pterocarpum,* planted as a hedgerow, produces a dense, closed canopy during the fallow period. This eliminates hard to manually remove speargrass.[21]

There are cases where delayed planting can reduce competition with the primary species. In West Africa, a dense cover crop of mucuna improved soils while eliminating weeds. Success, here, demands that the mucuna be planted a few weeks after the main crop.[22]

Intercrops

Selecting a companion species that crowds available niches or is allelopathic with common weed species is a useful weed control strategy. A high potential LER ensures that more essential resources are utilized and more niches occupied.

An abundance of roots in different soil strata can also block potential weeds. Ryegrass and white clover, both shallow rooted, help deter weeds in pepper and okra.[23]

All does not depend upon occupying belowground niches; physical blocking can also take a toll on weed numbers. As mentioned, large, slow-decaying leaves, as opposed to smaller, short-lived ones, work better. Large-leaf squash is added to maize or maize and beans, not for the minor yields but for weed and insects control (see also Plate 4.2).[24]

Hand Weeding

Not to be overlooked are the techniques of hand removal. Under the assumption that a species, upon growth, can overcome any immediate threat, weeding can be done a distance away from the plant. This requires a species that can handle weeds close to the plant. The

PLATE 4.2. Squash planted between tomatoes as a soil protecting, weed control measure.

large, sunlight-blocking leaves of squash and pumpkin excel in this role.

The opposite technique is one where weeding takes place near the stem. The assumption here is that the more distant weeds are not that strong a threat. A crop with a vertical root system accommodates this strategy.

If farmers have an intimate knowledge of crop and common weeds, they could remove only those species that exert a strong negative influence. This reiterates the belief that less competitive weeds augment biodiversity and block subsequent, and more competitive, weed establishment.

LARGER FAUNA

In some regions, mammals, birds, and other creatures eat crops. What complicates control is that some of these, mostly birds, are useful for insect control. Ill-considered tactics, as with the elimination of a bird species, can be counterproductive when herbivore insects thrive without bird-provided insect control. This is a case where sacrificing some of the crop to protect the whole can be a viable agroecological strategy. As such, the pluses and minuses must be considered before embarking on an eradication program.

A plethora of control measures is available for adoption, without having to resort to wanton hunting. A few are sampled here.[25]

Barriers

Fences are essential in farming. These serve a number of purposes, keeping animals in or out, and in demarcating boundaries.

A barrier could be a hedge, tree, or other crops. Beckerman (1983 and 1984) found circular, target-shaped plots in South America where each ring protects an inner crop. Within these rings were also plants to repel small mammals. This barrier strategy can also keep birds from crops. Some grain-eating species do not like wide-open spaces, and vulnerable crops can be surrounded by wide strips of nonsusceptible crops to deter them.

When hungry grazers border crops, fences do make good neighbors.[26] However, there are limitations: deer can jump tall fences, bison can charge through filmsy barriers.

In order to ensure that hedges and other living fences are not ravaged by the grazing animals, these are deterred with thorns or distasteful internal chemicals. There may be middle ground where a live fencing serves as a forage source of last resort. This requires plants that are edible but not preferred.[27]

The traditional English hedgerow is a prime example of a living fence. The alternative, characterized by wire or wooden fencing, is a dead boundary. This is some overlap between types; vines growing on a wire fence is a little of each.

Barriers can also be small scale. This is seen in Plate 4.3 which shows a rabbit guard around a fruit tree. This stops rabbits from gnawing the bark.

Repellant Plants

Another control measure, though seldom addressed, is the use of plants that can repel or make large fauna uncomfortable. Temperate gardeners have a short list of repellant plants from which to draw. These include the following:

- castor beans (repels gophers and moles)
- chives (rabbits)
- garlic (gophers)
- onions (rabbits)
- mint (mice)[28]

Other plant attributes discourage other fauna. Stem thorns, found with some palm species, deter climbing rodents. Birds may find it more difficult to get at varieties of wheat, barley, or other grains with long awns or bristles that stick out from each grain.[29]

Repelling Fauna

Even the largest of animals can be chased away. Ferocious dogs can deter many larger mammals as can aggressive geese. In this, size may not always matter. Even the diminutive honeybee can chase away elephants if the hives are well located.[30]

PLATE 4.3. An antirabbit barrier protecting the bark of a fruit tree.

For rodents and other small creatures, snakes, hawks, and owls can be the cornerstone of control. Select species of predator birds can be enticed to stay and hunt with a tailor-made habitat, including nesting sites and plenty of game.[31]

Repellant Compounds

Chemicals can be employed to chase away or deter crop-eating animals. Several possibilities exist in this technique of repelling fauna. At the large end of the spectrum, elephants can be dissuaded with a pepper spray. More common cattle and the like can be discouraged from eating the bark of fruit trees by smearing the stems with animal manure.[32]

Hunting

Often less sustainable is where animals are hunted for sport, food, and to protect crops. Much can be lost if this course is followed; without placing limits on the extent of hunting as a valuable resource, the fauna of a region, can be lost.

The large animals of the Mediterranean regions of Europe and Africa, for example, elephants, lions, and so on, were depleted by the Romans and are no longer part of the ecology and culture. In contrast, elephants and tigers still exist in India and South Asia. Although these creatures live free in comparatively small areas, they still have a cultural impact expressed in regional folk art.

More might be lost. The eradication of keystone species, those upon which others depend, can lead to a reduction or extinction of lesser species and the removal of an entire slice of a natural ecosystem.

Besides upsetting the balance of nature and losing potential predator-prey agroecological assets, this can also be a quality-of-life issue. Missing songbirds do detract from the enjoyment of nature. Where rice paddies are found, large reductions in area populations of birds and bats, fish and frogs can lead to a explosion in mosquito populations. Resulting mosquito-borne diseases adversely affect human health.

Most land users often do not worry about the niceties of conservation and will hunt animals to protect crops. With large animals, this is clearly effective, while it is less so with small rodents or other creatures that reproduce rapidly and are hard to catch.

In the case of the smaller pests natural predators can be more effective. Snakes, foxes, and small cats (e.g., the bobcat of North America, bush cats of West Africa, or even the barn cat) have rodents as their prime food source. These animals are less likely to attack domestic animals (the exception being the very young). Where attacks on domestic animals are likely to occur, nondestructive countermeasures may be easier to implement than rodent control.

The other alternative is to regard rodents as a human food source. Although alien to many cultures, this remains a distinct possibility.

Semihusbandry

More common than the scientific literature indicates, the sport or subsistence hunting patterns of many developed countries fall more under the heading of semihusbandry. This follows more closely the agroecological premise that some crops can be eaten, more so if this results in another consumable.

Semihusbandry considers wild animals as a renewable resource with a sustainable harvest. Any damage done to crops is a contribution to this resource. Farm layout and the purposeful planting of fruit trees can attract and promote this resource.[33]

Ideally, a natural forest or other reserve protects the breeding population. This area is not hunted; only the outflow is taken. A large protected area ensures sustainability. If there is no wild area nearby, close monitoring of animal populations prevents overhunting.

Some animals are better suited than others as a food source. They should be neither too large nor too small and the species not endangered in any way. These should also be

1. comparatively easy to hunt or trap;
2. not be capable of quickly destroying cropland; and
3. easily dissuaded from causing damage where not wanted.

The gains, in addition to more protein in the diet, come in not having to provide farm infrastructure (buildings, containment enclosures, and feed storage facilities).

An example of semihusbandry is deer hunting in Europe and North America. In the southeastern United States, the sale of deer hunting rights in tree plantations is a substantial income source. In South America, the capibara (a large rodent) is sought and is far more efficient than domestic animals at biomass conversion and reproduction.[34] In other regions, culture permitting, lizards and even snakes are nature-provided food source.

A study in West Africa found that species of rats, porcupines, and small deer make up about 80 percent of the animal harvest. In Peru, 41 percent of the harvest comes from four species.[35]

Chapter 5

Insects and Plant Diseases

In consuming the fruits of farm labor, insects have plagued humankind since the inception of agriculture. Since recent controls have proved environmentally damaging, insect management strategies—those sans insecticides—often front agroecology.

The distant historical record focuses on sudden, severe outbreaks, such as swarms of locust. The less dramatic and more mundane infestations of earlier times were most certainly handled by natural controls: an understanding of which was gained from closely observing and applying the lessons of nature.

Early agriculturalists, most certainly, came across plant diseases. Although the lessons of nature are harder to comprehend, some of what works against unwanted insects also thwarts plant diseases. Because of these parallels, these topics are jointly covered.

Due to large areas of single species, less interface with nature, and fewer mechanisms of natural control, bug infestation has become a major issue. The essence of the problem has also changed. The immediacy and severity of the solutions, for example, the application of harsh, less than environmentally friendly chemicals, have made this a sustainability question.

Rather than immediacy and harshness, agroecology is more a constant pressure pushing back these smaller crop-destroying organisms. For this, a step backward, examining both the approaches (more than one) and time frame, can be helpful.

Introduction to Agroecology: Principles and Practices

doi:10.1300/5553_05

MANAGEMENT

To eradicate insects or diseases in a sustainable manner, farmers should be willing to accept a certain amount of damage. However, if the threat becomes overwhelming, a saving response is required. Integrated pest management (IPM) was developed to assess threat levels and invoke a measured reaction.

The original form of IPM was to avoid blanketing fields with harsh chemicals upon finding a herbivore insect.[1] Instead, chemical use was to be localized and employed only where the insect populations were the largest. The aim was to reduce insecticide and application costs and, as a side benefit, to contain the environmental damage. Basically, this establishes an economic injury level (a minimal damage threshold) and implements control tactics, preferably before the minimal damage threshold.[2]

Rather than an identify-and-destroy strategy, more agroecological input adds more steps. For this, one must

- identify the pest or pests;
- find and exploit a cardinal control (a weakness in life cycle, a stage where predators attack, and/or a key predator);
- establish an economic injury level;
- look for alternative controls (one or more); and
- implement control tactics, before minimal damage occurs.

The first step, identification, pinpoints the most damaging pests and ranks these by the amount of havoc each creates. To keep these from reaching dangerous proportions, weaknesses in their life cycles are identified and, through field or crop management, these weaknesses are exploited. The fall or spring burning of crop residues may disrupt an overwintering organism. While a stationary egg, a harmful insect may be a tasty treat for a predator. These are maintenance defenses.

The initial containment may not totally eliminate a pest. Population dynamics, a common assessment in ecology, determine the expansion rate and when the per area density will become intolerable, and it is at this point that more immediate outbreak countermeasures are considered.

An improvement on earlier IPM, this strike-them-early and strike-them-late strategy is valid against an early emergence and a late population bloom. This is not the only course; agroecology offers more defensive depth and a wider array of options. These can be refined into a chain of anti-insect events, some in place and constantly active, others interjected, triggered by an IPM assessment.

BASIC INSECT COUNTERMEASURES

The countermeasures range from landscape-wide to local application and can be designed for broad range of pests or for specific scourges.[3] Most apply to insect dynamics; a few have application against disease spread. Briefly presented, these are as follows:

- Environmental factors
- Rotations
- Fire
- On-site predator insects
- On-site large predators
- Attractant crops and attractant biomass
- Decoy crops
- Trap crops
- Traps
- Barriers
- Passive repellant plants
- Active repellant plants
  —Traumatic release
  —Cut and carry
- Biowarfare
- Benign chemicals
- Harsh chemical

Although subjective, these are roughly ordered by their compatibility with the environment.

Environmental Factors

The condition in which plants grow can encourage or discourage unwanted organisms. Many factors fall under this imprecise heading, for example, tillage method, sowing date, and planting density.[4] For some of these environmental controls, the underlying control mechanism is not always clear.

Plant diseases and some herbivore insects prefer a certain microclimate. If not provided, this can help thwart a growth in insect populations. The type of tillage, as well as the type of fertilizers applied, encourages or discourages certain in-soil or on-soil fauna.

Robust plant populations, those supplied with ample essential resources, can better resist the attacking organisms. This boosting of defenses has been demonstrated when fewer pines per area and healthier trees increased resistance to pine beetles.[5]

On the contrary, less may be better. Grapes supplied with reduced irrigation water were not as attractive to harmful insects. The mechanism may have been in establishing water as the limiting resource. This decreased the nitrogen level in the leaves, degrading their palatability for insects. The fruit, not high in nitrogen, was not affected and showed no yield decline.[6]

Rotations

Common in agriculture are rotations to rid the soil of insects and diseases that shelter there, waiting for the return of their favored plant species. These organisms often cannot wait long. A lengthy delay, as with rotation or inserted fallow, or shorter one, as when a crop is planted late, are implementation strategies. A case in point, a short delay has been suggested as a means to reduce diseases in barley.[7]

Fire

Burning increases mortality of insects by killing them at a vulnerable, unable-to-flee stage in their life cycle. Commonly, this is an off-season measure when insects shelter in plant residues. The burning of crop residues has been found, by destroying eggs, to control the alfalfa weevil.[8]

This is not a universal tool. Fire destroys without discrimination and desirable predator insects may also succumb.[9] Therefore, burning is a preplanned event, aimed at certain insect species. The greater application may be with plant diseases that bide their off season in crop residues.

On-Site Predator Insects

Insect-eating insects can and do play a large role in the control of herbivore insects. The number of operative insects is impressively large. Although most of these regard herbivore insects as a convenient meal, others may achieve results though molestation, causing those that wish to dine quietly to flee elsewhere.

Common on-site predators include ants and spiders. As large groups, both offer numerous control options. Ants can be fearsome defenders and can even cause greater growth in some plants; use comes with caveats as some species devour large amounts of biomass. Ants also molest, driving mature insects to other locations, and will eat the nonmobile eggs and larvae of other insects.[10]

Spiders have equal potential and have a lighter touch as they do not consume plants, only other insects. Wasps also attack multiple species without crop harm.[11]

It is possible to commercially purchase predator insects. Ladybugs (eat aphids), trichogramma wasps (consume insect eggs), and lacewings (counter mites, thrips, and small worms) are among those that can be bought, shipped, and introduced into fields.[12]

On-Site Large Predators

Large predators figure into an insect control strategy. As many of these animals (e.g., birds and bats) are well adapted to a farmyard setting, this often requires a minimal introduction effort.

Natural wildlife can be encouraged by modifying field practices or by offering favorable habitats. Scattered trees can increase the effectiveness of predator birds, especially those that hunt from roosts. Habitat areas may also be part of an animal-based IPM. Not all large predators are wild; barnyard chickens, geese, ducks, turkeys, and hens love to snack on insects.

Attractant Crops and Attractant Biomass

Prey-seeking insects can be enticed to set up shop by planting preferred habitat species. Nectar-producing plants can host a variety of favorable insects, for example, ants are attracted to nectar and serve a deterrent function; vetch strips attract leatherwings that eat cucumber beetles; following an ancient Chinese practice, straw mulch near soybeans reduces damage; and dandelions lure ladybugs that feast on aphids. The list of attractant crops is conveniently large.

Decoy Crops

The purpose of a decoy is to keep insects and other creatures from eating the main crop. The long-term dangers of this are that insects will multiply and then, once the decoy plants are consumed, they will seek the main crop. As a result, decoy crops are employed only in specific situations and with other countermeasures. In this role, decoy crops become trap crops.

Trap Crops

Although seemingly counterproductive, the action of damaging insects may be encouraged. The immediate goal is to divert these insects away from a high-value crop. As with decoy crops, attractant crops are used in conjunction with another control method to concentrate populations of insects in an area for subsequent elimination. This is a "kill zone" approach to control.

Traps

Some insects can be enticed into physical and/or chemical traps. The entrapped insects are then destroyed. Synthesized pheromones as a lure may be the most potent weapon. Once enticed to a location, the targeted pests enter an imprisoning device or encounter a spot-applied, and therefore less intrusive, insecticide. As part of an ever-expanding list, insects susceptible to pheromone-based traps include apple maggots, bag worms, corn earworms, European corn borers, fruit flies, and peach tree borers.[13]

Not all snares need be high tech. In traditional farming of Southeast Asia, coconuts, each with a small hole, are buried. Unwanted ants enter and, once present in great numbers, the coconuts are removed; the ants are then killed and the coconuts repositioned to trap more ants.[14]

Slugs also fall victim to less sophisticated traps. Saucers filled with beer and placed at ground level attract and drown these leaf eaters. Melon peels, placed face down, also entice slugs for destruction later on.

Barriers

Barriers come in any number of forms. Besides stopping insects, these can halt diseases. Vegetative barriers are most common, but an unwanted crop species or even a strip of well-placed bareground can accomplish a specific task. Attractant species, for example, infested with plenty of ants, deter most other insects. Susceptible crops, protected by outer rings of nonsusceptible species, are also a barrier against disease and insect spread.

Passive Repellant Plants

There are literally thousands of plants that have the ability to repel specific insects (see, e.g., Plate 5.1). Effectiveness varies, and some are best against certain insect pests.[15] Positioned as part of an intercrop, their use can require a degree of plant-plant complementarity. Figure 2.1 shows the SIZ-determined placement of a repellant plant. A short list of temperate garden additions (along with the insect species repelled) includes the following:

- chives (repels aphids and mites)
- geranium (leafhoppers)
- horseradish (Colorado potato beetles)
- rue (Japanese beetles)
- sage (cabbage maggots and cabbage moths)
- spearmint (ants and aphids)
- thyme (cabbage moths)
- wormwood (cabbage moths and slugs)[16]

PLATE 5.1. A repellent species, in this application, insect-fending dill over broccoli.

Active Repellant Plants

Rather than placing repellant plants and hoping for the best, these can be positioned and their effectiveness increased through traumatic release. This is where cutting or damaging the leaves releases the volatile, insect-averting chemicals into the air. Being present in high concentrations, these should be more effective. This can be a periodic event or preoutbreak, when population increases threaten.

The other option is a cut-and-carry approach where banks of repellant plants are located nearby. When outbreaks threaten, the repellent biomass is cut and carried to where it will be most effective. The advantage of cut-and-carry over in-place repellant plants is that plant-plant complementarity is less of a concern, allowing greater flexibility in the location of the repellant plants and in the crops being protected. The disadvantage is in the monitoring and subsequent labor required to respond to an outbreak.

Biowarfare

As with plants and people, insects catch diseases, some of which are fatal. The idea is that these are introduced when outbreaks occur.[17] *Bacillus thuringiensis (bt)* is most prominent of the leaf-applied commercial insect pathogens. Another commercial product, milky spore disease (*Bacillus popilliae* or *B. lentimorbus*), lingers in the soil, finding and infecting future generations of in-soil grubs.

Resource-poor farmers can encourage biowarfare. Dead or dying caterpillars are collected, mashed into a diluted solution, and sprayed to spread the ailment. This anti-insect measure targets classes of insects while harming little else.

Insect diseases are not the only form of biowarfare. Pheromones sprayed around can cause confusion among specific insect populations, the hope being, in not finding mating opportunities, a reduction in their future numbers.[18]

Benign Chemicals

The application of some chemicals may not violate sustainability criteria. Some fairly benign chemicals are quite effective against insect and diseases. Environmental danger can be further mitigated if spot applied, that is, applied directly into a small area, not broadcast widely.

Much of this falls under the heading of home remedies. These are widely accepted countermeasures often utilized in noncommercial situations; the lists are long. Examples include a meal made from milkweed seeds that destroys nematodes and armyworms, and a solution of milk employed against cucumber leaf mold.[19]

Some have not been studied but seem to work well; others are quite well-known and studied. Diatomaceous earth functions by ruining the wax layer on beetles, causing death from dehydration. This list expands under home remedies, in this chapter.

Harsh Chemicals

Environmentally discordant agrochemicals are outside normal sustainability bounds. Besides risks to the health of workers, to local fauna, and in contaminating land and water, the spraying of toxic chemicals may not always have the predicted result. Chemicals are not selective in their action; these also kill predator insects, permitting the targeted insect to return in greater numbers once the chemicals subside. Poor dosing can cause menacing insects to develop insecticide resistance or result in other counterproductive side effects.[20]

Outside of broad applications, insular, contained use of a harsh chemical can be countenanced. As an internal component in a lure-and-kill trap, the environmental consequences, while taking normal precautions, can be kept within tolerable limits.

UNDERLYING CONCEPTS

Overcoming crop-seeking insects and diseases is made more difficult, not for the lack of options, but by an overabundance of control methods. Outside of the simplest, large-scale chemical spraying, the others require some understanding of the organisms involved. This is especially true when multiple controls, as with agroecological version of IPM, require some degree of cross harmony. It may be the lack of understanding that holds back the more environmentally friendly approaches. A number of concepts help sort through and put a friendlier face on this complexity.

Generalist-Specialist

Among the approaches, two aid in defining the problem and are the backbone for a series of defensive strategies. These are (1) a generalist approach where populations of all insects are kept in check and (2) a specialist approach which focuses on specific insect species.

A land user may decide to design agroecosystems and the landscape to encourage generalist predators, the idea being that these predators will go around eating all the herbivore insects encountered and, in the process, will keep crop damage well below the unacceptable threshold.

The opposite is where a problem insect stands out and measures to counter this insect species are put in place. An introduced predator that feeds upon the identified insect is one such measure. Multiple counters may also target, through an understanding of the life cycle, a threatening insect.[21]

Maintenance–Outbreak Defenses

The first of anti-insect defenses should be the most benign. These are the maintenance defenses that flatten population dynamics, keeping insect numbers at low base levels. Habitat management measures such as fire, crop rotations, intercropping, and/or encouraging predator-prey relationships are maintenance measures. These can be a generalist or specialist approach.

When an outbreak threatens, outbreak measures are quickly put in place or, if already present, enhanced. This can be done when the contained volatile chemicals of a placid repellant plant are brought out through traumatic release (leaf injury). Other outbreak defenses quickly brought to a site are insect-eating poultry or cut-and-carry repellant plants. Outbreak measures are usually specialist, relying upon IPM to determine the timing.

Predator Niches

How predators, birds, bats, or other insects, hunt crop-destroying insects is of great interest. This determines how these creatures are brought to bear and/or made more effective in their assigned task.

Bats, spiders (through insect-entangling webs), and some birds prefer to snag flying insects; some seek crawling or stationary prey; some attack early insect stages, that is, eggs, larvae, or caterpillars; owls and bats work by night; and some hunt in greater intensity at different times in the growing season, as when birds seek insects to feed large broods.

This approach is based upon niches, hunting and otherwise, that birds and other creature utilize. Proper alignment of these niches, as when insect populations peak, increases the performance of the predator and decides whether a line of attack is generalist or specialist, maintenance or outbreak.

Specific Interaction Zones

The concept of specific interaction zones puts a spatial and/or temporal dimension on the various controls. Besides addressing climatic gains, a SIZ marks out the effective range of an insect-repelling plant species, determining how far an insect-eating insect will travel seeking prey, or puts put physical dimensions on the hunting niches of insect-seeking fauna. For the latter, the niche may describe the preferred zone, that is, ground level, in flight, near or away from tall vegetation, etc.

In a cropping area, areas not covered or protected are at greater risk. Simple coverage may be effective where two hedges, each harboring predator insects, are in such close proximity that their zones of influence meet or overlap.

Better yet is whether the cropping area has different overlapping countermeasure, either temporal or spatial. This provides more protection, especially if the countermeasures are complementary (with predators, what one misses the other catches; with repellant plants, what one does not repulse, a second does). The planned reinforcement of protection (two or more working in tandem) is brought to the fore through prongs or tracts.

Reinforcing Prongs

Generally, SIZ-based countermeasures constitute planned side-by-side or overlapping defenses but are, in essence, separate. The same holds true with a series of defenses in which, as one is in danger of being overwhelmed, a second, possibly transitory, measure comes into play.

The concepts of prongs is where, acting in unison, two or more defenses are more effective together than if each is employed alone. There is an increased potency that comes when one defense heightens or reinforces the effectiveness of a second.

European literature from the 1600s indicates that fowl (geese, ducks, and chickens) roamed free through most crops. Besides directly eating insects, their movement would have caused some to flee. In taking flight, these would have been exposed to the second prong, the natural farmyard fauna that hunt flying insects (e.g., chimney swifts, barn swallows, and bats). This is also an example of niche feeding.[22]

In a vegetative version, a trap crop is employed to lure plant-eating insects away. In a push-pull arrangement, movement can further be encouraged by strategically placed repellent species, ushering unwanted insects away from the crop.

Such push-pull arrangements have been documented. In Africa, maize and sorghum are protected from stem borers by a repellent and a trap species. In this case, the push components include molasses grass, silverleaf, and/or greenleaf. The second prong, the pull component, is napier and/or Sudan grass.[23]

A third prong, and a further enhancement in a push-pull arrangement, might be a kill zone. This zone, well stocked with predator insects, is located near the trap crop.

Tracks

It is possible to stack countermeasures into a series of defenses. It goes without saying that, within agroecological precepts, these should start with the most ecologically benign. If an outbreak threatens, one proceeds on to those that are severe. Unless exceptional conditions warrant, most of the starting mechanisms should be generalist, going on to specialist measures if an outbreaks occurs. Each of the short-lived, specialist countermeasures is triggered by a separate, predetermined IPM threshold.

Within this context, the defenses or tracks form parallel sets of options. Figure 5.1 shows this diagrammatically. Each square represents a countermeasure and the down arrows represent a series of preplanned defenses. The upper blocks, in gray shade, are perma-

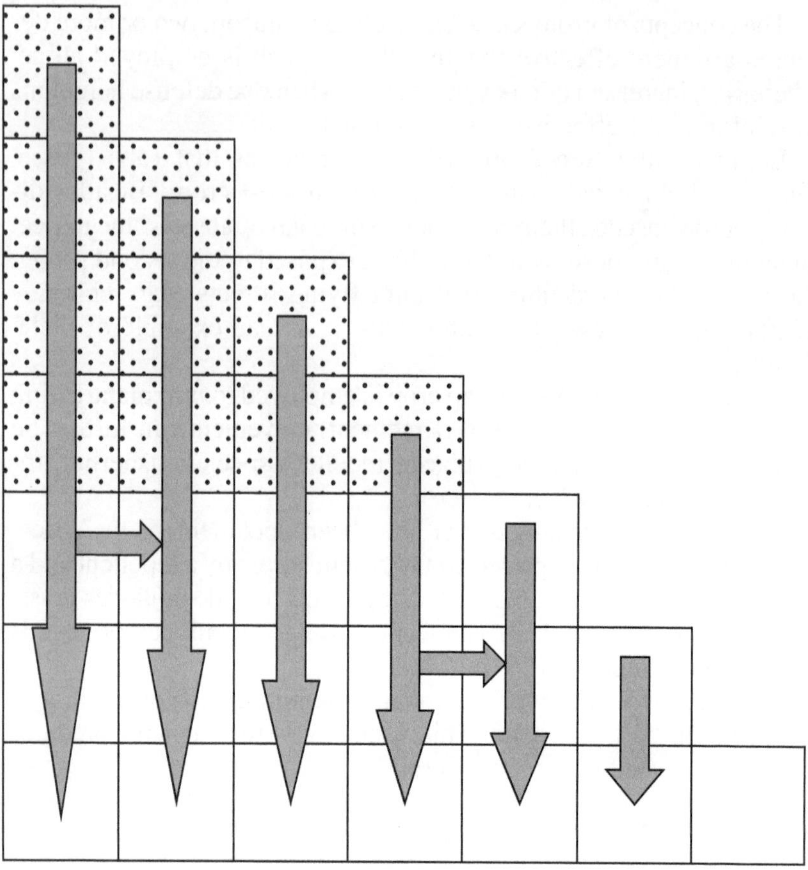

FIGURE 5.1. Stacked defenses, where each block represents one insect countermeasures, that are strung together into a series of reinforcing defenses (the vertical arrows). The upper, gray blocks are generalist, bent toward maintaining an insect status quo. As the situation evolves, the lower, temporary, specialist measures are bought to bear, including the switching of tracks (side arrows).

nently in place and formulated as generalist countermeasures. The lower blocks are latter applied measures, introduced as needed.

The tracks of the left involve more steps, each being environmentally less severe; those at the right are harsher, requiring fewer. The right-most mechanism represents an agrochemical application that kills all and leaves little; hence, no other remedy is required.

A multistep top-down track might include (as part of comprehensive strategy) the following:

1. nearby habitats for predator insects (generalist–maintenance)
2. movement corridors (generalist–maintenance)
3. cover crops or mulch to promote predator movement (generalist–maintenance)
4. repellant plants to relocate harmful insects closer to corridor kill zones (generalist–outbreak)
5. conveniently located cut-and-carry strips (generalist–outbreak, as needed)
6. introduced predator insects (specialist–outbreak, as needed)

A shorter version track with fewer blocks might include the following:

1. spread prevention barriers (generalist–maintenance)
2. bird and/or bat houses to retain unique predator fauna (generalist–maintenance)
3. trap crops to lure unwanted species to where predation is most severe (specialist–maintenance)
4. physical traps to capture those missed (specialist–outbreak, as needed)

As events unfold, a single track may not be the best course of action. It may be advisable to jump tracks (side arrows, Figure 5.1) to better address a specific and/or unanticipated insect threat. Once developed, jumping to other tracks, as an IPM-triggered event, may underwrite most insect control strategies.

Miniecosystems

Within larger agroecosystems exist smaller subsets. These internal miniecosystems can be devoted to insect control. Building one of these without a starting point is almost impossible; however, there are species upon which to construct a minisystem.

In ecology, there is the concept of the keystone species, the plants and animals upon which ecosystems exist. The consummate example is the beaver. These animals, in building dams, convert forest lands into wetlands and, in so doing, support a wide array of water-loving plants, insects, fish, birds, and mammals. These in turn support more fauna, those that subsist on or with the water-loving organisms.

Although nothing as dramatic occurs with disease and insect control, miniecosystems do exist at the behest of one keystone species.

The citrus ant, *Oecophylla smaragdina,* is the keystone for such an ecosystem. Found in China, this is not just a predator on prey, but forms a fairly complex mini- or subecosystem. Within this, herbivore insects are not entirely missing. In this case, the ants encourage and tend mealybugs. Although the mealybugs are damaging, this is not significant. The value of the ants is in their attacking other herbivore insects while not excluding smaller predators such as ladybugs. This miniecosystem is reportedly effective at containing all insect threats.[24]

A less documented case is the woodpecker. This bird provides a high degree of direct control (see avian control, later in this chapter) but, in attracting other birds, is a foundation for a miniecosystem. Chickadees, with their 70 percent insect diet and their liking for woodpecker-constructed nesting cavities, are a follow-on component.

Miniecosystems may be established not with insect-eating fauna but with plants. Often thought of as a scourge of agriculture, flowering weeds (or alternatively, a purposeful mix of less crop-competitive cover crops) can provide a habitat for a wide range of insect-attacking insects.[25]

EMPHASIS

The countermeasures provide ample options and prongs and tracks are effective implementers of multiple controls. Still, there are questions on how this is best done. For the sake of simplicity and to avoid unintended consequences (such as benign chemicals driving away insect-ingesting birds or repellant plants warding off beneficial predator insects), some advocate the keeping of emphasis within one group or category of controls.

The advantage of this strategy is that it requires less understanding of ecological interspecies relationships. Some categories seem to jump out as points of reference or emphasis:

- Repellant plants
- Home remedies
- Insect-on-insect dynamics (entomo-control)
- Birds (avian-control)[26]

Repellant Plants

Given the wealth of repellant plants, this can be the basis of control. Much of this falls under intercropping, pairing, and planting one or more generalist repellant species within crop fields. If needed, a nearby planting of a specialist control plant can be cut and carried when an outbreak occurs. A push-pull arrangement can be part of this strategy.

All-purpose, generalist, repellant species include the following:

- anise
- catnip
- garlic
- geraniums
- hot peppers
- marigolds
- nasturtiums
- petunias
- rosemary
- tansy
- tomatoes

Also documented are those specialist plants that fend off certain insects. Augmenting the list previously provided, additional examples (and the insects repulsed) are as follows:

- borage (Japanese beetles)
- chrysanthemum (nematodes)
- dill (aphids and spider mites)
- leeks (carrot fleas)
- mint (cabbage loopers)[27]

These are only a small percentage of the plants available for such duty.

Having a wealth of insect-parrying plant species opens other avenues. Given the choices and options, including temporal SIZs, weed and erosion control should be annexed into a comprehensive, anti-insect undertaking.

Home Remedies

There are instances where individuals, groups, or societies put emphasis on the so-called home remedies. These are readily available. Some are common household chemicals, many are utilized for other purposes, and others derive from native plant species. Most are fairly benign but with varying effectiveness.[28]

A generalist outbreak can be countered by burning leaves of tobacco under the affected plants. Another broad-based counter is a spray made by blending garlic bulbs, onions, and cayenne powder, adding soap to allow the mix to adhere to plant leaves.[29]

A host of specialist remedies are offered. Among these are the leaves of the skunk cabbage, placed in maize field as a decoy for cutworms. Slugs are countered by covering the ground with ashes or lime. Yellow or striped bugs are discouraged by spraying with water saturated with cow dung. Turnip or garden fliers are handled with sawdust, snuff, soot, or sulfur sprinkled on to a plant wet with dew. Plum curculio, an insect that strikes a range of fruit trees, is controlled by placing tomato vines under the affected trees. Banana peels counter aphids.[30]

Entomo-Control

Encouraging large number of predatory insects, with proper protocols, can be the basis for entomo-based control. There are cases where a single species sweeps away most of the damaging insects. The fire ant, *Solenopsis geminata,* is a generalist control in maize-squash agroecosystems of Mexico, effective at sweeping fields of most herbivore insects.[31]

Spiders, previously mentioned, can find a cropping environment a more than suitable habitat. Mixing spider species, by dividing the territory by plant species, intraplant elevation, or some other means, could make this a vigorous, multipronged defense.[32]

Of the categories, entomo-control has the largest aggregate of supporting literature.[33] Unfortunately, questions exist on implementing broad-spectrum ants and spiders. In contrast, the list for temperate zone specialist insects is quite long and includes the following:

- predatory mites (against a range of mites)
- snake flies (against aphids and caterpillars)
- assassin bugs (flies and caterpillars)
- big-eyed bugs (leafhoppers and spider mites)
- carabid beetles (cabbage worms, cutworms, slugs, and snails)
- parasitic nematodes (soil-dwelling corn rootworms and root weevils)
- mealybug destroyers (mealybugs)[34]

Of all the emphasis strategies, entomo-control, employing either a few or a broad mix of insect-eating insects, may be the most advocated and utilized. Introduced insects supplement, but may not strengthen, other ecological happenings. In contrast, the habitat needed to encourage a broad, permanent, predator-insect population can embody, through a well-planned ecosystem, weed control and soil retention.

Avian-Control

Clearly, bird species consume insects. Avian controls, either generalist or specialist, are often well within reach. Birds are niche feeders—what one does not consume, another will. This applies to bugs and unfortunately to agricultural output as well. The secret is in emphasizing one and not the other.

As discussed, barnyard fowl, chickens, ducks, and geese, partake of ground-dwelling insects. Those insects that choose to take flight are dispatched by shallows or swifts, birds that prefer to capture flying insects. This two-pronged attack is more of a generalist approach.

There are specialist birds that can be employed.[35] The housewren, with a 95 percent insect diet, has a taste for beetle eggs and larvae. With large broods and voracious appetites, these birds can be a component to control potato beetles. They can be attracted to site by properly designed and placed wren houses and a close tree-crop interface. Planting mulberry trees near nests is reported to keep wrens happy.[36]

(a)

(b)

PLATE 5.2. Habitat possibilities; *(a)* trees killed to house woodpeckers and *(b)* an agricultural birdhouse designed and positioned to entice and retain bug-eating wrens.

Woodpeckers have a 90 percent insect diet. These can be encouraged to stay around in larger numbers by providing large dead trees for nesting and by feeding through the winter. These are brought to an outbreak site by feeders that are operated daily but only for a few short hours. The key here is to whet, but not appease, their appetites. Once the feeder is empty, these birds then look toward the most immediate insect food source.[37]

Such birds occupy a ground-hugging niche. An in-flight hunter, the tree swallow, provides a second prong when an anti-insect avian undertaking needs reinforcement. With a 90 percent insect diet, tree swallows are drawn to fields with well-designed and placed birdhouses (see Plate 5.2) and by planting bayberry bushes as a supplementary feed source. Perches from which to hunt, for example, bushes, fence posts, or purposely placed T-shaped poles, may coax birds to outbreak areas.

Integrated Control

Under emphasis, cross-category controls are possible, but their use is limited to where in-category controls are not known or not effective. This can be to combat different specialist insects or to counter different threats by cross-assigning groups. For the latter, herbivore insects can be controlled with home remedies while, at the same time, raptor birds are encouraged to eat the rodents.

As emphasis strategies are entered into, particularly biodiversity in repellent plants or entomo- or avian-control with accompanying habitat, favorable cross-relationships begin to emerge. In their mature, developed form, integrated controls should underwrite disparate ecological tasks, only part of which resolve an insect problem.

In a case study of an organic vineyard, chickens ate the cutworms; owls and hawks, fostered in number with boxes and perches, ate gophers and moles. Covercrops, queen anne's lace, alyssum, and clover, prevented erosion while harboring spiders and ladybugs. The cover crops also fed sheep, which, in turn, kept the cover crop and any weeds from overwhelming the grapes.[38]

Chapter 6

Agroecological Concepts

The previous chapters have outlined a series of agrobiodynamic working principles. This chapter sets out to explain how these principles connect with practical applications.

The individual plant and the selection methods, through genetic change and an ordering of desirable properties, help in understanding the possibilities as individual agroecosystems are formulated. Also, a ranking of agroecosystems, based on ecological and other properties, positions agroecosystems and the crops contained where they are most advantageous.

Though all these help, it is the packaging of agroecosystems into neat catagorical blocks that manages the complexity inherent in agroecology. As economically and ecologically secure ecosystems, these blocks shape rural landscapes just as bricks form buildings.

AGRICULTURAL PLANTS

As one enters into the world of agroecology, a number of concepts emerge that focus on agricultural plants. The evolution of productive species is ongoing and is part of a larger picture, as is how species are selected and utilized in a productive (farm) setting.

Domestication

Domestication is the process by which a plant goes from its wild state to become an accepted and useful agricultural addition. Maize is a prime example. Although the wild parent, teosinte, does not resemble the plant found today, the earliest efforts, some 9,000 years ago,

Introduction to Agroecology: Principles and Practices

doi:10.1300/5553_06

produced recognizable maize.[1] All agricultural crops have undergone this process; some are close to their natural forms, while others have been subject to considerable domestication-related changes.

Human needs have been well served by improving upon wild plants. Aiming for higher yields, better nutritional content, expanded choice, tastier foods, easier propagation, expedited harvests, less risk, and so forth, humanity has acclerated this improvement. These changes paved the way for increased quantity of fruits per plant (as with an improved harvest index), less lodging (through shorter, stronger stems), ease of harvest (uniform ripening, dwarf tree crops with easy-to-reach fruits and nuts), and climatic resistance (through an accommodation with drought, flooding, winds, and temperature).

Domestication is a slow process in the hands of farmers. However, the results are quicker when research driven. In the mid-twentieth century, a concerted attempt was made to accelerate domestication. Through this green revolution, considerable success was achieved making common crops more amenable to conventional farming methods.[2]

Although headway through increased yields is undeniable, there are untended consequences. Most commercial crops are now far removed from their ecosystems of origin: many lack the ability to survive in or be part of a natural ecosystem. Unsuitable for life in the wild, most grain crops no longer throw their seeds; unnatural varieties of bananas, breadfruits, oranges, and watermelons are without seeds.[3]

Less noticeable are those undesirable characteristics that relate to competitiveness, that is, the ability to thrive, or even survive, in a competitive (biodiverse) environment.[4] As a by-product of genetic change acquired in a monocultural setting, higher yields may not transfer when the same varieties are intercropped.[5]

Heirloom Varieties

Over the past 10,000 or so years, agricultural domestication has proved exceptionally fruitful.[6] Transportation of species across diverse regions coupled with informal, but intentional, efforts and accidental events have produced considerable genetic variance and a multitude of common crop varieties. The movement of the potato throughout temperate South America, maize across temperate North America, and wheat in the Middle East, Europe, Africa, and Asia has led to hundreds of varieties of these plants.

More than 600 varieties of wheat have been described, and similar tallies exist for potatoes, maize, and other staple crops.[7] These, along with multiple varieties of tomatoes, beans, apples, peaches, etc., represent a human, quality-of-life resource, embellishing the diet in taste, vitamins, and in a host of other ways.[8]

As an offshoot of this process expanding the varieties, new and completely different food sources occasionally emerge. As variations of the same species *(Brassica oleracea),* cabbage, broccoli, brussels sprouts, cauliflower, kohlrabi, and collards feature different parts (leaves, buds, stems, and flowers) of the same plant. The same relationship holds with amaranth, a plant that can stress either eatable leaves or the grain.[9]

The evolution and expansion of useful plants is an ongoing process. A case in point, ancient Roman varieties do not resemble their modern counterparts.[10] Farmers in remote regions actively cross varieties, and the new varieties listed in seed catalogs shows this is ongoing in and away from the agricultural mainstream.[11] Along the way, this process produces many failures; some fall by the wayside, while others exhibit properties of interest and remain.

In agroecology, the multiplicity of heirloom or heritage varieties is an understated resource. Old, even unused varieties provide the genetic base, not only to improve the diet, but from which to seek ecosystem-friendly plants.

Despite the fact that humans have been active in agriculture for a long time, not all useful plants have been subject to exploitation. Foods continue to make the jump from their wild state to market prominence: macadamia nuts and kiwi fruits are recent examples.

There are species in semicultivation (gathered and known only in local markets) or in the wild that have not yet entered the domestication process.[12] Once noticed, these genetic reserves can enrich human experience and benefit future farmers.

DPCs

In regard to user needs, all plants have desirable plant characteristics (DPCs). Most have output attributes (food, fuel, and fiber) that encourage their use in an agriculture situation.[13]

Other plants are utilized not because of a useful output but for the service they offer. For instance, a large shade tree is so positioned because it protects people and domestic animals from the hot sun.

The lists of DPCs can be large and detailed. Usage starts with the fundamental properties, proceeding to the specialized lists.

Fundamental DPCs

Generally, if a plant species is to be welcomed into an agricultural environment, it must

- provide viable amounts of high quality output and/or strong facilitative services;
- resist diseases and insects;
- withstand adversity, for example, wind, drought, flooding, etc.;
- tolerate or overcome weeds;
- thrive on a range of soils, climates, and sites (especially for the sites in question);
- have ease of propagation in a horticultural situation; and
- not become a weed or nuisance.

Not all agricultural plants fit this somewhat idealistic model. Still, this list underlies much of the selection process.

Productive DPCs

The DPCs for productive species resemble the fundamental list with one prominent addition: first and foremost a productive species should yield marketable, useful, or interesting end-use output.

It goes without saying that if a food type is desired, there is little substitute; however, within a species, subdivisions become important. Culturally, the preferred ordering might be

1. taste,
2. texture,
3. appearance,
4. ease of cooking,
5. storage ability, and
6. alternative products (a useful stem, beautiful flowers, etc.).

The sweet potato is a staple food in New Guinea. With multiple varieties to choose from, people recognize subtle taste differences and have preferred breakfast, lunch, and dinner varieties. Across the world, the unheralded parsley is usually used as garnish, whereas the French, with multiple varieties, have regional or stipulated end-use classifications.

A commercial farmer in a less demanding market may see an end-use reordering. A commercial variety may be selected because the fruit transports and stores well (an end-use characteristic) and/or for ease of propagation (a fundamental DPC). Tomatoes grown for off-season markets, those that must be transported great distances, are often faulted for their lack of taste but pass the critical transportation test.

Facilitative DPCs

As with the shade tree, not all useful plants provide an end-use output. One plant, a facilitative species, can aid a second, productive species. As discussed in Chapter 3, this can be either through the transfer, direct or indirect, of essential resources, or through the mitigation of risk.

Beyond the fundamental DPCs, there is a sought after or standard facilitative package. Each purely facilitative species is expected to

- promote positive nutrient cycling;
- aid in nutrient capture (nitrogen fixation is common);
- help with insect control;
- assist in microclimatic modification, for example, wind-speed reduction; and
- suppress or eliminate weeds.

As with the fundamental list, not all can be found in a single species. In contrast to productive species in which the output is overriding, with more species to choose from, more elements of the standard package are expected. The actual ordering of these properties will vary with species and need.

In practice, productive species are often employed in facilitative roles. When this occurs, the difference between the productive and facilitative roles lies solely in the ordering of DPC. If production is first on the list, a species is generally not regarded as facilitative.

Specialized Lists

As needs advance, DPC lists change. Specialized lists subsist for any number of uses. Take the urban or household shade tree the in-use DPCs include the following:

- thick foliage (a high LAI);
- pleasing appearance;
- sturdy branches and trunk (to reduce branch fall or toppling);
- no unpleasant, drippy sap;
- fragrant, long-lasting flowers;
- useful fruits or nuts; and
- favorable habitat for songbirds.

Some trees clearly fall within these DPC parameters. In the tropics, many prefer mango trees; in North America, the American chestnut was once favored; and in middle Europe, the European basswood is found in large numbers.

Alternative use-oriented lists are possible and find application. Desirable soil characteristics is a version associated with the rotational alternatives and applies to the residual condition of soil after a crop is harvested.

Ranking

In selecting agricultural plants, land users implement some form of informal ranking system. These vary in manner and outcome.

To decide which variety is the best, the tie-breaking method is commonly employed. Species or varieties are compared on the basis of a few key requirements. The varieties that lack these properties are eliminated from contention. For those with the key attributes, second-tier characteristics break any tie. If more than one choice remains, a third selection helps to single out the best.[14]

From Uganda, once it is determined that a banana variety will grow on site (the first-tier decision), commercial farmers seek bunch size and crop maturation time while noncommercial growers look for stand longevity, taste, and maturation time (second-tier selection criteria).[15]

Generally, ranking is an unappreciated side of agrobiodiversity. This rises in importance when one considers the amount of available genetic resources, varieties that often go unnoticed, and the need to select from this vast pool.

AGROECOSYSTEMS

Rather than being just a collection of useful plants, agroecosystems have a regulated form and purpose.[16] The tendency is to keep to a well-trodden path, that is, to maintain current land use, simply because this is a well-demarcated course of action.

In going beyond mainstream agriculture into the sphere of biodiversity, the formulation possibilities expand exponentially. However, one challenge of agroecology is to weed out those systems that weakly address user needs while promoting those that serve best.

There is the conceptional view that where one agroecosystem is weak, another might be strong, where one leaves off, another begins. Placed end to end, these become a continuum of variations. The best one can hope, given the extensive list of intercropping possibilities, is that enough is known to define points on or sections along this agroecosystem continuum. These are the building blocks mentioned in the introductory part of this chapter.

Classification

Individual agroecosystems can be ordered or ranked in importance by a number of criteria. Put together, these help define ecosystems. The broad ranking criteria are as follows:

1. primary outputs (crops, trees, animals, fish, etc.);
2. revenue or profit potential;
3. landscape role (productivity, soil improvement, etc.);
4. economic orientation (see Chapter 7);

5. temporal reach (e.g., time in ground, time spent producing, etc.);
   a. primary crop (annual, biannual, or perennial),
   b. overall system rotation,
6. fundamental ecological principles (competitive production, facilitation, etc.);
7. user objectives; both in
   a. productivity—primary and secondary crops,
   b. the land use problem(s) resolved,
8. environmental benefits received (both tangible and intangible);
9. underlying ecological dynamic(s); and
10. spatial or temporal pattern.

Multiple classifications imply an ordering, one where the more important criteria come to the fore; those of least concern are secondary.

DAPs

Combining these criteria, one can derive and define various agroecosystems. As an ordered list, these become the desirable agroecosystem properties (DAPs).

The first desirable agrosystem property is its useful output. A farm is not complete unless there is a profitable mix of economically viable agroecosystems.

A fully yielding landscape is seldom the norm; productively inert systems are commonplace. These can be grass strips that hold soil or promote water infiltration. Bands of tall trees that retard surface-level air movement also qualify a listing of ecologically favorable DAPs when an agroecosystem

- resists soil erosion (water and/or wind),
- has strong infiltration or water retention ability (low runoff),
- does not yield to harmful insects, and
- resists unwanted plant additions (weeds).

Properties, such as sustainability or being insect unfriendly, are the result of positive amalgamations of ecological mechanisms (as discussed in previous chapters). To be precise, some of these properties stem from the DAPs of component species, others derive from the temporal and layout dynamics of ecosystems, and still others evolve from the resulting overall ecosystem (e.g., as accorded through ecosystem governance).

The susceptibility to any one danger may be determined based on one phase or an average across the life of a system. The erosion-susceptible bareground phase may decide if a system is to be placed on an erosion-prone site.

Ranking

Agroecosystems are ranked in the same way as useful plants: stronger properties are listed first and the rest appear in a descending order. As with DPCs, this is an imprecise process, one made difficult by undefined properties, numerous variations of agroecosystems, and an often unclear track toward an optimal end use.

Substitution

Changing the spatial pattern and the species composition of an agroecosystem changes the order of the DAPs. This brings about substitution, purposely altering an agroecosystem to reorder the DAPs. One reason for this is to accommodate changing goals. In essence, one slides along the continuum of possibilities seeking an improved or more agreeable DAP outcome.

In adopting agroecology, farmers frequently seek relief from costly and/or environmentally detrimental inputs. Substitutable inputs include the following:

- fertilizers,
- insecticides,
- herbicides,
- fungicides,
- manual weeding, and
- irrigation.

The agroecological ideal is where nature is assigned tasks and accomplishes these, such as replacing any or all of the above, through natural dynamics implemented through biodiversity. Many alternative systems utilize ecodynamics to supplant farm-external inputs. Inclusive in this is the movement of essential resources between plots.

With substitution, the ecosystem remains but in an altered form. This can involve swapping[17]

- one plant variety for another (e.g., one less prone to a resource shortfall and needing less irrigation and/or fertilizers);
- one input for another (e.g., replacing a costly fertilizer with homegrown biomass or manure);
- an input for more biodiversity (e.g., including one or more facilitative species and reducing insecticide use); or
- forms of biodiversity (e.g., replacing the annual planting of a facilitative species with more cost-effective trees).

Changeover

Substitution is not always the answer. When change becomes inevitable, adopting an entirely new agroecosystem may be warranted. This occurs if the current practice is

1. not in line with an upcoming economic or land-use situation (e.g., increased demands on the soil are draining the nutrient content or the market value of an existing crop is in decline), and
2. the cost of a changeover is compensated for by the increased yields of a new system (e.g., profitability is substantially higher with a different cropping system).

Changeover is fraught with danger. This is not the swapping of one monocrop for another and, unless the land user has had previous experience with the new system, this can be a fate-laden plunge into the unknown.

AGROTECHNOLOGIES

Rather than risking a fate-laden plunge, changeover can be a gentle step to a higher plane. Agrotechnology-based agroecology eases the transition by providing safe and solid steps.

An agroecosystem changeover is best done by finding superior points along the continuum of possibilities. Each point represents a standard agrotechnology; each point is, through study and observation, a solid landscape addition.[18] Many come with a high degree of biodiversity and with ecological safeguards.[19]

Although a concept in development, the cropping or productive agrotechnologies are a means to focus on the serviceable, but still ecologically distinct, land-use options. In essence, these agrotechnologies avoid the hassle of probing the full continuum of agroecosystem possibilities. Instead, the focus is on a few ecologically distinct points, each with a unique agroecological mechanism and clear track toward a goal. As user-safe havens, these offer a clear view of the connection between agroecosystem design (component species, spatial and temporal pattern, canopy arrangement, etc.) and outcome (the land-use problem addressed, economic expectations, etc.).

Subdivisions

For discussion, four independent agrotechnological categories are proposed with subdivisions:

1. Vegetative (spatial)
   a. Principal-mode
   b. Auxiliary
2. Temporal
3. Land modification
4. Mechanical

The vegetative (spatial) agrotechnologies can be principal-mode. Vegetative agrotechnologies are the workhorse plots that provide all or most of the farm outputs. Offering little or nothing as useful products are auxiliary agrotechnologies. These are separate but neighboring ecosystems that provide facilitative services, directly or indirectly, to the yielding systems. These can be a common, and often unnoticed, landscape feature.

Temporal agrotechnologies describe how vegetative systems function across time. This can be seasonal and, for some in the temporal category, the jump from one time period to a distant other.

Land-modification agrotechnologies are those physical changes in the land that are made to accommodate agricultural usage.[20] These may also go unnoticed as they underlay (as with terraces) or accompany (as with water channels) principal-mode systems.

The mechanical agrotechnologies employ machinery to overcome obstacles to production. Hydroponics, sprinklers, or drip irrigation are among those technologies that might be included.[21] As the focus of this text is more on accommodating nature, this group is mentioned only in passing.

Relationships

The main groupings (i.e., vegetative, temporal, and land modification) are not exclusive; an agroecosystem can fit within each of these three classes. Generally in describing a practice, one of the three aspects is focused upon. A rice paddy describes a type of land-use modification (i.e., the paddy), the contents (e.g., monoculture rice), and rotations (if any) are assumed.

Scaling

At the heart of each spatial agrotechnology lies a single ecological dynamic or a suite of effects. These can be the apportionment or transfer of one or more essential resources, a protection conferred, the substitution of an input by a natural process, or some other ecological happening. This occurs whether the plants are small or large.

With scaling, many of the seasonal intercropping agrotechnologies have ecological parallels with the larger, longer lasting agroforestry agrotechnologies. Clusters, based on efficiency in light distribution, and barriers, based on the crop protections afforded, are systems that scale upward or downward without loss of purpose.

Others have parallels but do not scale as well. Complex agroecosystems, based upon ecosystem governance, lack duration when replaced each season. The longer-term version, the agroforest, has more opportunity in advancing the underlying ecomechanisms.

For presentation purposes, an intercropping version is presented separately from the agroforestry variation if both are feasible and relatively commonplace. For some, the parallel version may be a rarity or offer less ecological and economic advantage. These are not described and must be extrapolated by the reader.

Understandability

A key rational for agrotechnologies lies in understandability.[22] Understandability refers to the ability to formulate a favorable starting design and to take assenting measures to insure an auspicious outcome.

With what may seem like underwhelming simplicity (at least in comparison to the other cropping options), monocultures are quite complex. Factors such as location, plowing method, the timing and density of planting, weeding, and the timing of the harvest are among the decision variables that help ensure an economically successful outcome.

An increase in the number of species (biocomplexity) causes an exponential rise in the decision variables and in the understandability hurdle. Set agrotechnologies come to the aid of those who do not have a profound knowledge of the local cropping environment and how to implement multispecies systems.

Each agrotechnology is categorized in order for farmers to know where and when it is best employed, what it requires (e.g., inputs), and what is expected (e.g., yields). When these agrotechnologies are presented with few formulations and suggestions on location, inputs, expected yields, etc., farmers can successfully utilize these without becoming enmeshed in what can be overwhelming biocomplexity.

Design Packages

Agrotechnologies gain strength from the inclusive application packages. This package focuses on the DAP properties. A design package is subdivided or grouped into the following:

1. Land-use problem:
    a. land relief (for hillsides, good erosion and infiltration properties and no system requirement for wind or light based directional orientation)

   b. soil sustainability issues (one or more of the component species should provide the soil-limiting nutrient, e.g., fix nitrogen)
2. Site and socioeconomic requirements:
   a. inherent soil fertility (richer soil favors multiple output systems, less endowed soils do better with a facilitative intercrop)
   b. seasonal rainfall (e.g., when in short supply, the proposed ecosystem should be miserly with water)
   c. soil pH (the ecosystem should be accepting of the soil pH and resulting nutrient balance) for the socioeconomic profile
3. Socioeconomic requirements:
   a. primary crop (whatever the ecosystem, it must be favorable to the preferred crop)
   b. secure land tenure (this is applied whenever the ecosystem is a long-term project)
   c. ample labor (the amount and timing of labor varies with each ecosystem and one, high or low labor, is selected accordingly)
   d. confinement of animals (systems can resist or fall victim to roaming animals)
   e. need for firewood (wood can be secondary output and selection plus)

If these and some still to be defined conditions are met, a standard agrotechnology might offer a complete package.

However, this is not always the case. A standard agrotechnology may provide only a starting point. An agroredesign to best fit the user situation may be in order (as in substituting the tree species), or an entirely different agrotechnology may be considered.

For most of the described agrotechnologies, the packages are understood but not carefully itemized. For a few, for example, alley cropping, many of the elements have been formally stated. For others, there is little to go by.

Some are more flexible than others. The agroforest, the most biodiverse of cropping systems and one of the most site agreeable, has a very concise package. The agroforest will address almost any land-use problem, has no site limitations except that the crops and trees will grow on the site, and, for the economic profile, offers abundant output of more than one product.

The monoculture, as an agrotechnology, also offers great crop and site flexibility. However, a monoculture alone, without further elaboration, addresses few environmental problems and may introduce many not present. With substitution, some of the immediate dangers and drawbacks (insect attacks, erosion, etc.) can be eased.

Agrotechnological Adjustments

For many agrotechnologies, the stated parameters or the standard design is not explicitly voiced or even known. It is these parameters that provide a starting point for modifications.

The standard design for hedgerow alley cropping has maize with leucaena hedges. The standard interhedge distance is 4 m; the hedge height upon cutting ranges from 0.25 to 0.5 m. Locationally, the hedges contour the slope.

Modifications help the system reconcile with the site, improve upon the environmental and ecological performance, or bring the system in line with local labor practices. In one alley cropping case study, farmers preferred ground-level pruning, employed fire, and changed the hedge species.[23]

Maize may not be the staple crop in all areas. This does not eliminate an alley system as the agrotechnology of choice. There are a large number of candidate species, some better than others, but whatever primary species is selected, it must be reconciled with the overall alley design including a compatible hedge species.

Agrobiodiversity Adjustments

Not all gains come from dimensions and management; component species are often interchanged. For crops, economic factors (i.e., market demand) may provide the impetus. For facilitative species, a better match with soils, climate, and/or a companion crop may trigger such a change.[24]

Rather than simply switching species, most agrotechnologies provide opportunity to add biodiversity. This is done without destroying the ecological essence of the system. Strips can be composed of more than one productive or facilitative species. The same holds true with hedges.

The options and opportunities of agrotechnologies are enormous. A few of these options are explored in subsequent chapters. The reader should keep in mind that greater biodiversity is always possible and, if well thought out, a positive step toward future success.

Chapter 7

Monocultures

In all regions of the world, the monoculture is the mainstay of agriculture. There are reasons for this. Monocultures are very good, but not always efficient, at producing a lot of one output.

Over the past 50 to 100 years, the desire for simplicity in research and teaching, as well as in practice, has focused much effort in this area. Since much of the now available seed stock, farm machinery, and research is for plots of one species, there is often the accompanying perception of economic superiority.[1]

In advancing this perception, refinement on the single plant system comes through precision in inputs and the championing of the super plant. The latter is where one species is genetically armored, increasingly through genetic engineering, against outside threats. Those threats not countered through genetics are answered through external, often agrochemical, inputs. These lines of development, more of an ex-nature approach, further the notion that the monoculture is the only viable commercial alternative.

The image of a boundless and genetically superior monoculture is ripe with fallacies. The main counterargument is that farms composed of field after field of a single species are seldom environmentally complete. There are dangers in offering potential crop-harming organisms such an inviting target. Also looming are the climatic threats.

In any ecological discussion of monocrops, less acknowledged mainstream perceptions must be challenged. Many see the monoculture as a single agrotechnology. However, considerable divergence is masked by preconceptions. This occurs when underlying genetic variation or seemly superfluous plants are taken into account.

Introduction to Agroecology: Principles and Practices

doi:10.1300/5553_07

Although with faults, monocultures do have agroecological application as agrobiodiversity-associated subsystems. Strip cropping can integrate monocultural strips, and multicanopy systems can contain an imbedded strata monoculture. Ecological or environmental harmony comes about when the single plant subsystem, strip or strata, is surrounded by or immersed in biodiversity. As renderings of biodiversity, monosubsystems are discussed in later chapters.

ECOLOGICAL UNDERPINNINGS

In nature, monocultures are rare but not unknown. A bamboo grove can represent the purest form, with all plants originating from the sprouts of a single parent. Colonizing plants, those that grow quickest, can populate an area in fairly pure stands.

Where sites are climatically challenged, only one species may find conditions acceptable. The resulting natural monoculture may endure for some time. Pure cacti or pure white birch stands represent a situation where climate eliminates all but one species. Cacti tolerate water shortages; birches survive extremely cold weather.

Generally, nature has other ideas, and considerable effort must be expended to keep the agricultural version unadulterated. With or without outside inputs, there are elevated threat levels. Seasonal monocultures go through a bareground phase where erosion peril is high. Hungary herbivore insects find clear expanses of one crop especially delicious and inviting. The battle against weeds is ongoing and constitutes another of the many drawbacks.

Ecologically, monocultures can be classified as a constrained fundamental niche where plants are free of interspecies competition but have constricted access, because of the rivalry with like plants, to the essential resources. Lacking much genetic expression, all plants take the same essential resources in the same proportion, and those resources not taken by one plant are available, in the same ratio, to another. In practice, essential resources are generally allocated on an area basis (density and spacing) with essential resources added as required.

As for classification, monocultures are species governed. Unencumbered by all but a few minor agrobionomic principles, the ecological thrust is one of simplicity. The sacrifices made, in terms of losses in potential productivity through increased threats, are large, but, by keeping to the lowest ecological denominator, monocultures allow resources (essential and management) to be focused in a way not possible with biodiversity.

ECONOMIC UNDERPINNINGS

The ability to direct, no matter how inefficiently, essential resources to a single output is economically attractive. The simplicity of the resulting ecosystem is also of interest. Lending and other financial institutions find these systems to their liking, merely because economic evaluation is so much easier (i.e., with one crop per area, revenue estimates are easily approximated from yield data).[2]

For commercial farms, markets drive simplicity. It is less difficult to seek lesser, rather than greater, biodiversity. For many products, crop differentiation is lost with processing and/or sale. Few supermarkets are willing to stock the 60 or more varieties of potato found in the Peruvian Andes. Even Andean farm markets tend to reduce this number.

With the strong push provided by suppliers of farm chemicals[3] and genetically "improved" varieties,[4] there is always the motive and coinciding advice to purchase more. This is not always a wise course of action.

As changes are made in an agroecosystem, these lead in one of two economic directions: (1) revenue and (2) cost orientation.[5] In the revenue orientation, increased costs result in increased yields and revenue. The accompanying proviso is that the increase in revenue must outpace the increased cost(s), otherwise profitability will decline or be lost.

Many commercial monocultures employ a high-revenue, high-input regime. With corresponding input levels, high-input intercrops can offer even greater revenue potential.

The other direction is cost orientation with a focus on using fewer costly inputs. As expected, yields (one or more) and revenue may also decline. To maintain profitability, cost declines should outpace any losses in revenue. Cost orientation is associated more with biodiversity, although monocultural applications are possible and economically viable.[6]

The ideal is where a reduction in inputs (costs) causes an increase in yields and revenue. This seldom happens but remains a hoped for possibility.

Strong, fixed commodity prices favor (as when farm price supports are in place) higher and more costly yields. Revenue orientation may have an advantage where fixed costs (e.g., land taxes) are high and must be covered. This discourages taking areas or seasons (as with a fallow) out of production.

Cost orientation may be the best choice when commodity prices are unsupported and fluctuate wildly. When prices are near bottom, low production costs allow farmers to eke out some profit.

Between a conventional revenue approach and organic cost orientation, European studies show, as expected, large yield differences but little change in overall profitability (revenue minus costs). The details for this are presented in Chapter 14.

MONOCULTURAL AGROTECHNOLOGIES

Looking at a single plot, not subsystems, three monocultural types can be observed. These monocultural types carry basic assumptions on ecologically countering adversities. A diversity in monocropping can be with seasonal or perennial systems, for example, tree crop, or silvicultural plantations. The breakdown is as follows

1. a single, genetically identical species that prospers through superior plant characteristics and outside inputs (pure monocultures)
2. a single species with varietal diversity or plants with a single genus that have a like output and the trait differences provide an solid ecological footing (varietal monocultures)
3. multiple species, but with a very strong emphasis on single plant and output (perceived monocultures)

Under the multiple species heading are the perceived monocultures. These are commonly viewed as monocropping, but by all measures, including ecological dynamics, perceived monocultures are biodiverse systems.

Pure Monocultures

The purest form of the monoculture has plants of a single undiluted genetic parentage growing together. These may be clonal or closely bred to remove any genetic variation. Despite movement toward purer forms, most monocultures, those grown from seeds, contain considerable genetic likeness but are not genetically identical. In most cases, these minor differences are agroecologically unimportant.

The pure form operates under the assumptions that (1) a single or a few superior varieties exist and best serve the overall needs of farmers and (2) any shortcomings (in DPCs) are surmountable through management or other interventions. Any site can be employed. Farmers are cautioned to avoid those prone to erosion or other damage, unless an appropriate land modification agrotechnology is also employed, for example, terraces.

The pure monoculture addresses no land-use problem except for an acceptable level of outputs. The strong link between input and outputs, the unqualified simplicity, and the ability to machine plant and harvest make monocropping a commercially attractive option. This view spans cultures, societies, and peoples.

Patterns too have been simplified; most monocultures employ a row pattern. Disarray is apparent with grains, (e.g., wheat). This is only a planting convenience; it is cheaper to broadcast seed rather than individually place them in rows.

Since the focus is on the individual plant, the desirable attributes (DPCs) are key to success. The best DPCs, in order of importance, are

1. high productive potential;
2. the ability to resist common, and not so common, plant-attacking insects and plant diseases;
3. being attuned to a wide range of sites and climatic zones;
4. resistance to normal on-site climatic variation;
5. tolerating or ignoring unintended competition (e.g., weeds); and
6. a sought after and/or marketable output.

Superior varieties can be bred or genetically enhanced. Missing an overlooked or unanticipated attribute can result in spectacular failure, especially if the susceptible variety is widely planted. What is lacking genetically is made up through outside inputs. This often makes these revenue-oriented systems.

Varietal Monocultures

Rather than selecting plants with minor or no genetic difference, land users can go farther afield, pick, and interplant disparate varieties of one crop species (see, e.g., Plate 7.1). Most common, and many not so common, crops have tremendous genetic variation, much of which remains underutilized. These are often labeled as heirloom or heritage varieties.

Ecological Assumptions

Varietal monocultures can be designed to thwart insects and diseases by exploiting the resistance of one variety that is lacking in another. In China, multiple varieties with the same plots counter fungal rice diseases, doubling the yields from what would have been expected if a single variety was grown. In Syria and the mountains of Pakistan, interplanted varieties of wheat combat a range of threats. In Mexico, mixed maize varieties exist for the same reasons.[7]

Other gains come from exploiting varietal differences through placement, the matching of the resource needs of the variety with the site resource profile, for example, a drought-resistant variety on a normally dry site or the most insect-resistant variety on the most susceptible tract. The result is field subdivided by multiple varieties placed in response to differences in soil, microclimate, etc., across a plot.

As an Alternative

As an agriculture alternative, varietal monocultures require certain assumptions for success. These are that

1. suitable varieties exist that can address various sites, resources, and/or threats;

2. there exists enough understanding so that the DPCs can be effectively utilized and any shortcomings are surmountable;
3. any intervarietal differences in planting, harvest, and management can be reconciled; and
4. the output (product) differences (if any) are locally acceptable or can be profitably marketed.

The output can be nondifferentiated (as with feed gains), close enough to mix without a price penalty, or the varieties can be mar-

PLATE 7.1. A multivarietal lettuce planting.

keted separately. In cultures that prefer multigrain breads, mixing wheat varieties may incur no market penalty. When consumers like a homogeneous product, the varieties should not have divergent output. Cotton fiber may be less distinguishable, but this is not the case with vegetables.

Orchards or tree crop plantations can fall under this heading. Some product differentiation may be a plus especially when varieties ripen at different times. This can deintensify the harvest, requiring less daily labor over longer periods. Reducing the daily harvest while extending the product stream also benefits processing mills.

Perceived Monocultures

Some monocultures are such only in name. These contain other species but are either managed as if the other plants do not exist or the biocontributions of these plants are acknowledged and not expanded upon.[8]

The accompanying species can be unwanted weeds in which case the damage done (in terms of lost yields and revenue) is far less than the cost of removal. Plate 7.2 shows a weed-infested monoculture, most likely not weeded because the damage done to yields does not outweight the weeding costs.

If the unwanted additions offer something or a service of value, they may be encouraged.[9] In an advanced form, the secondary species are purposely placed or are those that naturally regenerate. As an example, a cover crop may be employed but not to full effect, that is, the system is still viewed and managed as a monoculture.

In an orchard or tree crop plantation, the "weeds" may be a permanent and pervasive feature. Except for mowing and the acknowledged erosion control service provided, the biodiversity is ignored. Agrochemicals substitute for the natural ecological processes, such as insect control, that could flourish if the system were not viewed and managed as a monoculture.

Some perceived monoculture may be more purposeful. Often a few widely scattered stalks of a different species may exist in an otherwise pure monoculture, for example, several maize plants may rise above a field of potatoes. The same occurs in tree crop plantations

PLATE 7.2. A biodiverse monoculture, in this case a maize crop, with a large and tolerated weed population.

where fruit trees of a different species may be intermingled. The reasons are similar, the effect on productively being almost negligible; without major interference, these are regarded and managed as monocultures.

Chapter 8

Seasonal Intercropping

Intercropping is where two or more short-term productive species are grown in close physical association and with a high degree of interspecies interaction. No attempt is made here to overturn this perception. The defining parameter is that all crops provide a useful output. In practice, this line is fuzzy, as when a second species can have more of a facilitative and less of a productive role.

Going beyond joint productive-facilitative plants are those plants that are purely facilitative, such as cover crops, that aid crop growth. These have enough ecological overlap to be jointly discussed with the more traditional multioutput possibilities.

PAIRINGS

There are two forms of nonwoody or nonagroforestry intercropping. These are the

1. *productive intercrop,* that is, two or three herbaceous species growing together with the intent of obtaining multiple yields, and
2. *facilitative intercrop,* that is, combining one or more nonproductive plant species with one or more productive species (see Plate 8.1).

Productive Intercrops

For the productive intercrop, there is little problem in finding two or more crops that are niche compatible, have a high LER potential, and share a common climate. The number of worldwide examples at-

Introduction to Agroecology: Principles and Practices

doi:10.1300/5553_08

PLATE 8.1. Two sample intercrops: *(a)* a fully productive version, beets and carrots, and *(b)* a facilitative version, maize with an understory of leguminous canavalia.

tests to this. Some bicultures commonly associated with temperate gardens are as follows:

asparagus + parsley
asparagus + tomato
basil + pepper
bean + beet
bean + celery
bean + cucumber
bean + eggplant
bean + tomato
beet + kohlrabi
beet + onion
cabbage + celery
cabbage + tomato
carrot + leek
carrot + onion
celery + leek
celery + tomato
cucumber + radish
cucumber + tomato
eggplant + radish
garlic + onion
garlic + potato
garlic + tomato
horseradish + potato
kohlrabi + lettuce
leek + onion
lettuce + onion
lettuce + pea
lettuce + radish
lettuce + tomato
mint + radish
muskmelon + radish
onion + pepper
tomato + watermelon

Bicultures of staple or commercial crops are common or found in various regions:

- Maize is found with
  cassava
  cotton
  cowpea
  cucumber
  groundnut (peanut)
  mungbean
  muskmellon
  pea
  pigeonpea
  pumpkin
  sorghum
  soybean
  squash
  sugarcane
  sunflower
  sweet potato
  safflower
  chili
- Sorghum is found with
  alfalfa
  groundnut
  maize
  millet
  oats
  chickpea

- Cassava is mixed with

banana	mungbean
cowpea	pumpkin
groundnut	

- Sugarcane with

maize	sunflower
soybean	sweet potato

Some other mixes with commercial potential are as follows:

banana + coffee	oats + pea
barley + oats	oats + rye
cabbage + soybean	pigeonpea + sweet potato
cauliflower + rapeseed	radish + sunflower
cotton + garlic	rice + sugarbeet
cotton + groundnut	rice + tobacco
cotton + cowpea	ryegrass + wheat
cotton + sisal	sorghum + groundnut
jocote + pitya	soybean + tomato

In addition to bicultures, triple mixes (tricultures) have been suggested:

asparagus, parsley, + tomato	carrot, flax, + potato
bean, maize, + squash	cotton, cowpea, sweet potato
bean, onion, + summer savory	garlic, lettuce, + tomato
beet, celery, + onion	millet, sorghum, + cowpea
broccoli, nasturtium, + squash	millet, sorghum, + groundnut
brussel sprout, cauliflower, + thyme	plantain, cocoyam, + cassava
bush beans, eggplant, + marigold	yam, cassava, + maize
cabbage, mint, + potato	

Among the listed four-crop mixes are

cassava, yam, cocoyam, + maize
millet, sorghum, groundnut, + cowpea
mungbean, maize, potato, + wheat

Even a five-species mix has been documented:

yam, cassava, cocoyam, maize, + pigeonpea

Those given here have been formally documented;[1] many more undoubtedly remain undescribed in gardens and farm fields. The tally shows that seasonal intercropping is not limited to a few combinations with well-established complementarity. However, as expected, some species intercrop better than others, for example, beans have wider complementarity and can advantageously share a plot with more species than can watermelon or cucumber.

Facilitative Intercrops

With purely facilitative intercropping, there are long lists of possible companion species. Those utilized in California include[2] the following:

barley
cereal rye
annual ryegrass
purple vetch
annual bluegrass
rose clover
crimson clover
bur clover
wimmera 62 ryegrass
cacamonga brome
blando brome
wimmera 62 ryegrass
black medic

Cover crops from California or any other area have important DPCs. These species

- do not ascend or climb on taller plants;
- are shorter than the crop;
- repel herbivore or attract predator insects;
- produce ground-level heavy shade through high biomass or large leaves;
- may provide forage or an alternate product (e.g., beans);
- are fairly resistant to climatic threats (e.g., drought and frost);
- fit well within the planting, harvest, and growth sequence of the accompanying crop;

- are best if they possess allelopathic properties to retard the germination of weed seeds; and
- offer a below canopy microclimate that facilitates decay and recycling of nutrients.

As with any DPC list, the ordering and ranking is use dependent.

ECOLOGICAL UNDERPINNINGS

As theory suggests, the options are greater on resource-rich sites. In well-tended backyard gardens, rich in nutrients and moisture, one can expect to find closer spacings, high LER values, and pairings beyond those that are documented.

For these better sites, light may be the limiting resource and species with light compatibility should do well. A taller, light-demanding, low LAI crop over one that is shade resistant fits this need. Other light pairings are possible: vertical versus horizontal light or two crops of equal light competitiveness.

For subsistence farmers and low-input commercial farmers, there is a tendency to stick with recognized pairings as, in having wider ranging complementarity, these extract the most at the least cost. The best combinations are those that transcend the site and grow well together when resources are abundant or when one or more are limiting. Maize and beans are grown together throughout most of the Americas and these seem to, with limits, transcend site. Still, below set rainfall levels, resources cannot support a maize-bean coplanting.[3]

The real chore is in matching a site with the species mix. In a study of maize, soybean, and oat strips, the strip-internal rows experienced higher yields; those at the edge, that is, those bordered by a dissimilar species, suffered under dryer conditions. When ample rains excluded water as the limiting resources, the edge rows experienced the higher yields.[4]

As with the strip example, pairings are essentially resource dependent, and unexpected pairings may thrive given the right mix of resources. Until the science has advanced, intercropping is mostly trial and error, where established combinations, as listed in this chapter, guide the process.

Forming and clouding the line between the purely productive and purely facilitative species are the dual-purpose plants where the output may or may not be harvested. These plants have the DPCs to be an excellent facilitative addition irrespective of the output possibilities. Squash (for weed control) and legumes (for nitrogen fixation) can fall into this nonharvest option category.

ECONOMIC UNDERPINNINGS

The point of having a complementary or a facilitative crop is an increased LER . This is certainly the case. A scan of recorded LER values (shown in parentheses) for some common intercrops are as follows:

- groundnut and millet (1.26)
- bean and maize (1.53)
- maize and pigeonpea (1.67)
- barley and fava bean (1.85)
- cocoyam, groundnut, and maize (2.08)
- cassava, groundnut, and maize (2.51)

One can expect cost increases, mainly in planting and harvest, while experiencing reductions in weeding. These are revenue-oriented systems where the high costs are offset through high output levels.

For facilitative-crop pairings, yields can increase or decrease. If on the decline, one expects decreased costs. The facilitative species hairy vetch, planted with maize, has an LER of 0.95, but weeds were reduced by 76 percent. For red clover intercropped with maize, the LER was 0.97 and weed reduction was 72 percent. Both of these mixes are clearly cost-oriented.[5]

The need to replant at periodic intervals places some practical limits on the degree of spatial involvement: complexity can be costly and, unless amortized over a long period, not worth the effort. This keeps the number of agrotechnologies from multiplying to the degree found with the to-be-discussed agroforestry variants.

As a class of agrotechnologies, governance differentiates the simple intercrop from the more complex versions. Simple intercrops, bicultures, and tricultues are heavily weighted toward plant-on-plant governance. Greater biocomplexity and longevity are needed to bring system governance fully to the fore.

There are some intercrops that, when times are good, provide abundant multiple yields, and when times are bad, these systems still manage to eke out some yield. Although this form of suppression has not been documented,[6] the idea is to intercrop two species in which one is drought resistant and the other is not. If rainfall is high, both thrive and yield. If rainfall is low, the drought-resistant species produces while the other species dies. Although not offering yields in parched years, the suppressed species still provides facilitative services including providing biomass for the nutrient cycle in subsequent seasons. This has an unquestionable application where seasonal rainfall amounts vary greatly.

SEASONAL AGROTECHNOLOGIES

The classifications of intercrops are far from exact. The few representatives presented here are based on spatial pattern and purpose. The seasonal category of agrotechnologies resolves into

- simple intercrops where multiple species are more or less evenly mingled (simple mixes);
- having different crops in adjoining strips (strip cropping);
- employing lines of seasonal plants to protect seasonal crops (barriers/boundaries);
- purposely planting a ground-hugging species beneath a seasonal crop (cover crops);
- grouping seasonal crops so as to maximize light efficiency (clusters); and
- a large number of seasonal species mixed together in a more or less disarrayed pattern (complex cropping).

Simple Mixes

In accordance with theories of intercropping, the gains from complementarity are best captured through a high degree of plant-plant interface. An individual pattern or some other close pattern does this best. See Plate 8.1 for examples of productive and facilitative intercrops.

When crops are hand planted and harvested, there are few restrictions on this pattern. When farm machinery is employed, intermingled crops are somewhat problematic,[7] the exception being where outputs, upon harvest, can be mingled, sold together (as with mixed grass species, or later separated or sold without value loss.

Possessing a high degree of interface, productive versions of these systems offer the highest potential LER. Equally important, the mixing of species brings about many ecological happenings.

In practice, this is catch-all category. On occasion, these are scaled-down versions, in size and duration, of the specific ecosystem dynamics found in agroforestry. A case in point, if one plant is under another, being protected from the full intensity of sunlight, this is a shade system in miniature.

Strip Cropping

Instead of mixing crops, cross benefits occur when species are located in close proximity. Strip cropping with a different species in each strip can be ecologically rewarding; it can accommodate farm machinery and avoid much of the hassle associated with crop intermingling.

As with most agrotechnologies, there are design options. Strips can be productive—each strip providing a direct economically significant output—or facilitative, where there is a movement of biomass (e.g., green manure) from the facilitative strips to crop strips.

Productive

Production intercropping with or without farm machinery can be a difficult proposition. One solution is strips where different crops are raised in a crop-alternating strip pattern. In cross section, this can be *aaaabbbbccccbbbbaaaa, aaaabbbbccccddddaaaa,* or some other strip arrangement that promotes favorable interspecies interactions. These follow along the standard agroecological practice of locating plants with a high degree of complementarity next to each other or employing a buffer species to close a wide interspecies interface.

The width of each strip is determined by planting and harvesting equipment with the provision that each strip is not too wide. In order to promote cross-species interactions, 3 m wide strips seem, in ecological terms, to be a good starting point.[8]

Seasonal rotations (Chapter 11) promote favorable soil conditions. If there is any drawback, it is the erosion-prone bareground phase. Dangerous on hilly sites, this can be countered by leaving some strips in fallow, by inserting permanent facilitative strips, and/or by adopting contour layout.

Facilitative

In a facilitative form, crop strips alternate with facilitative strips. These can be passive, serving as barrier to insect or disease spread, and act as a conduit for the movement of beneficial insects. Formulated to control erosion, strips have reduced soil loss by 90 percent.[9]

Facilitative strips can be a very active facilitative component where biomass is transferred. Mowing machinery can cut and blow this into the neighboring strip at a comparatively modest cost.

Biomass movement can be

- for a soil cover and erosion control (transfer made after the crop has emerged and/or at the end of the season);
- to provide nutrients and/or augment soil moisture holding capacity (transferred before the crop strip is plowed);
- an insect control measure (when insects are most threatening); and/or
- to control weeds (soon after the crop is established).

The design of the overall system, including the strip contents, offers a range of alternatives. The vegetative content of facilitative strip is generally dictated by the facilitative service provided, for example, a repellant species when insects are to be kept at bay, allelopathic when weeds need to be controlled. By adding greater biodiversity, a facilitative strip can be a mix of facilitative species, each addressing a different task.

Biomass movement between strips expands the options. Figure 8.1 shows the biomass flow options with permanent biomass strips. Normally, these have purely facilitative strips alternating with productive strips. This design is machine friendly (i.e., strip widths are such that

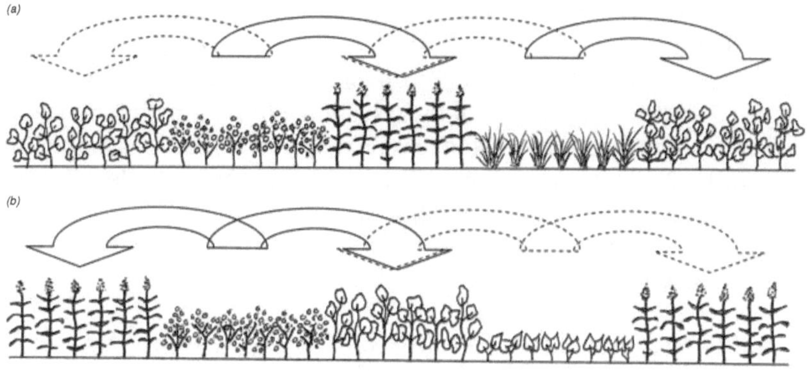

FIGURE 8.1. Shown in cross section, two biomass transfer strategies that pair a type of biomass with a crop. The solid arrows represents biomass movement, either *(a)* seasonally or as *(b)* the immediate situation demands.

the biomass can be cut and blown onto the crop strips and the biomass is of a type that accommodates frequent cutting).

The option exists to pair the nutrient content of the biomass with the crop needs (e.g, high nitrogen biomass with nitrogen-demanding species). As in Figure 8.1a, the direction of the biomass delivery shifts as the crops rotate with the crop strips; each crop received the designated biomass.

Another design alternative has biomass strips for nutrients and others for insect and/or weed control species. These are transferred right or left (as in Figure 8.1b) as the situation and crops demand.

There is yet another dimension that both complicates and adds additional avenues of ecological gain. These are the rotational sequences. The facilitative species, as a strip fallow planted or otherwise, sets the stage for cropping the next season or can be part of a cropping sequence that takes advantage of residual soil conditions. The biomass strips are not permanent and, generally, no biomass movement takes place. This does not preclude these being a cut-and-carry component in an anti-insect defense.

Many land users tend to treat each strip as a discrete cropping entity without regard for the facilitative possibilities. At the very least, one should have the crop strips parallel along the contours of the land.

The design of a strip might also incorporate structures to reduce water flow; ditches and/or in-place vegetation can help in this.

A second low-tech option is to locate species or varieties where they grow best. This means taking advantage of richer, high moisture soil at the bottom of a slope and less fertile, drier conditions higher up on a hillside. This can mean multiple varieties at each elevation to take advantage of changing conditions.

Barrier/Boundary Systems

Although seasonal barriers may seem, for cost reasons, outside the sphere of intercropping, there are applications when a temporary barrier is appropriate. These might be with barriers or surrounding boundaries that constrain insect spread and/or are the smallest component in a windbreak network.

In their least intrusive form, seasonal barriers are usually one or two plants wide. These are of a species that repels herbivore insects or attracts insect predators or provides some other form of protection.

An effective insect barrier is buckwheat that surrounds cucumber and guards against cucumber beetle.[10] As a windbreak, a tall annual species, such as sunflower, sorghum, maize, sugarcane (see Plate 8.2), or Jerusalem artichoke, if the interbarrier distance is not too great, is used.[11] As for specific pairings, documented are sunflower with tomato and grass strips as wind protection for cotton.[12]

The rule of thumb regarding a windbreak SIZ is that these be 30 "heights" apart. A "height" is the actual height of windbreak expressed as a vertical distance. There is more on this topic in Chapter 12 under permanent wind structures.

Figure 8.2 shows the basic layout. Plate 8.3 is a seasonal windbreak which, as a cost savings, spouts anew each season. Timing is-

FIGURE 8.2. A seasonal windbreak design where the boundary strips provide most of the protection, augmented by the taller, in-field, crops.

PLATE 8.2. An antiwind barrier: sugarcane protecting potato.

sues, especially in the early season when crops are more susceptible, do limit their usage.

Windbreaks are not the only type of barrier. Border strips can be a formable obstacle to crop-eating insects and birds. Most encircle field as a single line of vegetation. In large fields, encircling strips can be many meters wide where the outer rings are of a crop type not susceptible to the threatening organism but at the center is the delectable crop. These wide crop barriers are less documented but, as concentric circles, have been described in regions of Peru.[13]

Cover Crops

Under the heading of facilitative intercrops comes the cover crop combinations. These commonly have one of more primary and productive species with ground-hugging nonwoody understory (see Plate 8.3 and Figure 8.3). The overcrop can be annual or perennial species.

Facilitatively, these plants contribute in any number of ways. Foremost is to protect bare soil during the postplowing, preplanting phase.

PLATE 8.3. A complex seasonal garden protected from wind by sunflowers and Jerusalem artichoke.

FIGURE 8.3. A cover crop–primary crop biculture.

In addition, nitrogen fixing and weed and insect control are high in the list of expected facilitative burdens. Two lines of attack are possible perennial and seasonal cover crops.[14]

A perennial cover species is ecologically better but, unless highly complementary, can be crop competitive. Clearly a perennial cover crop must survive being quashed when the primary crop is harvested and must regrow when plowed under. This is less an issue with chisel (no-till) plowing.

A seasonal cover crop is another option. Cost issues aside, this offers the option to change the cover species as part of larger rotation.

To avoid competition, the cover species can be planted after the main crop. As well as forgoing protection during the bareground phase, there is the cost of a second planting. This may prove economically advantageous if weed control is the primary objective. In a case from Africa, soil-protecting and weed-excluding mucuna is planted well after the maize, which avoids mucuna-maize competition.[15]

In contrast, simultaneous planting may require a cover species that is slow to establish. If the primary species has a high degree of complementarity with the cover species, the primary species is planted at normal or lesser density (compared with the monocultural optimum). If not, it may be best to encourage the primary species with closer than normal spacing. This suppresses the cover crop, rendering it less competitive, while still accomplishing weed control and other functions.[16]

Clusters

In intercropping, light is often the limiting resource. By increasing the gathering ability of component species, this limitation slackens. Some species prefer more vertical, midday light, while others like more horizontal, early morning and late afternoon light. In accommodating these needs, light use can be made more efficient, permitting closer spacing, and increasing the chance for high LER values.

The pattern utilized is a cluster or clump where tall plants, those that seek vertical light, are in the core and those that want horizontal light are at the edge. In application, the clumps become fairly biodiverse and disarrayed, especially of uneven age and height. This feature may augment system light collection efficiency.

The planting of short-statured plants, those that seek vertical light, between the clumps (as in Figure 8.4) can upgrade the agrobiodiversity. Light efficiency, and the relative height differential, may be further improved upon by positioning the vegetation on mounds.

Complex Cropping

Some household gardens mix annual species with enough density, diversity, and disarray so that ecologically these reach the thresholds for ecosystem governance although, in strength, governance may be abbreviated (see Figure 8.5).

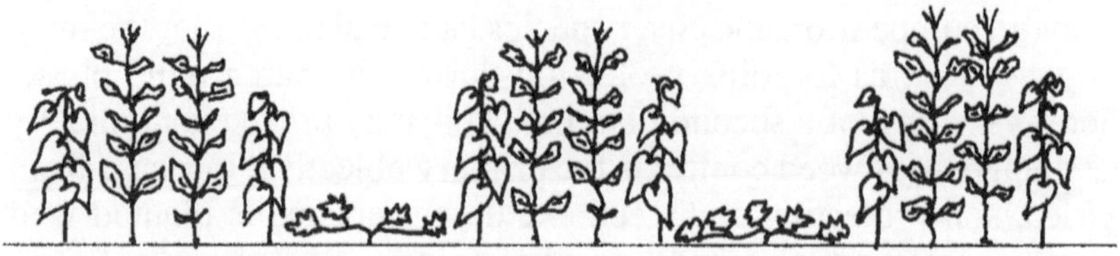

FIGURE 8.4. A clustered assemblage designed to maximize the collection of light.

FIGURE 8.5. Density, diversity, and disarray can be utilized, as shown, as a basis for seasonal gardens.

A mix of annual and perennial species, even without multiple canopies, resolves much of the dilemma. Given sufficient levels of biodiversity, most of ecological gains will be captured by the system.

Complex gardens many be more commonplace than the literature suggests. There are examples from Nigeria and the South Seas, both with over ten species.[17]

Others are historic and still active remnants of early European farming, often a feature next to farm dwellings in France and Spain. Besides herbs and species, these produce a range of medicinal plants and minor foodstuffs.[18] In southeast Asia, gains have come, and disarray realized, from taking advantage of in-field microenvironments, that is, strictly matching crops with the microclimate and microsite soil properties.[19]

Beauty can be combined with functionality and can be the basis for disarray with biocomplexity. Mixing flowers, as well as crops, can produce are aesthetically pleasing garden.

Chapter 9

Facilitative Agroforestry

Rather than replacing or replanting the facilitative species each season, the option exists to employ woody perennials in various facilitative roles. The resulting agrotechnologies fall under the heading of facilitative agroforestry.

PAIRINGS

Throughout the world, there are thousands of tree and shrub species. If only a small percentage were to find service as facilitative additions, the active list would be quite long. Since comparatively little effort has gone into developing facilitative systems, this remains a largely untapped source.

Most of work done has focused on high or superior biomass plants, those that supply nutrient-rich leafy plant matter to help the productive component. For the tropics, a short list of common superior biomass species found worldwide includes the following:[1]

Acacia angustissima
Acacia auriculiformis
Acacia saligna
Albizia lebbeck
Calliandra calothyrsus
Enterolobium cyclocarpum
Flemingia macrophylla
Gliricidia sepium
Leucaena diversifolia
Leucaena leucocephala

Introduction to Agroecology: Principles and Practices

doi:10.1300/5553_09

Senna siamea
Sesbania sesban

ECOLOGICAL UNDERPINNINGS

Since crops are ecologically supported and the site is protected by trees, marginal lands can be better yielding when nutured by trees. This statement is a corollary of agrobionomic theory where facilitative systems are advantageous on nutrient-poor soils, less so on richer sites. This is because the opportunities for facilitative gains diminish if more essential resources are on hand.

Trees, often thought of as a component of forests, do exhibit favorable niche associations with nonwoody plants. Some cohabit temporal phases; others overlap in their growth cycles. In the ecosystems of western North America, lodgepole pine often grows in combination with grouse whorteberry or pine grass. In natural successions, white pine can facilitate the establishment of subsequent poplar and perennial shrubs can aid the subsequent growth of oak trees.[2]

Given the number of available nonproductive species, farmers should strive for a very close match between the tree and crop and a resulting agroecosystem that includes a full complement of standard facilitative gains. As with any intercrop, the key is in possessing the right DPC mix. As a source of nutrients, it is helpful if the leaf nutrient content is in tune with the crop, for example, high nitrogen leaves are directed to a nitrogen-demanding crop.

Rather than selecting any tree species, superior biomass trees are often looked at. These are fast growing, can be continually pruned, and, over a short period of time, produce a considerable volume of leaves. These have multiple uses: animal forage, green manure, and, if possessing the right DPCs, weed and insect control.

More important on the list of DPCs is that these provide high quality leaves rich in those nutrients desired for certain crops. A case in point, the tree gliricidia, in addition to offering two to three times more biomass per area than natural regrowth, is higher in leaf-contained nutrients. The breakdown is nitrogen (4.25 percent), phosphorus (0.27 percent), and potassium (3.22 percent). Correspondingly, the leaf nutrient content for ordinary biomass has nitrogen at 2 to 3 percent, phosphorus at 0.2 to 0.3 percent, and potassium at 1 to 3 percent.[3]

Not be forgotten are the less observed but equally potent nutrients from fine root decay. Plowing or root pruning, in addition to reducing immediate tree-on-crop competition, provides the transfer mechanism. This is not insignificant; in one case, sorghum yields were 33 percent greater when the tree roots under the crop were cut and naturally recycled.[4]

Still, despite the advantages of and focus on the superior biomass tree, these are not the only choice. The DPCs for any one application may require something different. In this regard, one must pay attention to the end use requirements. A case in point, shade trees need not be superior biomass species.

ECONOMIC UNDERPINNINGS

The mechanisms of facilitative agroforestry overlap greatly with those of facilitative intercropping. Since the role of the trees is to benefit the crops, a good mix of DPCs, well directed, should have a large yield impact. LER values as high as 3.69 have been recorded.[5]

In contrast to intercropping, where the potential for high yields is a selling point, woody perennials remain across seasons. Without the need for replanting, trees can confer standard ecological benefits at a lesser cost. Cost containment is often the driving force in adoption.

The disadvantage lies in these being temporally fixed systems. The land user must be assured of stable markets and farm prices, as well as land rights, before considering a long-term investment.

The Cost Equivalent Ratio

The LER and RVT, with their focus on yields, are measures of revenue orientation. When costs are of chief concern, or are the driving force for adoption, cost advantage must be demonstrated. The cost equivalent ratio (CER) does this, the operative equation being

$$\text{CER}_{(\text{RVT})} = (C_a / C_{ab})(\text{RVT}) \qquad (9.1)$$

where C_a is the cost of monocultural production, and C_{ab} the cost of intercropping with a facilitative species. As with the LER and RVT, results above 1.0 are good.

This equation, and the rationale for facilitative trees, can be demonstrated comparing monocultural with tree-shaded coffee. In a study from Central America (Perfecto et al., 1996), the yield for coffee grown alone is Y_a, while Y_{ab} is the coffee yield with a shade tree. C_{ab} is the management inputs with the trees, and C_a is the costs without trees. With coffee prices (P_a) added, the

$$\text{RVT} = (P_a Y_{ab} / P_a Y_a) = 314/1397 = 0.22 \tag{9.2}$$

The revenues, through the RVT, are quite low (0.22 < 1.0). Analysis with the CER produces a more favorable result,

$$\text{CER}_{(\text{RVT})} = (C_a / C_{ab})(\text{RVT}) = (\$1740/\$269)(0.22) = 1.42 \tag{9.3}$$

The crude CER equation (specifically, C_d / C_{ab} or, in numbers, \$1740/\$269) shows costs that are almost 6.5 times lower with a shade system. This translates into a CER of 1.42 when the revenue (i.e., RVT) figures are factored in.

The numbers explain the economic interest in and usage of facilitative agroforestry by resource-poor farmers. Facilitative agroforestry also has commercial applications: when prices are low, profits are still possible. This has not gone unnoticed; facilitative trees are commonplace throughout the world.

The Nonharvest Option

The nonharvest option, mentioned in passing in the previous chapter, is more of economic force when the facilitative species is a perennial. This is because, in contrast to intercropping, fruit trees can be a low-maintenance facilitative addition and may be utilized in this role despite fruit yields.

Under normal circumstances, secondary species, fruiting trees included, are purely facilitative and any output is not harvested. Harvest occurs only when the secondary product

1. suddenly attains a market value;
2. only small percent of the output has use (as with household spices, medical plants, and where only a few fruits of the highest quality are marketable);
3. for subsistence farmers, this is the only food available, the primary crop having failed; or
4. it supplements what is normally produced (when pasture grasses wane, a forage-cover crop may plug a shortfall gap).

The nonharvest option, leaving fruits where they fall, may seem counterproductive. Outside this cultural opinion, the subsequent decay and recycling of nutrients may be the wise economic choice; nutrients may eventually end up in a higher valued future harvest.

Outside a changing market or climatic emergency situation, the nonharvest option can be part of more ingrained economic strategy. One version trades away the secondary output, for example, the unpicked fruit in exchange for labor inputs such as weeding, harvesting the primary crop, and so on. If labor is not needed or goes elsewhere, the secondary fruit is not harvested.[6]

AGROTECHNOLOGIES

Finding a favorable tree species, although important, is only part of the overall design; other factors intercede to make one tree-crop mix better than another. Spatial pattern is the visual element that separates these agrotechnologies and, in most cases, gives them their unique ecological dynamics.

The ecological and social gains (and the resulting agrotechnology) stem from

- a single tree offering habitat or shelter to sought-after fauna or shade-stressed domestic animals (isolated tree);
- scattered groups of trees protecting and encouraging understory crops (parkland);
- perennial strips as anti-insect, antiwind, or antierosion obstructions (protective barrier);

- rows of short-trimmed trees that, while not blocking light, do facilitate nearby crop growth (hedgerow alley cropping);
- having wide rows of trees between crop strips such that the crops are not overly shaded, but are wind protected and ecologically encouraged (strip cropping);
- shrubs or pruned trees residing beneath taller crops or tree crops and, without blocking light, these confer the benefits of association (crop over tree);
- a vine or tree is physically supported by a sturdy plant (physical support systems); and
- having a nonproductive tree overstory that provides multiple benefits to the shorter crop (light or heavy shade systems).

Isolated Tree

One or a few scattered trees in a large field provide little in the way of useful output but can offer facilitative benefits. This follows along natural lines. For example, the savannas of Africa are visually defined by a few scattered trees.

The gains come in different forms and can depend partially upon the tree species utilized. There does not seem to be a preferred tree species in any given area. This suggests that any ordering of DPCs is possible. As these trees are widely spaced, the immediate effect on crop yields and tree-crop complementarity is normally not an issue.

The DPCs may vary with need. In dry regions with large cattle populations and a hot midday sun, heavy shade may top the list of DPCs. Another usage follows from these same conditions. In the arid zones of South America and Africa, forage can be stored in tree canopies, out of the reach of hungry animals.[7] Besides shading animals, trees provide a place for workers to rest.

It can be beneficial to encourage birds that hunt from roosts. These can be insect or rodent seeking.[8] A nonvegetative alternative is found with the traditional German practice of placing T-shaped roosting poles in fields.[9]

Although set at a wide spacing, the trees may help with water infiltration where rains are brief and intense. For maximum effectiveness, trees are placed in water-collecting depressions, their roots punching drainage holes in soil substrata.

As many isolated trees systems are less facilitative and more productive, this topic is addressed further under productive agroforestry. Figure 10.2 illustrates either the facilitative or productive version. It is difficult to say which version is more common.

Parkland

If one or a few trees are good, more trees may be better. The parkland system has a number of trees widely spaced across a field. Another defining characteristic is the presence of a single tree species (see Figure 9.1).

Parklands with short crops, grasses, and grains visually liken to natural savannas (grasslands with scattered trees). Questions remain as to the extent that agricultural parklands ecologically resemble natural savannas.

Besides the many advantages of the isolated tree systems, additional gains are possible. Tall trees may slow wind speed enough to increase the deposit of airborne nutrients. Also belowground nutrients may be brought to the surface by the deep tree roots. Despite the potential, some report yield increases, while others find no gains.[10]

Yield facilitation may be only a small part of a larger picture; other returns come in countering some of the standard threats. As with isolated tree systems, rodent-eating birds, those that shelter, rest, and/or wait in trees, are a positive aspect. Trees can have

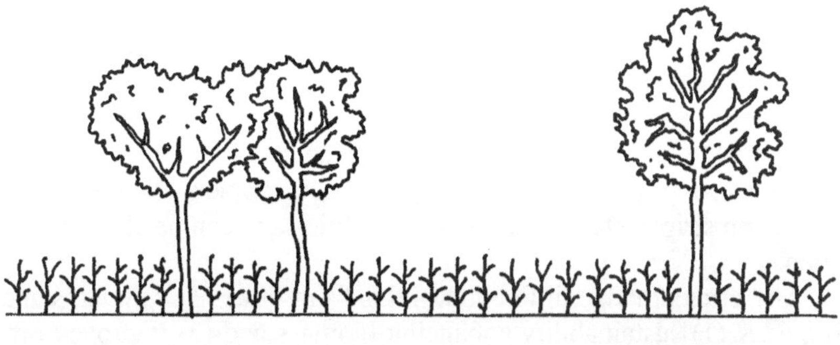

FIGURE 9.1. A classic parkland system showing a few disarrayed but widely spaced trees. As a defining characteristic, the trees are of the same species.

enough presence (density and/or size) to obstruct wind microbursts that can flatten grain crops or interfere with fruit setting and growth. In addition, trees harbor bugs that pollinate or prey on unwanted herbivore insects.

Far from an impromptu system, there is a rationale why a single tree species dominates. The overriding DPC is tree-crop compatibility, lessened somewhat if the trees are pruned prior to crop planting.[11] Because of this, the list of usable trees is very short. Most often, there is a strong regional specialization in one or two species. Worldwide, the species found include the following:

Prosopis cineraria (India),
Nothofagus obliqua (Chile),
Faidherbia albida (sub-Saharan Africa),
Baobab—*Adansonia digitata* (sub-Saharan Africa),
Shea tree—*Vitellaria paradoxa* (sub-Saharan Africa),
Balanites aegyptiaca (sub-Saharan Africa),
Hyphaene thebaica (sub-Saharan Africa), and
Oaks—*Quercus* spp. (parts of North America and Europe).[12]

With grazing parklands and with a wider range of available grass and forage crops, the number of serviceable tree species expands slightly. Plate 9.1 shows both a cropping parkland with pruned trees and a grazing version.

Alley Cropping (Hedgerow)

On the premise that with increased management, trees can be more effective facilitators, hedgerow alley cropping was developed. The basic description for alley cropping has rows of heavily pruned trees (as hedges) placed between crop strips (see also Plate 9.2).

Alley cropping is not an entirely new development. The medieval European strip systems share commonalities and can be classified as such.

The pruned hedge differentiates this from other alley systems and provides (1) sustainability enhancing biomass and (2) improved off-season erosion control. Studies have shown maize alley cropping to decrease storm water runoff by 40 percent and soil loss by 48 percent.[13]

(*a*)

(*b*)

PLATE 9.1. Two parkland systems *(a)* a cropping version with pruned *Erythrina glauca* and *(b)* a grazing version with *Pithecellobium dulce*. Both photos were taken in Nicaragua.

PLATE 9.2. A variation of hedgerow alley cropping where the hedges are more biodiverse and taller than with the standard design.

To ensure crop yields, the hedges are normally pruned prior to the planting of the crop. This eliminates competition for light while the biomass (green manure) enhances the soil. This is a temporal and spatial repositioning of nutrients: those the trees capture during the fallow (or off season), those that might have escaped during a previous planting season, or those the trees locate in deep soil strata.

Design Package

Alley cropping is a well-stated agrotechnology with fairly well-developed design package.[14]

1. Land-use problem:
    a. Sloping site with off-season erosion danger and/or
    b. Soil sustainability issues[15]
2. Site situation:
    a. Generally fertile soils
    b. Adequate seasonal rainfall
    c. Soil pH greater than 5.5
3. Socioeconomic profile:
    a. Maize as the primary crop
    b. Scarce land requiring hillside cultivation
    c. Secure land tenure
    d. Ample labor
    e. Confinement of animals
    f. Need for firewood

Standard Design

The design parameters for standard hedgerow system has 4 m wide strips bordered on both sides by a single or double row of trees pruned to a 0.5 m height. Figure 9.2 shows standard-design alley cropping in cross-section.

Listed, the DPCs for hedge species are[16]

- nitrogen fixing;
- ease of propagation;
- a high aboveground biomass production;
- a high and well-balanced nutrient content in the leaves;
- nutrient-rich, quick-decaying, fine roots;
- prunes well without dieback;
- burns well (depends on use);
- no toxic substances in the foliage (for a secondary use as forage);
- a large ratio of leaves compared to woody stems;

- small leaves when rapid decay is needed or larger, slower-decaying leaves when erosion or weed control is sought;
- leaves that shed easily;
- root and moisture compatibility with crops;
- attracts predator and/or repel herbivore insects;
- does not harbor, but discourages, plant diseases;
- does not sprout from the roots; and
- leaves that are highly, moderately, or not palatable to forage animals (depends on use).

For Africa, some of the species that have proved practicable are as follows:

> *Leucaena* spp. (*Leucaena diversifolia* or *L. leucocephala*)
> *Gliricidia* spp.
> *Calliandra calothyrsus*
> *Senna siamea*
> *Enterolobium cyclocarpum*
> *Acacia auriculiformis*
> *Sesbania sesban*
> *Albizia lebbeck*

As one might expect, some tree species are better with certain crops than with others.[17] Most of the work done is with hedge-maize compatibility, but other crops benefit from this system. There are also the traits for individual trees that must be accommodated, for example, pruning height.[18]

A number of practitioners go beyond the standard design. Strips may be seasonally rotated (one in fallow while the next is farmed). Hedge height can be lower (at ground level as with a minihedge)

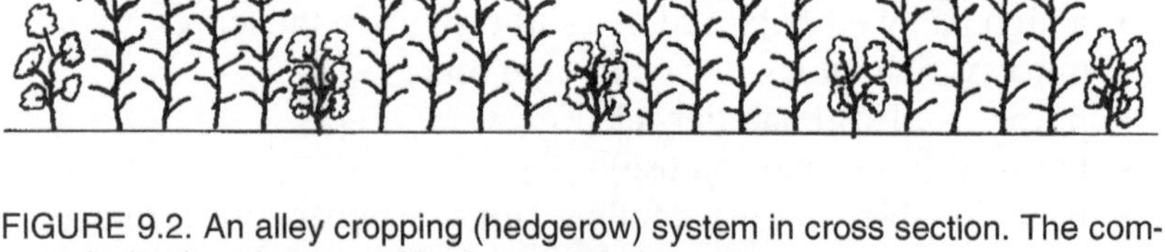

FIGURE 9.2. An alley cropping (hedgerow) system in cross section. The common design has the pruned hedges spaced as shown.

where the trees are machine pruned (i.e., mowed).[19] As always, greater biodiversity is possible. Facilitative hedge species are not immune to insect attack or climate stresses, and as such, the mixing of hedge species is a viable strategy.[20]

Protective Barrier

A land user always has the option for some between-crop barrier. These have the same ecological roles as their seasonal counterparts but remain in place much longer. Among the functions are to stop the spread of insects and plant diseases, block winds, halt erosion, and/or to promote water infiltration. In contrast to auxiliary systems, these are considered part of the ecosystem being protected.

Strips

Rather than hedgerows, strips of facilitative woody perennials may serve best. As with intercrop strips (Chapter 9), tree-crop strips alternate across a plot. The facilitative strips function in much the same way as hedgerow alley cropping and strip intercropping, that is, biomass is transferred to crops (see Figure 9.3).

If the intent is mechanized agriculture, a mower can cut the biomass and directly blow this onto the crop strips. This does not eliminate woody species but does require a tree that can survive severe and frequent ground-level pruning.

The design options are much the same as those described in Chapter 8. The biomass from a nutrient-favorable tree species can be guided to a crop that likes the nutrient mix provided (as in a seasonal directed nutrient flow—see Figure 8.1) or where one strip provides nutrient-laden biomass, another insect-repelling biomass (see Figure 8.2).

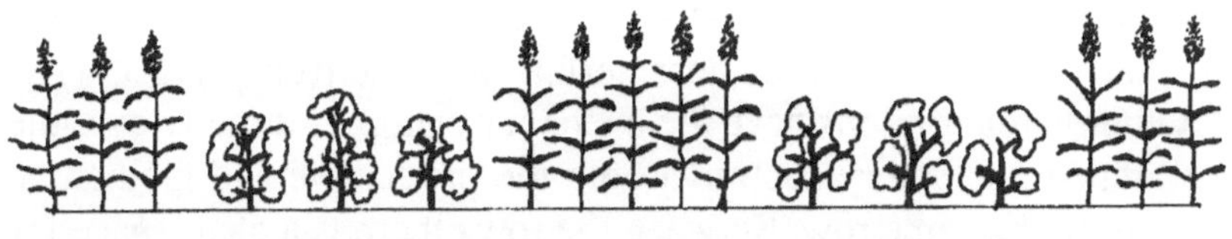

FIGURE 9.3. Strip cropping where, in this cross section, strips of permanently in place, pruned trees alternate with seasonal crop strips. The biomass, cut from the trees, is carried and applied on the crop strips.

The facilitative strips need not be uniform in composition. A mix of facilitative species can offer cropping flexibility, for example, mixing predator-harboring as well as biomass-providing plants.

The strips usually run along the land contours but, where the topographical relief is slight, light needs or wind direction may help determine the strip orientation. When possible, the biomass is transferred uphill.

Counter patterns are also found. These have the rows in the pruned tree strip perpendicular to the overall strip orientation. Utilized where the biomass is hand cut and hand carried, the counter row layout, with labor movement along, not across, the rows, makes the transfer easier.[21]

Crop Over Tree

With a few exceptions, most of the facilitative agroforestry systems have trees above the crop. One of the exceptions has crops over the trees.

As with all tree-based polycultures, gains are expected on a number of ecological fronts (offering a more or less standard facilitative package). The primary reason for this ordering is light apportionment, which is not a factor as the facilitative species is shorter than the crop species.

Despite the potential gains and the elimination of one point of interspecies contention, applications are limited and mainly associated with tree crops that can rise above pruned trees. Documented examples include plantain over flemingia, and jackfruit over *Calliandra calothyrsus* or *Acacia Angustissima.*[22]

One of the drawbacks is the need to harvest, and between-row facilitative trees do get in the way. Figure 9.4 shows this in an uncrowded, ease-of-entry, row design where space is provided so that the facilitative species does not interfere with the fruit harvest.

Other layout options have the overstory canopies directly over the facilitative understory such that both the productive and facilitative tree are in the same row. Between the rows there is a clear space for harvest and maintenance. Although not as ecologically potent, there being fewer facilitative trees, any fruit- or nut-bearing tree species needing a facilitative assist can be aided by this design.

FIGURE 9.4. A crop-over-tree design where fruiting palm trees are the overstory for biomass-producing, pruned trees. The wide spacing expedites the harvest of the fruit.

Physical Support Systems

In nature, vines have evolved to climb on trees. Although this lesson may have been lost, nature can still be the model. Vines, seasonal or perennial, are common in agriculture and physical support is a major cost. By substituting tree support for a trellis, cost savings of the more or less standard facilitative gains are an expected result.

Perennial Vines with Tree Support

Employing trees to support vines has a long history. Documented by the Romans and in middle ages, this practice, now out of favor in Europe, has been discovered by other societies.[23] Of the perennial crops needing support, grapes, black pepper, passion fruit, vanilla, kiwi fruit, and guaraná qualify.[24]

In Bolivia, grape vines grow inside a tree canopy. Aggressive tree pruning and ample sunlight ensure high yields. This approach is not universal; some opt for vines above the canopy. A short-statured supporting tree species, especially those with umbrella shaped canopies, may be the best choice. Following ancient Roman guidelines, elm was suggested for grapes and an umbrella-shaped variety does exist *(Ulmus glabra).*

In and above canopy vine positioning are not the only options. Vines may be raised below the supporting tree canopy where the trunks of trees prop up the black pepper vines. For this, a high tree canopy and wide tree spacing will ensure ample understory light. A tall, clear tree stem (the type often associated with a vertical light preference) can have both current facilitative (support) use and future wood value.[25] These three support options are illustrated in Figure 9.5.

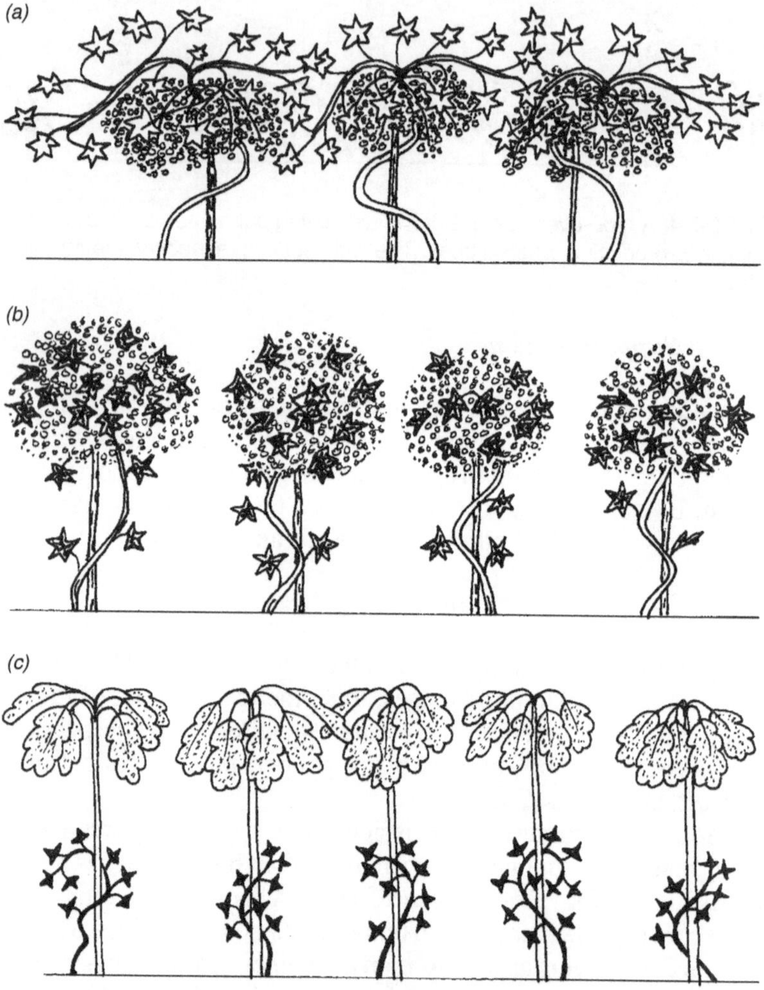

FIGURE 9.5. Physical support systems: *(a)* vines over shrubs or trees, *(b)* vines within the canopy, and *(c)* trunk-supported vines.

Seasonal Vines with Tree Support

Seasonal vines can also be accommodated with trees (see Plate 9.3). Seasonal vines include climbing beans, yams, cucumber, and sometimes hops.[26] The same three positioning options exist (i.e., above, in, or below the supporting canopy). In addition to perennial support (as found with perennial vines), seasonal crops can be supported with impermanent trees.

A common version is where young trees, 1 to 2 m in height, are killed or suppressed by girdling. The vine grows on the now leafless or sparse canopy. After the vine crop has been harvested, the tree can either recover, resprout from the lower stem, or be replanted.

PLATE 9.3. Vine-over-tree system where pepper vines cover an unidentified tree species. This example is from Costa Rica.

Root Support Systems

On sites with loose or shallow soils, tree crops with a narrow root base are in danger of toppling. Loose wet soils and winds can be causal where orchard or tree crop species lodge.

A second species may be added to stabilize the situation. The facilitative species can be shorter or taller than the tree crop depending upon light requirements for the primary species. The key is having facilitative plants with a strong, spreading root system that can tie together everything belowground.

Shade Systems

In contrast to row systems where vertical light enters the gaps between the rows, with shade systems light penetrates through the canopy. This is a visual clue that qualifies a shade system, subdividing it into light or heavy shade.

For an agricultural shade tree,[27] the specialized DPC list includes

- low tree-crop competition;
- low propagation and establishment costs;
- smooth bark that allows predators to easily locate herbivore insects;
- self-pruning with good bole form (branches do not interfere with agriculture);
- limited maximum size;
- strong branches and stems (blow-down resistance);
- small leaves to minimize raindrop coalescence and drip damage or large leaves to suppress weeds;
- for deciduous trees, rapid flushing of the leaves to regenerate the shade;
- wind resistance (to prevent stem or canopy breakage);
- valuable or useful secondary outputs; and
- high salvage value (high wood or firewood value).

These DAPs apply to either light or heavy shade systems.[28] Plate 10.1 shows a side-by-side comparison of two shade systems. Plate 10.1a is a fully productive ecosystem. Plate 10.1b is facilitative with a branch pruned, but productively inert overstay tree.

Light Shade

The defining characteristic of light shade is the comparatively large amount of sunlight that penetrates through the overhead canopy. In most common facilitative designs, the overstory is of a single species with open low LAI canopy. Trees that gather vertical light serve well and are spaced to make up an unbroken, single canopy strata. In this way, the light received by the shorter species is filtered and reduced in intensity (Figure 9.6).

Except for the protection from intense light, the ecosystem initiates a more or less standard package of facilitative effects. The overstory provides nutrient additions, wind protection, a more humid microclimate, favorable predator-prey dynamics for insect control, and opportunity for harvest index gains.

The intent is revenue orientation. Because the yield gains may not be strong enough to bear the cost of maintaining a purely facilitative overstory. Cost-based designs are rare. When found, the trees are low maintenance and/or with nonharvest potential. Most light shade systems are of multiple output (see Chapter 10).

Heavy Shade

A dense overstory, either as single species or a mix as found with natural forests, produced a unique agroecosystem. Heavy shade mag-

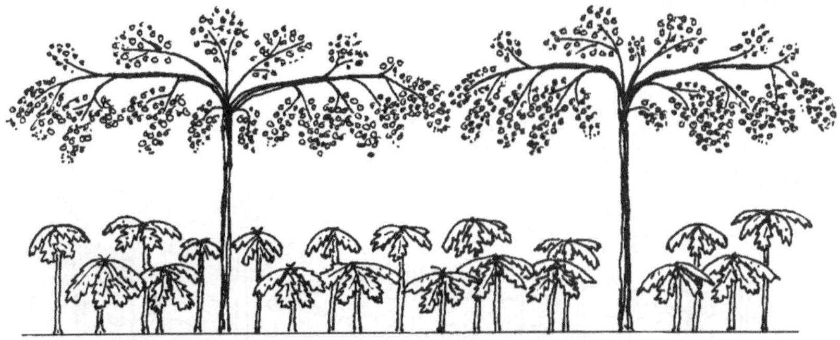

FIGURE 9.6. A light shade system where a tree species, one that overtops the crops, does not intercept much light. The overstory tree can be purely facilitative but more often is an yielding and economically contributing member of the agroecosystem.

nifies the facilitative gains in an environment where light is the controlling force.

The ecological gains include constant microclimate, nutrient facilitation, positive insect dynamics, and, through shading by two canopy layers (overstory tree and understory crop), the elimination of weeds.[29] The CER example given earlier in this chapter, that is, $CER_{(RVT)} = 1.42$, is for heavy shade over coffee.

Economically, these are very much cost oriented where the cost reductions exceed any lost revenue. These are found more on large, resource-poor farms or in remote areas away from the main sources of labor. Figure 9.7 is of a heavy shade system with a nonproductive overstory.

As most of the light is captured by the overstory/understory agriculture is limited to shade-resistant crops. As light is the limiting resource and the system recycles ample nutrients and holds water in sufficient quantity, these tend to be self-contained.

The DPCs for shade trees are given in Chapter 6. Although detailed, these only touch upon the topic. At the front of any debate is whether a fast- or slow-growing overstory species is best.

With fast-growing trees, a system will reach maturity quickly. High value, wood-producing trees are often slower growing, so compromises must be made. On the positive side, slower growth can mean strength (falling branches are less a problem) and, at the end of the rotation, farmers can reap a large monetary reward when the trees are cut and sold.[30]

FIGURE 9.7. A heavy shade system where shade-tolerant crops are raised below a light-intercepting tree canopy. The trees are often only facilitative, reducing costs rather that directly or indirectly contributing to increased yields.

A variation favored by environmentalists is an overstory of naturally occurring forest species. These canopies harbor migrating birds and provide an agreeable habitat for many native species. As opposed to monocultural coffee, shade systems with diverse overstory have 94 to 97 percent more bird species. The trees can be planted or the crop placed within an existing forest. The product, often coffee or cocoa, can be sold, at a premium, under an environmentally friendly label.[31]

There is a user benefit from a mixed species–mixed height overstory. Since the economic focus is on controlling costs, trees that prove troublesome can be removed without handicap or notice from the jumble of overlapping canopy stratum.

Chapter 10

Productive Agroforestry

Agroforestry, when fully productive, has both the trees and accompanying crops providing marketable or useful yields. Commonly, fruit trees or some tree crop are raised above a seasonal or perennial crop species. Other versions have short-rotation, wood-producing trees above a seasonal or perennial crop.

PAIRINGS

Worldwide, the number of noted tree-crop pairings is quite large.[1] As with intercrops, many more have undoubtedly escaped scientific notice. The list includes the following:

rubber + coffee
coconut + cacao
papaya + macadamia nut
coconut + pineapple
rubber + cacao
olives + opuntia
cloves + banana
laurel + cacao
rubber + cassava
citrus + cucumber
breadfruit + jackfruit
grape + olive
grape + almond

ECOLOGICAL UNDERPINNINGS

With the exception of the time frame, fully productive, multiple-output agroforestry has much in common with seasonal intercropping. The parallels are in how essential resources are divided, used efficiently, and in the single ecological dynamic that makes each system unique. Also, all the resulting agrotechnologies save one, the agroforest, are species governed.

Introduction to Agroecology: Principles and Practices

doi:10.1300/5553_10

In combining trees with seasonal crops, the greatest pairing difficulty lies in the size disparity and light apportionment between the trees and the crops. These are resolved with pairings and patterns much the same as found with facilitative agroforestry: a light-demanding tree is positioned over one that is shade resistant, plants are matched whether they collect vertical or horizontal light, or by pairing two crops of equal height and light competitiveness.

Since the trees have less of a facilitative role, the land should be very fertile and capable of high levels of productivity. Marginal sites are generally less favored unless the nutrient extraction rate is low. Even on good sites, there are sustainability concerns where the sheer volume of a harvest overwhelms the ability of soils to naturally recover. External nutrient inputs, either chemical or organic fertilizers, may be required. The nonharvest option may be part of this.

ECONOMIC UNDERPINNINGS

Facilitative and productive agroforestry are more than a site-accommodating strategy. The general rule is facilitative systems for marginal sites and poor soils, fully productive mixes on richer sites and soils. These also have strong economic leanings that can transcend, where possible or feasible, a site. The exceptions are the agroforests which can combine revenue and cost orientation (a reduced CER and an increased LER) (see Plate 10.1).

With purely facilitative systems (as described in Chapter 9), costs may be low, and, when selling prices are low, the user may still eke out a profit. This is the cost-orientation approach. Cost systems tend to be favored where labor inputs are costly or not available.

With multiple outputs, when the selling price of the principal crop is low, the user can apportion the costs among the various outputs and sell more of the alternative or secondary crops. These measures cushion the economic impact, allowing a systemwide profit. This is the revenue-orientation strategy.

These economic strategies do take into account climatic and other risks. Well-chosen, facilitative trees should protect the primary crop in all kinds of ways. With productive agroforestry, the trees are gener-

(a)

(b)

PLATE 10.1. Two coffee shade systems: *(a)* revenue oriented with a productive overstory and *(b)* cost oriented with a productively inert overstory.

ally less protective, offering instead an alternative output if the primary crop fails.

AGROTECHNOLOGIES

The breakdown of productive agroforestry into separate agrotechnologies is generally based on light apportionment. The breakdown is as follows:

- widely spaced trees with open, crop-filled expanse around each tree (agroforestry intercropping)
- one or two, often productive species, with a large plot (isolated tree)
- a close spaced, in-row planting with wide between-row spacing which is open to the sun (treerow alley cropping)
- multiple rows of short or nonspreading trees alternating with crop strips (agroforestry strip cropping)
- groups of species internally ordered to maximize light collection (perennial clusters)
- a single touching layer of low LAI overstory trees over a perennial understory (light shade)
- spaced, high-LAI trees with a shade-resistant understory (heavy shade)
- closely spaced, disarrayed, tree mixes with multiple canopy layers (agroforests)

Often thought of as an agrotechnology, silvopastoral systems are actually variations of other agrotechnologies where grasses or forage are substituted for seasonal or annual crops. Irrespective of this, special-purpose, fauna-inclusive systems do exist. Along more exotic lines are the feed systems which include aquaforestry, entomo-agriculture, and avian-agriculture. Here, domestic and nondomestic mammals along with fish, lizards, frogs, insects, or birds are the primary crop. Their distinctiveness lies in specialized trees which feed whatever creature is being raised. For discussion purposes, these are placed under a separate feed system heading.

Agroforestry Intercropping

Agroforestry intercropping has spaced trees with plenty of intertree, unimpeded space. Between the trees are shorter crops. This can also describe an orchard with scattered taller species. These differ from other scattered tree systems in that there are more overstory trees per area, the overstory has a greater economic presence, and the system is often ordered.

Figure 10.1 shows agroforestry intercropping. This drawing represents a cross-section from any view as the system is symmetric. Among those combinations encountered are citrus trees with cucumber in Mexico and, in Kenya, cashew and coconut over papaya, banana, and tomato. An orchard-style intercrop is found in Brazil where a few widely spaced brazilnuts rise above rubber trees.[2]

Isolated Tree

The positioning of one or two isolated trees in a large expanse can be facilitative (see the preceeding chapter) or productive. Productive trees are more the norm as the area taken is small and the competitiveness of the tree of lesser concern. Many times the system is accidental: a land user is loth to cut an otherwise useful tree and it remains. Figure 10.2 shows a lone tree in a crop plot.

The facilitative gains generally transfer from the purely facilitative version of this agrotechnology without loss. In Mexico and other tropical regions, mango trees, with their dense canopies, can be seen occurring singly in farm fields.[3]

FIGURE 10.1. Agroforestry intercropping showing spaced trees with interspersed crops. Because of the importance of the LER and RVT, the trees are economically important and complementary with the crops.

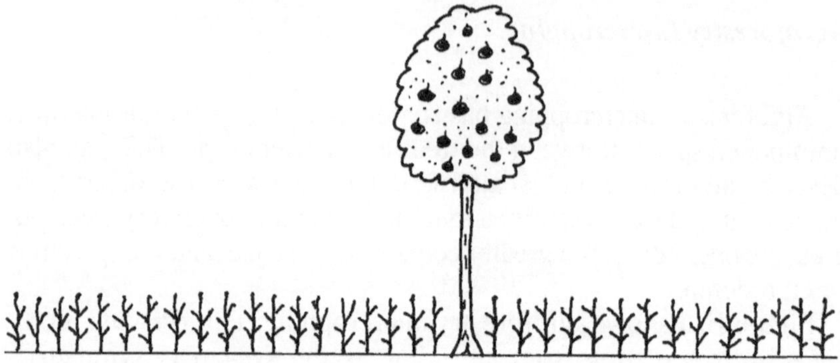

FIGURE 10.2. A productive isolated tree system which has a single fruit tree located within a large cropping area.

Alley Cropping (Treerow)

Instead of being widely spaced, trees are closely spaced, often touching, along rows. Between the rows, the clear, interrow alley is planted with one or more shorter crop species. In a cross section, this is shown as *aaaabaaaabaaaa,* where species *a* is the crop, species *b* the alley tree. The visually distinct alley comes about (1) by selecting a tree species that does not spread branches laterally, (2) through pruning, or (3) as a brief temporal stage before the canopies spread and the system transforms into a light or heavy shade design. This is shown in the cross-section in Figure 10.3.

Because of the added maintenance and crop yield penalty from loss of areas and tree competitiveness, the trees are usually productive. The high LER or RVT compensates for these costs.

Specific tree species help keep overstory maintenance costs low. Species of palms and cloves have the requisite nonspreading canopies.[4] Another tree, cinnamon, is also suitable because the bark is continually harvested by cutting the branches (see Plate 10.2). This holds back branch spread.

Visual distinctness produces an equally pronounced design package with the following elements:

1. Economic needs
   a. two harvest-compatible crops
   b. need for high output levels

2. Site situation
   a. a flat or slightly sloping land relief
   b. generally fertile soils
   c. adequate seasonal rainfall

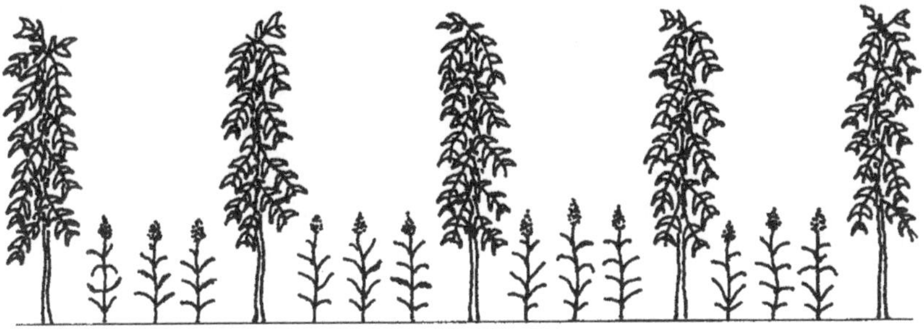

FIGURE 10.3. A cross-section of the treerow form of alley cropping where horizontal light enters through the interrow gaps. In most cases, the trees are a productive, economically augmenting component.

PLATE 10.2. Treerow alley cropping with cardamom under cinnamon. The alley is maintained by harvesting the branches from alternate rows of the cinnamon trees (as shown, left center).

The row-strip pattern is the only arrangement employed with this agrotechnology.[5] This system is better as a flatland system for good reason. To better accommodate light requirements, a north-south row orientation may be needed, that is, the alleys are positioned such that midday light shines down the open strips.

Although this is a good starting point, it may not be the optimal use of light. Some crops respond better to early morning light and afternoon shade. If this condition is diagnosed, the row orientation might be, in the northern hemisphere, southeast-northwest or, in the southern hemisphere, northeast-southwest. This has the advantage that the morning dew protects crops from excess drying. Because of moisture stress, this orientation protects from direct sunlight in the late afternoon.[6]

If there is some gradient, the rows can follow the land contours (any of the layout options in Figure 2.9 are possible). However, where land relief determines row orientation, light may place an upper limit on crop yields.

Wind may play a role in the system disposition. Winds could be problematic if the rows are oriented such that winds accelerate down the rows. If light dictates row orientation, a wind problem may be better resolved with an external barrier.

Another option has crops only within every other strip. This arrangement is utilized where overstory is continually yielding and harvest roads are needed to remove the output. This is often the case with high-yielding oil palms.

Agroforestry Strip Cropping

Alternating strips of trees and crops or different types of trees results in the agroforestry version of strip cropping. With tall trees and shorter crops, this is a variation on treerow alley cropping where the row orientation and the design package closely follows the treerow guidelines.

The difference lies in wide, rather than narrow, tree rows. In a cross section, this appears as *aaaabbbbaaaabbbbaaaa*. The wider tree strips can be a basis to lessen the negative tree-crop dynamics or as a means for greater productive exploitation of the tree strip. With fewer rows per area, the tree-crop interface distance is less a factor. This alternating design is shown in cross section in Figure 10.4.

Dwarf fruit trees in alternating strips with various crops exemplify this type of system. One practical benefit is that, given that the crop is harvested first, the now open strips expedite fruit harvest.

Other design options are apparent. If the tree species utilized is not highly complementary with the crop, the wider strip allows the insertion of a buffer species. The reasons, as previously stated in Chapter 2, are to

1. prevent canopy overhang;
2. block root spread;
3. reduce weed growth in tree strip understory;
4. allow normally competing species to be grown with a lesser interface distance; and
5. reduce the amount of unused space, increasing the LER.

In cross section, this layout is of the form *aaaacbbbbcaaaacbbbbcaaa,* where *c* is the buffer species. Figure 2.8 demonstrates the sought-after effect.

Clusters

Light efficiency clusters (see Chapter 8) can be upscaled and applied to groups of woody plants (see Figure 10.5). As with their seasonal counterparts, trees that seek vertical light are in the core and those trees or shrubs that wish for more horizontal light are at the edge. In more biodiverse situations, the clump or cluster pattern is ordered as a minimum interface design.

FIGURE 10.4. Tree with crop-strip cropping. This cross section, looking down the strips, shows alternating strips of productive trees and seasonal crops.

FIGURE 10.5. A light efficient cluster pattern. Species that prefer vertical light at the top center, horizontal gathering species at the edge, and shade tolerant species in the internal core. This is a scaled-up version of an intercrop cluster (as in Figure 8.4).

Often fairly biodiverse, these can have various combinations of shade, forest, and fruit trees with coffee and banana at the edge and in the understory.[7] These differ from light shade in that each plant receives some direct, unfiltered light. This differs from agroforests discussed (later in this chapter) in that they are species rather than ecosystem governed.

Light Shade

At times, economics seems to favor productive rather than facilitative light shade systems. As with the facilitative designs, a single overstory canopy filters light, protecting the understory from the negative impact of direct, intense sunlight. The overstory provides strong facilitative services but can be more of a cost and less of a cost savings. A productive overstory fills this revenue-cost gap. The design is represented in Figure 9.6.

The problem is finding a productive overstory species with the right DPCs, a low light area index canopy being paramount. Although few such species have been identified, light shade systems are among the most common, mainly because palms (notably coconut) have the right DPCs. Under palms are found shade-resistant cocoa or slightly light-demanding species.[8]

Heavy Shade

In contrast to the facilitative version, heavy systems with a productive overstory are relatively uncommon. As with their facilitative counterparts, the principal reason is cost control. Under these conditions, there is little reason to accept a higher upkeep productive overstory and risk upsetting the economic balance. Figure 9.7 is of a heavy shade system.

Although uncommon, a few species with a productive overstory have been documented. In Haiti, breadfruit has been raised as overstory for coffee and cocoa. A triple-canopy system from Brazil has nature brazilnuts over rubber trees which, in turn, are above cocoa.[9]

Agroforests

The agroforest (also referred to as a forest garden, forest farming, polycultural agriculture, or tropical home garden) represents a class of systems with extreme environmental affability. Millions of farmers throughout the world utilize these low-input, high-output systems. Although seldom studied, these are ubiquitous in tropical regions and, in some areas, few farms do without. Agroforests also exist in a few temperate countries.[10]

Of all the agriculture and forestry alternatives, the agroforest is unique in combining environmental friendliness, high productivity, low maintenance, and a compatibility with natural ecosystems. If there is a disadvantage, it is having to accept not much of one thing, but a mix of outputs. For many subsistence farmers, the resulting diversity in diet is a blessing. Because of the highly favorable ecological properties, the agroforest epitomizes the agroecological ideal.[11]

Visually, these are characterized by vegetative density, diversity, disarray, and duration. The defining characteristic is ecosystem governance incorporating a large percentage of tall woody perennials.[12]

Basic Premise

The best way to illustrate the agroforest is to fashion a hypothetical example. If one wanted a temperate agroforest, a wide range of agricultural species, including a large percentage of trees, are interplanted.

For trees, one might combine multiple varieties of apples, pears, peaches, plums, walnuts, butternuts, and mulberries along with a few forest species. For shrubs, quince, gooseberry, current, raspberry, blueberry, and blackberry could be among the included species. Vines are part of a natural forest system and add to the productive biodiversity; therefore, grapes are a good candidate for inclusion. As with a natural forest, these are put together without any apparent spatial or temporal pattern.

Forest ecosystems often contain a large aggregate of herbaceous vegetation, and agroforests follow this model. Rather than wild plants, crops assume this role. Maize, squash, beans, tomatoes, and a host of other garden species might be considered. Many farmers add decorative plants, and an abundance of native and exotic perennial flowers will increase the aesthetic worth of this construct.

Outside of natural forests, these exemplify ecosystem governance. For biodiversity and agrobiodiversity, up to and exceeding 300 species are possible with associated high values of biodisarray, biodensity, and bioduration. Most individual gardens have a lesser number of plant species but well above threshold value for ecosystem governance.[13] Figure 10.6 is representative, in cross section, of the mix of species commonly encountered in this agrotechnology.

FIGURE 10.6. An agroforest in cross section showing biodiversity, biodensity, and biodisarray. Along with disarray, canopy patterns are still possible; in this drawing is a difficult to discern midpoint design.

Management

Handling the biocomplexity takes less effort than might be supposed. Nature has a large say in things and takes charge of the facilitative tasks. However, without guidelines or rules of thumb to focus or direct resources toward useful outputs, management can be unfathomable and at the total whim of natural forces. To surmount this, suggested guidelines are as follows:[14]

1. if a plant output is needed and the plant is producing well, leave it; if not, alter the competitive environment;
2. if its production is not needed, neglect it;
3. if it is negatively affecting a more desirable output, prune it or, as a last resort, remove it; and
4. if space exists and essential resources are unused, often as measured by the amount of light striking bareground, plant or let something grow.

Given the modicum of existing knowledge on placement and on how species interact in complex ecosystems, rule-based management can be quite effective.[15]

Because these guidelines are efficient over time, agroforests are a good launching ground for new and unknown species. Once familiarity is acquired, subsequent use of these same species can be in simpler environments where more understanding of the underlying agricultural requirements is needed.

Subcategories

There are three subcategories of agroforest. These subdivide by both location within farm landscape and by types of plants contained. These are as follows:

1. Home gardens
2. Shrub gardens
3. Forest gardens which divide further into
   a. intense forest gardens
   b. enriched forests

c. forest farming
d. mixed crops and forests

Far from being uniform, agroforests of all types are often internally fragmented or subdivided. This is a lesson taken from natural ecosystems where transition zones (i.e., ecotones) between systems often contain high levels of biodiversity. It is these internal and edge zones, often indistinct, of varying species mix, and having a different light collection profile, that are formed and exploited for greater agrobiodiversity.[16]

Home gardens. Those systems closest to dwellings are the home gardens (see Plate 10.3). In temperate zones, these can be a mix of trees, shrubs, and trees. The outputs are spices, herbs (garden or medicinal), minor fruits, nuts, and vegetables, and other interesting and useful products. These are briefly mentioned under intercropping. However, full-blown examples with a large percentage of trees fall under an agroforest heading.

In humid tropical regions, most often these resemble, visually and ecologically, forest ecosystems.[17] Often surrounding houses, these provide a mix of spices, herbs, fruits and other outputs that have household use. As a tropical side benefit, these shelter houses from the sun and the heat of the day and provide an ecologically prudent means to recycle organic household waste.

Shrub gardens. The shrub garden is a less encountered version of the agroforest. These qualify as agroforests as the species contained are primarily short-stature perennials, and the outputs are fruits, nuts, and other treeborne products. Nonwoody, useful perennials, short nonwoody fruits (such as bananas), and short (such as papaya and rose apple) or dwarf fruit varieties are intermixed with a few taller trees (see Plate 10.3).

Forest gardens. A number of variations are associated with forest gardens. These may be characterized less through annual outputs and more for their overall facilitative effects. Of the four versions, (1) intense forest gardens, (2) enriched forests, (3) forest farming, and (4) mixed crops and forests, the intense forest garden is generally located at the forest fringe. As with all agroforests, mixed outputs are expected. These, however, offer more of the lesser utilized products or those that require infrequent harvesting. This includes wood

(a)

(b)

PLATE 10.3. Two forms of agroforests: *(a)* a tropical home garden surrounding a house and *(b)* a temperate shrub garden composed of dwarf fruit trees (apples and peaches are visible) and other perennial crops (horseradish is discernible).

and timber trees, products not overly associated with the household versions.

The enriched version is where forests at the farm fringe are economically augmented by adding species or through attrition of the less useful trees. Here, timber trees share the ecological stage with those producing fruits, nuts, and other products. An early European version of the agroforest would have been enriched with oaks or beech, both supplying high-value wood and eatable nuts.[18]

Forest farming takes away the forest understory and replaces this with a crop. This is a variation of heavy shade system except that the shade and facilitative effects result from a canopy of mixed tree species. In North America, a number of outputs have been suggested; goldenseal, ginseng, saw-palmetto, slippery elm bark, elder flowers, and eucalyptus leaves are botanicals that can be produced in this type of environment. Mushrooms are always a candidate for inclusion.[19]

Mixing crops with enriched forest fragments or forest gardens puts an environmentally sound agroecosystem in a facilitative role. These utilize coarse patterns and ecological spill over (through SIZ effects) from the dynamically rich agroforest to benefit nearly and closely connected crops. Figure 10.7 has this type of arrangement where crops are found in clusters or strips with biodiverse setting.

In an auxiliary role, the agroforest is not economically inert, and this adds some output to the expected facilitation. Where risk is para-

FIGURE 10.7. A forest farming arrangement that has crops, either in clusters or strips, within an agroforest. This drawing shows mostly a minimum-interface spatial pattern within the tree strip.

mount, the agroforest can supply alternatives outputs, including life-saving foodstuffs when crops fail.[20]

Feed Systems

Domestic and nondomestic fauna of whatever type, aquatic creatures included, are found in or near forest ecosystems. Through agroforestry, natural habitats can be duplicated or improved. Because various animals can be ecologically ingrained within and contribute to the economic outcome of an ecosystem, feed systems constitute a grouping of agrotechnologies. These can overlap with other agrotechnologies when grass replaces understory crops or overstory trees feed hungry animals.

There are pastoral systems. Cattle eat grass and pastures provide a lot of it. It is also possible to put a pasture below a tree canopy, and tree-pasture systems (silvopastoral) constitute what may be the most common form of agroforestry. Grasses and forage, being a large class of plants, have many species and varieties that are shade tolerant and thrive beneath a canopy. A list would be extensive, but, as a sample, most forest plantations can support some form of grazing, and nut trees (coconuts, walnuts, macadamia, etc.) can be employed as can a wide range of fruiting species. These also qualify as light shade systems.

Keeping with the grazing theme, grasses and other herbaceous vegetation work well in some climates but not so well in others. For cattle and the like, a mix of trees and grasses may offer forage year round. The grasses are eaten during wetter periods, and drought-resistant trees are consumed as forage when the other feed sources are not available. Pruning may be required to get the forage down to ground level.

Going beyond domestic grazers, agroforestry offers an eclectic mix of systems supporting a range of creatures.[21] Much of this borders on semihusbandry.

The simplest are the nonpastoral applications. These take whatever food type these creatures seek and provide more of it. Pigs thrive on fruit and nuts and can be fed and sheltered in an ecosystem where fruit and nut trees produce in abundance.

Some of these can be purposely designed. The dehesa systems of Spain support grazing and pigs through grasses and oaks at 20 to 49 trees per ha.[22]

The traditional streuobst of Germany are more mixed species orchards employing the nonharvest option. Grazers keep the grasses down, and pigs, geese, chickens, and so on convert fallen fruit to soil-enriching manure.[23] This may be a temporary abode as all the animals would have a primary ecosystem of residence, cattle in pastures or fallow fields, pigs in the surrounding forest, and geese and chickens in the agriculturally active plots.

Similarly, avian-agriculture or avian-forestry is where birds are raised in mixed tree-based systems. Chickens and turkeys are birds that, in their natural habitat, forage at ground level and shelter in trees. As with other feed systems, forage comes both from trees and understory vegetation. This, along with other forest-inclined domestic animals, is found in agroforests and similar agroecosystems.[24]

Placing water under or near trees produces a different outcome and other possibilities. Aquatic animals, for example, fish and frogs, find water-based systems to their liking and, as with terrestrial feed systems, the accompanying plants can be direct or indirect feed source. As a direct feed source, fish that eat leaves (e.g., carp) can be fed with above-pond trees.[25] Fish that eat fruit find solace when above-pond fruit trees shed plenty of fruit into the water.[26] Insect-eating frogs and fish like trees that harbor and dump insects into the water. These can be provided. Some species of willow serve well and are utilized above salmon ponds in southern Chile.[27] Aquaforestry is depicted in Figure 10.8 (see also Plate 10.4).

FIGURE 10.8. An aquaforest feed system where the fish eat leaves or fruit falling from the overhanging trees.

PLATE 10.4. Aquaforestry: in this example, a self-sustaining pond with the tropical fish species tilapia.

A rare version, often a more culturally oriented agroforestry practice, is that of entomo-agriculture. Well documented are silkworms raised on tree leaves, specifically mulberry leaves. The commercial versions are exclusively cut-and-carry, taking the mulberry leaves to the young silkworms.

Other insect-based systems have not yet evolved along commercial lines. This includes those insect species that are part of the diet in less squeamish cultures.[28] When consumed, these can be raised in a welcoming agroforestry environment. Most examples of harvestable edible bugs fall under the heading of supplementary additions rather that being the driving force for a new or separate agrotechnology.

Chapter 11

Temporal Agrotechnologies

In the proceeding chapters, static agroecosystems are discussed. The perspective taken is one of a moment in time or where the component plants are temporally insular.

This is not always the case. Many interplant or interspecies relationships have a forceful temporal dimension. This is often exploited for productive or environmental gain. The possibilities are strong enough such that a grouping of agrotechnologies can be based solely upon the temporal plane. Temporal strategies also underwrite some interpretations of alternative agriculture.

ECOLOGICAL UNDERPINNINGS

As an alternative agricultural approach, rotations can assume a lot of ecological responsibility. Rotating crops can

- contribute to soil retention and nutrient buildup;
- control insects and weeds;
- interrupt the buildup of residual chemicals (mainly allelopathic compounds but other toxins, for example, decay retarding wood extractives and in-leaf insect repelling chemicals); and
- through the addition of moisture holding in-soil biomass, help in countering drought.

Carryover

Each season, the soil on a given site contains a mix of essential resources (as in Figures 2.1 and 3.3). Theory states that this essential re-

Introduction to Agroecology: Principles and Practices

doi:10.1300/5553_11

source profile is best utilized by the crop species or variety that is least contained by one or more limiting resources. After harvest (and the removal of some of the elemental resources, that is, N, P, and K plus trace elements), an elemental resource profile remains. These residual minerals, as the basis for a new soil mineral profile, carry over to the next season (the arrows in Figure 11.1).[1]

The carryover is augmented by mineral gains. These come in the form of windborne, rain-carried, and/or internally released nutrients. Combined with light and water, a new essential resource profile is born.

Temporal Complementarity

The fundamental rule of temporal complementarity is that two species, those that normally grow well together, also sequence well. This is because those minerals that are residual at the end of one season (and available for the next) are the same that are exploited for intra-

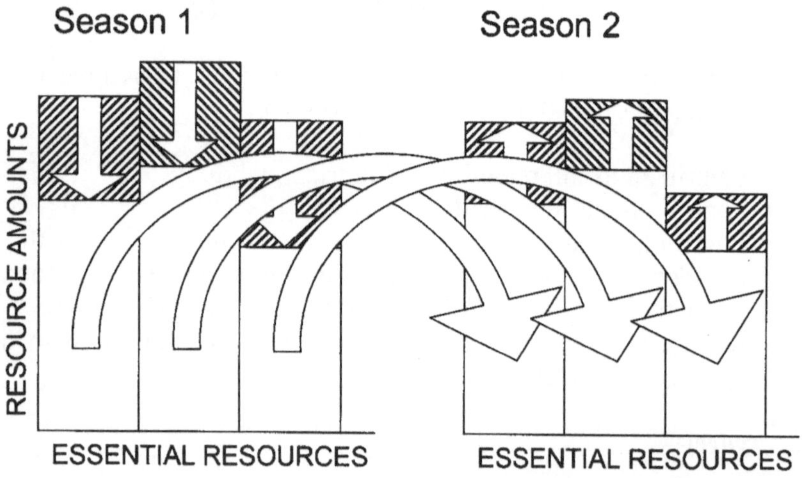

FIGURE 11.1. The carryover of nutrients from one season to another. In season 1 *(left),* some of the mineral nutrients are lost or harvested (the losses are stripped) and others carry over *(arrows).* In the second season *(right),* those carried over, augmented through site gains (the stripped additions), are the basis for an upcoming season and crop.

season complementarity (as with Figures 3.3 and 3.4). In other words, instead of two species occupying the entire soil resource profile, the unused portion (as after a monoculture) remains and can be utilized in a subsequent season by a second species that is complementary with the first (as in Figure 11.1).

The reverse is not true and two species that serve well in rotation do not necessarily thrive when grown together. This is due to the transitory resources, light and water. These may not be abundant enough for the two distinct species to share without incurring a low LER.

Although water capture mechanisms have been demonstrated (e.g., hydraulic lift), increased water capture in intercrops is not high; tallies of –6 percent to +7 percent have been mentioned. This does not include indirect mechanisms as when in-ground biomass boosts the moisture-holding ability of the soil.[2]

The majority of established rotational sequences support the first contention, that is, those that grow well together sequence well. This intercrop-sequence affinity traces back to early agriculture; for example, it has been long noted that barley both intercrops and sequences well with clover.[3]

Other long-established cropping sequences are as follows:[4]

maize → bean
maize → bean → potato
tomato → onion → squash
oats → clover → ryegrass
soybean → groundpea (peanut) → sunflower
wheat → maize → millet
wheat → clover → alfalfa → ryegrass
clover → alfalfa → ryegrass

Compared with the lists of familiar intercrops (see Chapter 8), a high degree of correlation is evident.

To take full advantage of a site, a preestablished ordering may be required. This occurs when one species is facilitative, gathering some of what a second needs. The gathering species will, by necessity, be first in the sequence so the second can garner what was gathered. Commonly, a nitrogen fixing plant will precede a species that demands this element.

Temporal complementarity follows spatial complementarity with regard to planting densities and interface distance. Among the reasons for an *aabaabaa* arrangement is a relative abundance of those nutrients that species *a* requires, less than those for species *b*. If sunlight and water are not the reason, this translates into uneven rotation where species *b* has less of a rotational presence. That is, species *b* may be planted every third, rather than every other year had an *ababa* spatial sequence had been the site norm.[5]

Temporal SIZs

Take the case of an upcoming potato crop, one which might be plagued and productively sapped by in-soil nematodes. Nematode-countering plants, for example, varieties of marigolds,[6] might be a secondary species, not with potato, but in the pre-potato rotational sequence. The result is that soil is better situated to receive the potato crop.

A temporal SIZ introduces considerable pairing flexibility and is especially useful if a facilitative species is competitive with the recipient crop but not in the precursor rotation. To be a valid option, the SIZ effect must linger in the soil and be exploitable in one or more subsequent seasons. Figure 4.1 applies; the only change needed is to relabel the lower x-axis from distance to time.

The possibilities extend well beyond nematodes and may be inclusive to many of those detrimental or positive organisms that reside fully or part-time in the soil. Temporal SIZs, and the resulting flexibility, can also be applied in nutrient situations. Again, the facilitative species is a predecessor to, but not concurrent with, the benefit-receiving plant.

Successional Approaches

Nature has definite ideas how things work across time. This can be true with temporal complementarity where one productive species naturally follows another. This occurs when a facilitative species establishes the right soil conditions for a productive species. There are two ways to implement these principles: (1) using established crops with recognized cross-season complementarity, or (2) select untried successional species based upon known DPCs (or in this case, redesignated as desirable soil characteristics) taking into account both preplanting and postharvest soil properties.

Species Based

Many regions have, through trial and error, established rotational sequences. Rotations from medieval Europe are indicative of this. These early farmers understood which crops should follow in sequence, utilizing a three-course system: (1) winter grains (wheat or rye); (2) spring grains (barley, oats, or buckwheat) or, at times, nitrogen fixing crops (peas, broad beans, or vetch); followed by (3) a pure fallow or fallow-pasture mix.[7]

If, for economic or other reasons, the known sequences are not viable, then one can devise a sequence based on side-by-side complementarity. The lists given in Chapter 8 are a good starting point. If these do not temporally provide the output mix that markets demand, the DPCs come into play.

DPCs

The basis for temporal DPCs is that each species desires certain preexisting soil conditions and, upon harvest, leaves the soil in a set condition. By matching the pre- and postsoil DPCs (or desirable soil characteristics) for each species, a rotational series is formulated. Any gaps can be plugged by external inputs (fertilizers) when a precursor crop or an inserted fallow cannot fill the gap in mineral resource needs. A case in point, beans may follow wheat, not because this is a recognized sequence but, through a postwheat soil analysis, conditions are ripe for beans.[8]

An expanded list of soil DPCs (both pre and post) should span the full range of soil properties. Mineral nutrients; moisture-holding biomass; weed-controlling preconditions; in-soil, insect-killing chemicals; and beneficial in-soil flora and fauna are part of this.

Management options can heighten any cross-seasonal effect. Inclusive in these are the (1) frequency of planting, (2) timing of plowing and sowing, and, if required, (3) types and timing of outside inputs. As an example, yellow dwarf virus and some weeds can be controlled by the late drilling of barley.[9] When these threaten, this management variation becomes part of an unfolding but not set-in-place plan.

Agrotechnology Progressions

For many temporal progressions, especially those with trees, each interseasonal stage may also be distinct, spatial agrotechnology. For example, a rotational sequence may pass from a monoculture to a monoculture, from a seasonal intercrop to a seasonal intercrop, or from a monoculture to a seasonal intercrop.

For others, the progression may be more elaborate. The spatial advancement for an extended taungya (a temporal variation) is often an agroforestry intercrop followed by a treerow alley system proceeding on to light then heavy shade systems. Plate 11.1 shows two agrotechnologies as associated temporal stages, that is, being points along a temporal sequence, these share a primary species (in this case poplar).

When the progression involves an interchange of spatial agrotechnologies (as in Plate 11.1), best results may come by stringing together, across time and with ecological dynamics in mind, a series of optimal designs. This involves species pairing complementarity and a more-or-less standard design for each phase. Most keep the shared, cross-seasonal plants as is; other designs modify the planting densities. For the latter, a smooth transition is required, that is, it is easier to remove than to add trees.[10]

Contributing Factors

As mentioned, management inputs, such as sowing and plowing, have broad implications. There are pros and cons and a number of considerations. This can be a complex topic. Enough is presented here to broach, but not resolve, the options.[11]

Plowing loosens soils, allowing root penetration and, by deeply burying many weed seeds, offering a brief reprise from unwanted vegetation. The disadvantage is that soils, for at least a short period, are highly exposed. For root crops, there is a postharvest bareground phase. As with a postplowing period, soils are exposed, both to wind and water erosion, and the leaching of nutrients (nitrogen figures highly in this).[12]

(a)

(b)

PLATE 11.1. An extended taungya sequence in a poplar plantation: *(a)* the postplanting, light-shade stage with sugar beets and *(b)* an intermediate, alley phase where currants are the understory. The photos are of a commercial system in southern Chile.

There are measures that reduce this danger. Windbreaks and strips provide security, as can the simultaneous or delayed planting of a quick-growing cover crop. As an offshoot of the discussion on cover crops (Chapter 8), the latter helps in two ways, with direct soil protection and by outcompeting weeds.

A temporary, nonintercropping cover is a counter possibility, one that has the cover crop planted after the primary crop is harvested. This holding crop naturally expires on subsequent plowing and/or the sowing of the next primary species. In cold temperate zones, winter rye serves well, surviving a cold winter and reaching maturity in late spring. The rye is not usually harvested. Yet another possibility to address the bareground dilemma are nearby cut-and-carry strips where imported biomass is quickly transferred after plowing and sowing or postseason, for example, when a root crop is harvested.

Skipping the bareground phase may be possible. For this, no-till agriculture provides a continual ground cover that mitigates much of the erosion and leaching danger. Ground-level protection is through residual cropping debris. Unless well formulated, weeds often emerge as the main no-till threat. With this tillage option, weeds can be combated with cover crops that survive and regrow after a no-till sowing. Other counters include burning, mulch, and/or some animal control.

Tillage and cover crop strategies to engage weeds and erosion are only part of the picture. Also entering the equation are damaging insects and nutrient flows. For each crop and cropping system, the goal is to reach an accommodation, compromising as little as possible on each agroecological issue and outcome. The study briefed at the very end of Chapter 5, that of an organic vineyard with multiple cross-purpose ecological features, engages many of these elements, but in a less risky, more forgiving perennial setting. Because the options multiply, the implementation difficulties magnify as various crops come and go over a temporal cycle.

Fallows

A fallow (unfarmed period) stands separate in the scheme of agrotechnological classification employed here. The fallow is, in essence, the temporal equivalent of the auxiliary system. In and of themselves, these generally produce little or no agricultural output but serve to augment the productivity of a later planting season. Fallows vary in length from a single season to 50 or more years.

Further confusing things is the productive fallow. These yield comparatively little as the need to build up soil nutrients is of paramount importance. Still, the opportunity for some output cannot be overlooked, especially when a fallow can last many years.

Most productive fallows are long term. After the main crop has been harvested, a multiyear sequence is entered into. The initial years are somewhat productive, while harvests in the latter years are reduced, allowing nutrients to build in anticipation of cropping.

Where trees are involved, the roots, if not frequently cut, grow in diameter. As a practical concern, this can seriously hamper plowing after a long fallow period. If a fallow is part of strategy, the tree should have DPCs such that this is not a problem (as with a deep-rooting species).

Slashing

At the end of a long-term fallow, the woody vegetation must be removed. The leaves are nutritionally valuable whereas the woody stems and branches are low in essential nutrients and can even uptake minerals in the process of decay. Furthermore, the stems and branches often impede planting, adding to labor costs. There are two options in resolving these issues: (1) slash and mulch and (2) slash and burn.[13]

Slash and mulch is where, after cutting, the leaves remain but the stems and branches are manually removed. The mulch shortens the erosion-prone, bareground phase and the on-soil biomass can smother weeds.[14] On the negative side, the nutrient demands of the crops and the release of nutrients through leaf decay may not be optimal.

Slash and burn is where the standing vegetation is cut and burned and crops planted. Besides quickly releasing nutrients, an intense or hot burning destroys weed seeds lurking in the soil. The drawbacks are in the nutrients lost and enhanced erosion danger from the burned-over bareground.[15]

There are numerous variations that take advantage of specific fallow vegetation. One employs a perennial root stock, such as bamboo, as the fallow species. This is cut and burned, and later resprouts. In seed-based regrowth, the burn germinates seeds remaining from fallow, for example, with kudzu, but the regrowth is slow enough to allow the crops to yield before being burdened by kudzu-on-crop competition.[16]

ECONOMIC UNDERPINNINGS

A byword of agroecology and the overriding temporal question is long-term sustainability. Outside of the what is, the remaining questions are in how sustainability is gauged.

LER offers insight, either through a temporal LER (TLER) or an LER-based statistical trendline. The first of these, the TLER, is derived as follows:

$$\text{RLER}_1 = (\text{LER}_a)_1 + (\text{LER}_b)_2 + (\text{LER}_c)_3 + (\text{LER}_d)_4 \qquad (11.1)$$

For the above, the RLER is for a rotational sequence of species *a, b, c,* and *d,* a four-season cropping sequence.

$$\text{RLER}_2 = (\text{LER}_a)_1 + (\text{LER}_a)_2 + (\text{LER}_a)_3 + (\text{LER}_a)_4 \qquad (11.2)$$

LER_a is for a single crop (a) sequence across four seasons. The temporal LER (TLER) is calculated as

$$\text{TLER} = \text{RLER}_1 / \text{RLER}_2 \qquad (11.3)$$

Time-based RVTs and CERs can be similarly derived.

The statistical method provides stronger indication of sustainability but requires a longer data stream. For the actual calculation, the LERs for the time periods under consideration are put through statistical regression analysis. An upward trending regression line is good, and downward is bad. What makes this easier is that no direct yearly comparison is required and unplanned events, insect attack, drought, and so on, if random, are generally factored out.

In addition, other methods indicate sustainability while addressing the financial issues. Monetary return is important to most farmers and part of long-term decision making. Net present value (NPV), applied to yearly returns, can be a measure of sustainability.

One must follow NPV guidance with some trepidation.[17] Putting too much emphasis in the initial years is unwise as long-term damage is masked. On the flip side, NPV can promote greater agrobiodiversity. A case in point occurs with plantations and orchards in the initial postplanting period. Here costs are high and output low. Adding a crop as an income source during this early, high-cost period can effectuate a favorable NPV.

Straight financial analysis, analyzing the revenue stream with or without NPV, is a gauge of fiscal sustainability, remembering that, if a system is not profitable over time, overall sustainability may not count. Without long-term profitability, a system will not be chosen.

Examples abound where profitability trumps sustainability. Take where the cost of fencing is a grazing determinate. Adding trees to the contained pasture (as with a light shade system) can reduce grass production, lessening the per area stocking rate and/or reducing the grazing period. If not already depreciated, amortizing the cost of new fencing against a lessened stocking rate or a shorter grazing period can make the tree biodiversity option financially unattractive.[18]

TEMPORAL AGROTECHNOLOGIES

To fully implement rotational ecodynamics, the temporal agrotechnologies play a large role. As with nontemporal versions (previous chapters), this can confer gains without becoming enmeshed in complexity of unguided ecological happenings. Covered here are the following:

- Time periods without any planned interseasonal dynamics (discrete rotations)
- Rotations where each season is a distinct event, but with an exploitable ecological overlap (series rotations)
- Cross-seasonal plantings where the duration of one overlaps with others, leading to a series of intercrops such that the system remains, but in planned change (overlapping cycles)
- Intercrop plantings based around the tree growth dynamics of plantations or orchards (taungyas)

- Ecosystems, that once started, have no planned end (continual or enduring)

Single or Discrete Rotations

The simplest form of implementation carries the view that each season or crop is a single, discrete temporal entity. There is no carry-over or regard to previous or future land-use alternatives. Any soil deficiencies or lurking in-soil dangers (insect larva, weed seeds, etc.) are overcome through appropriate external inputs (fertilizers, introduced predator insects, insecticides, etc.). Figure 11.2 shows a sequence of like crops.

This series offers the greatest flexibility in what to plant each season. Economic flexibility, in being able to respond to shifting market demands, is traded off against the ecological gains and the decreased costs that normally occur with well-planned successional dynamics. Existing evidence suggests that this is not a wise trade (see section on Economic Gains, Chapter 14).

Series Rotations

The planned rotation sequence is well established in agriculture and few doubt the merits. This is the easiest of sequences; another crop is planted at the end of one cropping sequence, either directly or delayed by climatic event (i.e., the onset of winter or a dry season). In either case, the sequence is such that as much as possible is accomplished with cross-seasonal crop pairings. In accordance with the theoretical underpinnings presented earlier in this chapter, this can be done with temporal complementarity or based upon residual or temporal DPCs (or desirable soil characteristics).

A fallow can be part of temporal planning and a preplanned rotational series. As the land rests, ecologically good things happen, that is, insect pests, those awaiting a suitable crop, become disappointed; some plant diseases cannot lie dormant forever and can be curtailed with time; and soils continue collecting nutrients. Economically, it is more cost-effective to harvest high, rather than low per area yields, and a fallow, inserted in a sequence of crops, will increase the per area harvest and reduce the per unit cost of the harvest. Figure 11.3 shows a series rotation incorporating a third season fallow.

Overlapping Cycles

There are plenty of examples where temporal differences in cropping can be exploited. Some involve short-term crops, other systems are longer term. A classic example is lettuce interplanted with tomatoes; when the lettuce is harvested, the tomatoes remain. Clover or winter rye, overlapping through a late planting with root crops, prestarts the second rotation.[19]

Most cases are more complex, as is shown in Figure 11.4. A documented case from El Salvador can be shown diagrammatically, across time (left to right) as[20]

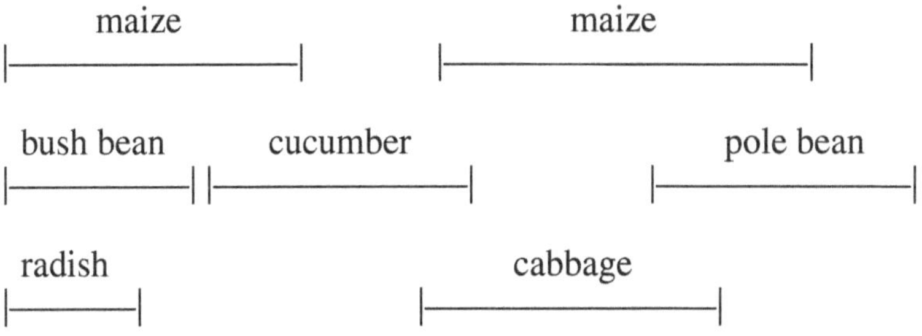

Overlapping cycles apply to perennial systems as long as the system has a beginning and an end. Orchards containing trees of varying ages also qualify.

Taungyas

Taunyga is a forestry term that originally denoted the planting of crops with trees during the tree establishment phase.[21] The concept has been expanded with four taungya categories: (1) simple, (2) extended, (3) multistage, and (4) end stage.

Employed mostly for instituting forest tree plantations or agricultural orchards, these, in their more advanced form, offer considerable cropping possibilities. These differ from trees in overlapping cycles in that, with a taungya, the tree is the primary species whereas, with

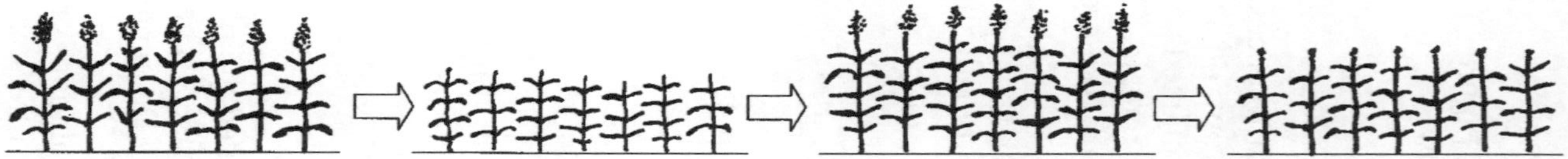

FIGURE 11.2. As shown in this four-season string (left to right), single rotations are a series of like or ecologically independent species planted across seasons with little or no planned, ecologically significant, temporal crosslinks.

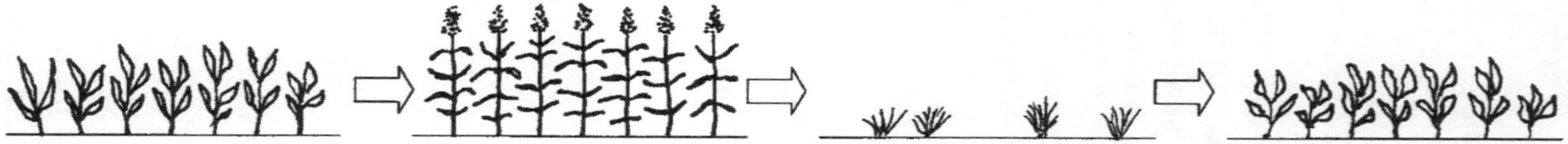

FIGURE 11.3. Series rotations where, for soil improvement or other ecological gains, crops are changed seasonally in a predetermined sequence. A fallow, shown in the third season, can be part of a series rotation.

overlapping cycles, the relative importance of each component varies with time. Ecologically, these take advantage of an ever-changing essential resource situation with a plantation or orchard.

With taungyas, obscure tree planting options come to the fore. The planting of mature trees, as tall as 4 m high, can initiate a sequence. This is done for good reason, large material plantings ensures a good quality plant; if fruit bearing, quicker yields; if for wood, a straight, clear stem; and if in a pasture setting, the height keeps the foliage away from hungry animals.[22]

Simple

In a newly established plantation or orchard, the trees are small, draw few resources, and offer considerable opportunity to plant crops between the trees. Ecologically, the crops replace weeds or other scrub vegetation that would normally colonize the open ground. Rather than bearing the cost of weed removal, the crops become an economic plus.

The simple taungya occurs for only one or two seasons. As the tree become competitive with understory, the taungya phase ceases and the system resorts to the intended plantation or orchard form. Figure 11.5 is of a simple taungya with cropping in the first year. Because of the versatility, any tree plantation or orchard may initially host almost any short-duration understory crop.

Extended

The extended taungya starts life much the same as simple taungya. Instead of ending, the understory relationship continues after the trees become competitive and an economically viable understory extends well into or for the full life cycle of a plantation or orchard. This can be with a single species or planned sequence of understory crops. Figure 11.6 shows, for the life cycle of an orchard or plantation a series of understory crops.

Examples include, in South America, orange trees planted with maize and then papaya. In North America, walnut trees have been coplanted with Christmas trees.[23] Often, shade-resistant grasses and grazing complete the cycle.

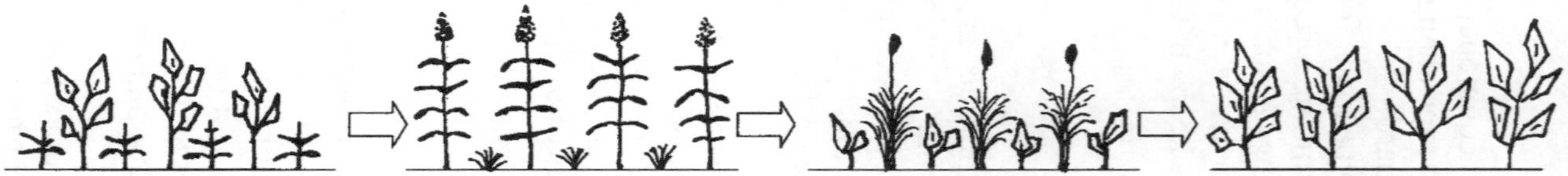

FIGURE 11.4. Overlapping cycles is where some productive plant species cross, overlap, and interact between normal seasonal plantings. In this illustration, each crop takes two seasons and temporally overlaps.

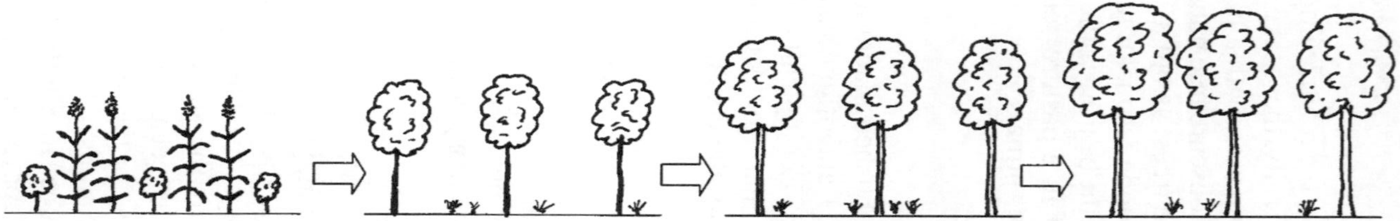

FIGURE 11.5. Shown through four different time periods, a simple taungya has intercropping in the first season. This serves as a prelude for a multiyear orchard, tree, or tree crop plantation.

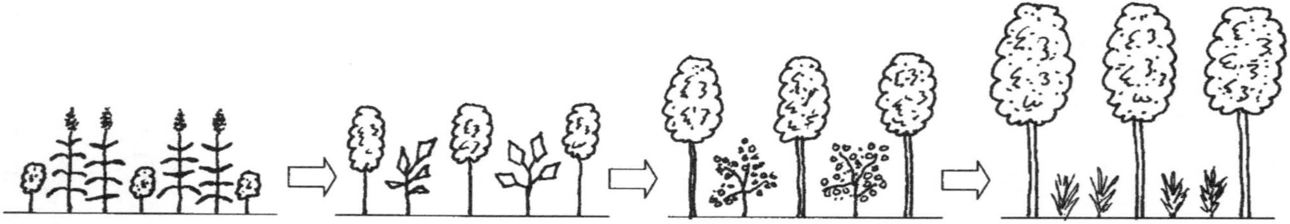

FIGURE 11.6. To take advantage of surplus resources and accommodate the changing essential resource situation, an extended taungya has series of secondary species under a long-term, multiyear plantation.

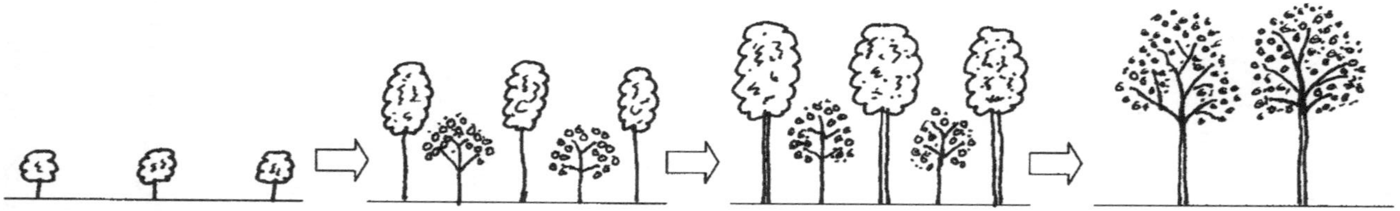

FIGURE 11.7. The multistage taungya features a switch in primary species across the life of the plantation. This is similar to an orchard or plantation version of the overlap sequence (see Figure 11.3).

Extended taungyas can be difficult to implement as each crop has to be complementary with the primary species in an ever-changing, essential resource situation. Practical and market considerations add to the complexity. This may preclude an active understory in orchards or tree crop plantations (e.g., coconuts or oil palms) where multiple harvests must be coordinated.

Despite the difficulties, these find use. In a Chilean poplar plantation, maize or sugarbeets precede oats or currants. A shaded pasture completes the 22-year cycle. The photo sequence (Plate 11.1) shows the first two: single-season sugarbeets followed by multiseason currants.

Multistage

Even more complex are the multistage taungyas. These start life as simple taungyas but, as the situation evolves, there is a midstream switch in the primary species.

The simplest form is an overlap of two plantations: planting the second rotation before the first has been harvested. Figure 11.7 illustrates a switch in primary species where the species that starts the sequence is not present at the end.

Other permutations exist where a transposition occurs with the secondary, overstory species. From Brazil, banana and cocoa are interplanted, the banana providing shade for the young cocoa thereby ensuring survival. As the banana matures, fast-growing shade trees are planted. These replace banana as the facilitative species and the shade tree-cocoa mix continues.

Also possible with coffee, cocoa, and other shade-tolerant crops is to replace the fast-growing shade with a slower growing, higher valued tree. This occurs a few years into the sequence before the faster-growing trees become too large. As an added facilitative effect, the fast-growing trees can guide the slower ones, producing tall, straight, higher-valued logs in anticipation of the final tree harvest.[24]

End Stage

Plantations, especially forest trees, are not always harvested in one swoop. If taken at different times, there are canopy opening, surplus essential resources, and opportunities for cropping. Figure 11.8 shows

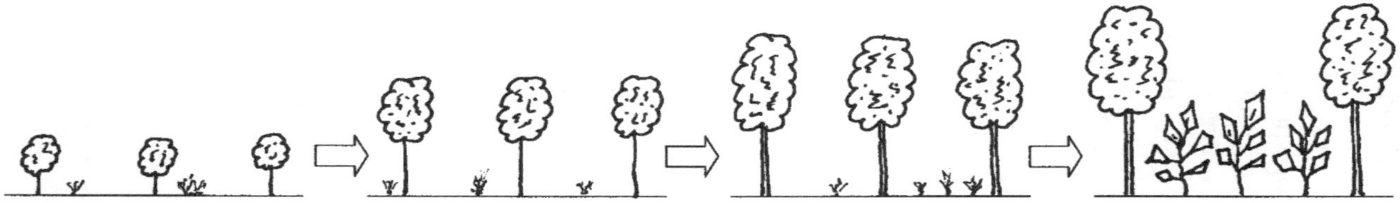

FIGURE 11.8. The end-stage taungya takes advantage of thinning late in the life of a plantation where now surplus essential resources permit a brief, latter stage cropping cycle.

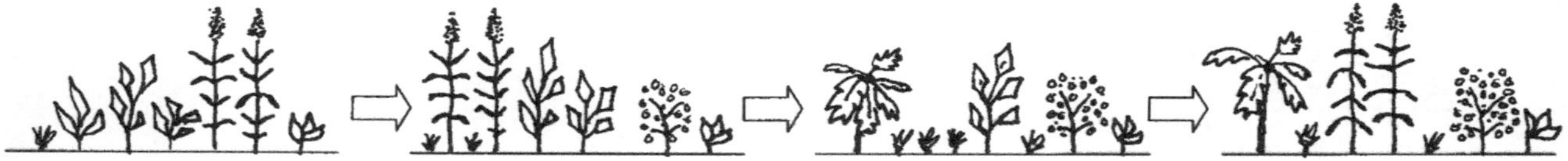

FIGURE 11.9. A continual agroecosystem that, across time, is ever changing, but has no definite end. In contrast, the overlapping cycles (Figure 11.3) do end and therefore contain fewer perennial species.

the partial harvest of the trees where, in the last phase, rows or gaps occur in the canopy. These are exploited for crops or grazing.

A few have been studied. Clason and Robinson (2000) describe a pine plantation in the southern United States where the trees were thinned after 20 years. The open space was fertilized and planted with forage grasses. The grass yields were reported to be 50 to 75 percent of what would be normal if the trees were not present. The fertilizer increased tree growth 30 percent. Combined with the grazing output, the anticipated LER is between 1.8 and 2.1.[25]

Continual

With the continual sequence, the individual plants change but the agroecosystem remains. Common with silviculture in natural forests, agricultural versions exist. The most prominent are pastures where the individual plants die and are replaced but, because this is staggered, the ecosystem endures. Other complex perennial ecosystems, for example, agroforests or mix tree-crop gardens fall under this heading. Figure 11.9 illustrates continual change in a nonending ecosystem.

Continual systems are mostly ecosystem governed. Allowing nature to run its course with only minimal inputs is cheaper than trying to direct successional dynamics through active management.

Chapter 12

Auxiliary Systems

There are two main categories under the heading of vegetative agrotechnologies: (1) productive (principal-mode) and (2) facilitative (auxiliary). The auxiliary systems differ from productive systems in that these are not expected to provide any directly useful (e.g., sellable) outputs. Their sole, or overwhelming, role is as a facilitative landscape feature, often located near productive systems, positioned and purposed to enhance the productive potential of those systems that do produce. Any direct output that may result is highly secondary to the system's facilitative role.

ECOLOGICAL UNDERPINNINGS

All ecosystems have positive specific interaction zones that extend outside the bounds of the immediate ecosystem. The assumption behind auxiliary systems is that, when individual principal-mode agroecosystems cannot protect themselves against all or some threats or when the internal mechanisms are weak and can be overwhelmed, specific interaction zone protection can be obtained from dedicated, neighboring ecosystems. As auxiliary systems are devoted to one or more ecological tasks, they can be very efficient at delivering the needed services.

Although formulated for certain tasks, there is a more or less standard package of ecological gains associated with auxiliary systems. Elaborated upon, auxiliary systems can

- inhibit the movement of plant diseases and damaging insects,
- provide a haven for insect-eating insects and insect-eating fauna,

Introduction to Agroecology: Principles and Practices

doi:10.1300/5553_12

- harbor pollinating creatures,
- control winds with related benefits,
- slow water movement and capture escaping essential resources,
- modify site microclimate (above- and belowground), and
- prevent or encourage the spread of specific influences between neighboring plots.

Of course, not all auxiliary systems offer this complete package. If a system does make inroads into the aggregate, the primary task or design purpose still determines the frequency and location of auxiliary structures in the overall landscape.

Figure 4.1 graphically illustrated the SIZ overreach and its spillover effect to a nearby agroecosystem. As with individual species, values above one represent the positive effect of the SIZ. These occur, for example, with predator insect dynamics, windbreaks, and waterbreaks. With riparian buffers, downstream water purity is a form of SIZ.

Because the individual species at the edge or interface of an auxiliary system may not be complementary with the bordering crop, measures must be taken to keep the LER high. A favorable buffer species between the two systems can improve upon an unharmonious plant-on-plant interface.

Auxiliary systems produce facilitative SIZ effects. This does not mean that these die or fade away at some predetermined distance. The targeted productive system may not have the strength of presence to create the facilitative influence but may have the inherent ability to stretch or widen the SIZ of the neighboring auxiliary system.

Examples of SIZ expansion abound. Predator insects may not breed in a farm field but may, upon entry, find the condition favorable for extended travel. Similarly, a crop plot may be defenseless against strong winds but, because of isolated or parkland trees, may be able to counter breezes or windbreak-evading downdrafts.

If the DAP of the targeted ecosystem is not favorable for SIZ expansion, smaller internal structures, specifically placed, can help convey the facilitative effect. Predator insects are often delivered from an auxiliary habitat to the core of a farm field through corridors (see Figure 12.1).[1]

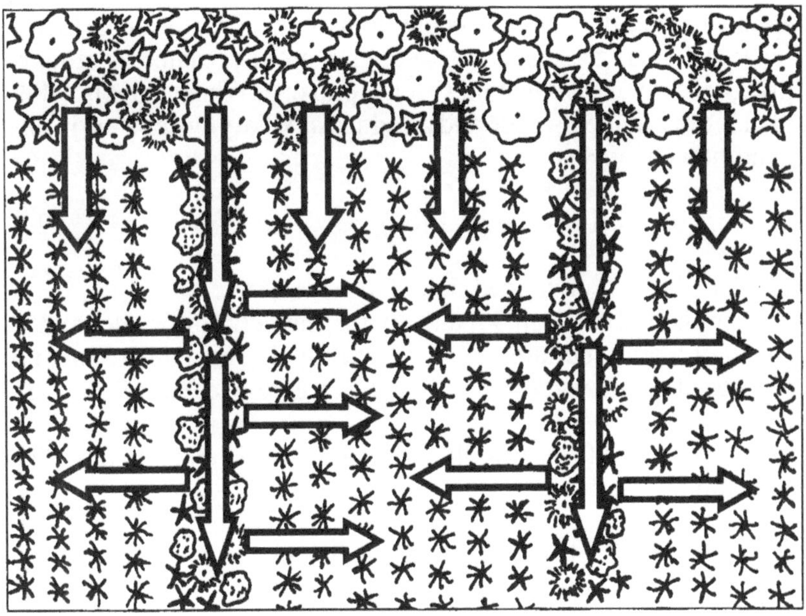

FIGURE 12.1. A system of corridors designed to convey chosen predator insects deep into farm fields. The arrows show hoped-for predator movement from the larger reservoir structure (top) into the vegetative corridors and the crops.

The relationship may be stronger as when plot-internal corridors, besides being avenues for flora and fauna movement, also offer habitat possibilities in their own right. This is an SIZ augmentation.

ECONOMIC UNDERPINNINGS

For these nonproductive landscape structures the economics are not complicated. The simplest calculation is through, the landscape LER (LLER). This is determined as

$$\text{LLER} = a(\text{LER})_1 + b(\text{LER})_2 + \ldots + n(\text{LER})_n \tag{12.1}$$

where the individual plots or ecosystems are $(\text{LER})_1$ through $(\text{LER})_n$. The actual areas for systems 1 through n are formulated as ratios (a through n above) such that these add up to one. When a plot is subdivided into a principal-mode and an added auxiliary system, the resulting add-on change has

$$c(\text{LER})_3 = p(\text{LER})_{3a} + q(\text{LER})_{3b} \tag{12.2}$$

such that, for the subdivided field 3, $c = p + q$.

When fields 1 through 4 make up, respectively, 25, 25, 40, and 10 percent of a farm (buildings and the surrounding area utilized are not counted), the ratio values (a, b, c, and d) are, respectively, 0.25, 0.25, 0.40, and 0.10.

For the unmodified farm, the original LLER is equal to one. This calculation is

$$\begin{aligned}\text{LLER} &= 0.25(1.0)_1 + 0.25(1.0)_2 + 0.4(1.0)_3 \\ &\quad + 0.1(1.0)_4 = 1.0\end{aligned} \tag{12.3}$$

Adding a nonproductive auxiliary structure onto plot 3 (which takes 25 percent of plot 3 or 10 percent of total farm area) increases the yields from plots 3 and 4. The result is

$$\begin{aligned}\text{LLER} &= 0.25(1.0)_1 + 0.25(1.0)_2 + 0.3(1.90)_{3a} + 0.1(0)_{3b} \\ &\quad + 0.1(1.3)_4 = 1.20\end{aligned} \tag{12.4}$$

Similarly, income (LRVT) and cost (LCER) can also be determined with the equations below where

$$\text{LRVT} = a(\text{RVT})_1 + b(\text{RVT})_2 + \ldots + n(\text{RVT})_n \tag{12.5}$$

and

$$\text{LCER} = a(\text{CER})_1 + b(\text{CER})_2 + \ldots + n(\text{CER})_n \tag{12.6}$$

For a complete picture, financial accounting must be utilized. Here, the benefits received, in monetary terms, should be greater than the yearly costs of operation. Risk and sustainability would also be included in the analysis.[2]

AUXILIARY AGROTECHNOLOGIES

As a landscape feature, these agrotechnologies have a long history of use. Early hunters and gatherers modified the landscape to improve their natural take, even before the idea took hold that things could actually be planted.[3]

Far from new, the number of systems under this heading is varied in form and role (see, e.g., Plate 12.1). The recognized agrotechnologies are the following

- those that store or foster forms of useful biodiversity, either flora and fauna (reservoirs/habitats);
- those that encourage the spread of biodiversity (travel corridors);
- wind-countering structures (shelterbelts and windbreaks);
- strips of vegetation that divest moving water of impurities (riparian buffers);
- control structures that prevent scouring and soil loss during flooding (waterbreaks);
- strips that halt or impede the spread of wildfires (firebreaks);
- areas that arrest the spread of productively negative cross-landscape influences (buffer zones);
- structures that hold back animals or people (living fences);
- areas of useful plants that are cut and carried to other ecosystems (cut-and-carry); and
- areas devoted to collecting water runoff (catchments).

What makes this topic interesting is the considerable overlap between the different systems, so much so that the primary purpose can slip from view. A few cases in point: windbreaks can also be travel corridors; shelterbelts can be reservoirs and habitats. The peril is when one sees only the windbreak but not the habitat. Without proper analysis on the effectiveness of the windbreak, an attempt to change the design may unknowingly destroy an effective habitat.

The other danger comes in the opposite direction, that is, in viewing these as multipurpose systems. The primary purpose may be lost or weakened attempting to do too much with a single system. In placing a windbreak to ensure effectiveness as a habitat, this may render the windbreak as less than optimal as a wind control structure.

(a)

(b)

PLATE 12.1. Auxiliary-type structures: *(a)* a cabbage field laced with travel corridors and *(b)* a wind-protected pasture.

Reservoirs/Habitats

Areas of land can be set aside for predator insects, birds, bats, and other organisms of control.[4] Such creatures (e.g., insect-consuming birds) may utilize these for breeding and shelter, exiting to seek food for a growing family. In addition, the habitat might shelter a core population (e.g., of predator insects); whereas the offspring seek their own territory outside this zone and within crop plots.

Blocks of natural or less intensely managed vegetation serve this purpose and, as such, are a store of native plant biodiversity. Habitats are best if connected to like areas. The options are to combine reservoirs/ habitats with agroecosystems such as agroforests, forest plantations, and other perennial blocks.

In order to function optimally, the areas are enriched with food and habitat needs. For example if woodpeckers are desired in neighboring fields, plenty of dead nesting trees should be included in nearby habitats. For other sought-after birds, specific designs of birdhouses can be a useful addition.

Corridors

The installation of corridors across an agricultural landscape may seem counterproductive, taking land without providing much in return. This is far from the case. Corridors serve base functions such as

- facilitating the movement of predator insects (see Figure 12.1);
- providing a habitat for pollinating insects and animals;
- the possible augmentation for a distant windbreak;
- allowing native plants, small animals, and even microbes to colonize in receptive ecosystems;[5]
- erosion control; water infiltration strips;
- insect and disease containment barriers; or as a
- rest zone for far-ranging creatures (including birds) that hunt in-field insects or rodents.

In practice, corridors come in many forms. In one study, 1-3 m wide grass corridors subdivided larger fields and served both as a self-contained habitat or, when connected, as a path to attract and spread beneficial insects or birds from larger breeding grounds (i.e., reservoirs or habitats).[6]

In most cases, these extend parallel to crop rows. As with strips and similar patterns, these allow for efficient forward travel by tractors and draft animals. Less used, but possible, are counterpatterns. These corridors extend perpendicular to crop rows. For this, farm machinery inactively passes over the corridors to complete the run.

The density (frequency) of corridors is a function of SIZs and placement according to the anticipated interactions with cropping areas. The width and contained vegetation are formulated for the types of organisms that are expected to utilize these natural highways.

Wind Structures

In agriculture, the windbreak reigns supreme is the premier auxiliary technology. Where needed, an ecologically complete landscape will come equipped with windbreaks, often set out in league with their larger cousin, the shelterbelt.[7]

Although windbreaks and shelterbelts are formulated differently, they both serve to protect crops and animals from prevailing or sudden damaging winds. For crops, the danger is that

- constant, drying winds rob soil of moisture;
- stronger winds cause stems, branches, and leaves to rub and, to repair the damage, resources are subtracted from growth and yield;
- strong gusts can cause sandblasting, again with external damage and yield reductions;
- brief, sudden bursts can lodge crop species with harvest losses;
- pollination efficiency is reduced;[8]
- airborne contaminates are contained in the area of origin;[9] and
- in wind unprotected plots, temperatures tend to fluctuate with accompanying growth and yield stress.[10]

These are not minor effects. Unabated leaf rubbing can decrease maize yields by 28 percent, wheat yields by 11 percent, and bean yields by 45 percent; windbreaks eliminate this problem. Similarly, water use in soybeans was reduced 20 percent through windbreak establishment.[11]

For animals, comparable hazards lurk. Although animals are seldom blown over, when unprotected and exposed to weather extremes, they do (1) gain weight slowly; (2) experience greater mortality; and (3) breed less frequently.

Outside simple wind protection, these structures have other applications that may change placement and design. Subdividing plots into very small units to maintain a humid local climate when drying endangers crops. Where frosts are prevalent, overhead trees in a nearby windbreak can help retain daytime heat. Along these same lines are those windbreaks that channel cold air flowing downhill away from sensitive crops.[12]

Established guidelines suggest windbreaks and shelterbelts at 40 to 50 "heights" apart and interplaced windbreaks at 30 "heights," a "height" being the measured altitude of the wind-interfering structure.[13] Among the factors to be considered are the relative sensitivity of crops and the severity of the winds. As an example, garlic is not considered to be wind susceptible whereas strawberries and cherries are.[14]

Shelterbelts

In contrast to windbreaks, shelterbelts tend to be wider, sometimes taller, strips of trees that cross the landscape a wide intervals. The spaces in between are subdivided by narrower windbreaks. Shelterbelts are usually are more than three trees wide and, as a result of internal root and mutual canopy support, are less prone to stem breakage and toppling than the windbreaks. These can be located where soils are loose or shallow.

Being wide, shelterbelts have strong secondary productive uses. Figure 12.2 shows two shelterbelt designs: one for forestry (wood) outputs, the other more suited for tree crops (fruits, etc.). In North America, the suggested species for a productive shelterbelt are, along the edges, hazelberts, Manchurian apricots, plums, juneberries, jostaberries, bush cherry, gooseberries, basket willow, and currants while, in the core, are found, pawpaw, Asian pear, persimmon, black walnut, and mulberries.[15]

Windbreaks

Windbreaks are not as wide as the shelterbelts; common versions are one, two, or three species wide. The narrower windbreak, with

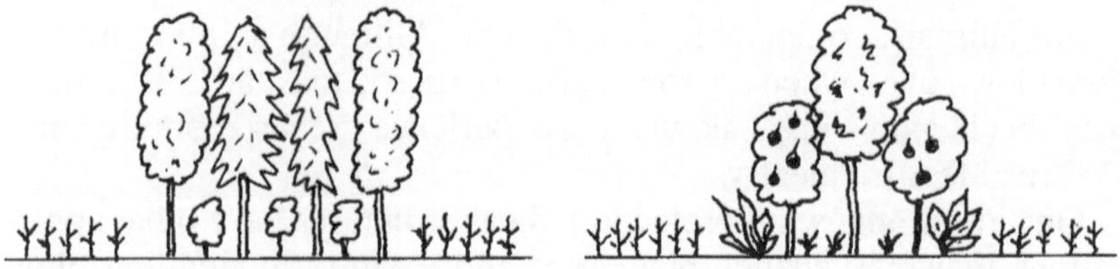

FIGURE 12.2. Two shelterbelt types: forestry *(left)* and tree crop *(right)* designs. The closed shaded internal portion of the forestry design is tailor-made for growing wood as clear, high value tree stems. The open edge of the tree crop design is better suited for fruits, nuts, or other light-demanding tree crops.

less internal support, can have more difficulty resisting, without toppling, the push of strong winds. Where winds are lighter, these can be stand-alone structures; where winds are stronger, they may be better integrated into a shelterbelt-windbreak mix.

To counter winds, a key factor is ground level density, more open at the upper reaches. The ground-level density requirement eliminates strong winds at the crop level. The open upper structure keeps wind from breaking or pushing over the trees.

Often farmers find one tree species that is dense lower down, open higher up. Trees with vertical canopies fit well. If not, a dense shrub may be the understory for a single row of open canopied trees. If necessary, three species may be employed.

Zhang Fend (1996) offers a brief study of an integrated shelterbelt-windbreak system located in a hilly and windy part of Mongolia. The shelterbelts were in downhill bands connected crosswise by anti-erosion windbreaks. Poplar was the main species in the 5 to 10 m wide shelterbelts. These reduced wind velocity by 32 to 38 percent, increased soil moisture by 3 to 6 percent, lowered summer temperatures by 0.1 to 0.7° C, and increased winter temperatures by 0.5 to 1.6° C. These seemingly modest changes increased maize yields 64 percent and millet yields by 70 percent.

Riparian Buffers

The need to protect water sources, streams, rivers, ponds, lakes, springs, and even wades from soil-laden or nutrient-contaminated runoff precipitates the installation of riparian buffers. The placement of the buffers plays a large role in their effectiveness.

Streamside buffers, those that line the banks of streams, rivers, ponds, and lakes, are the last line of defense, filtering any water flow before it reaches streams or ponds.[16] Figure 12.3 shows this type lining the banks of a diagonal river.

On the theory that defenses must be strongest where water flows are greatest, there are finger buffers. These are strips of vegetation

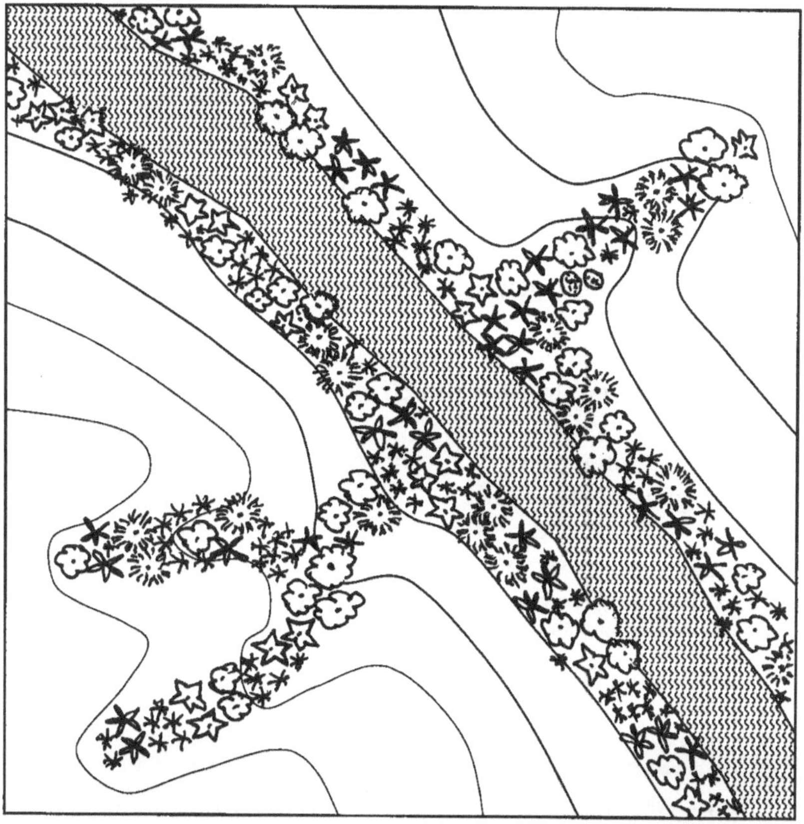

FIGURE 12.3. A bird's eye view of various riparian structures. Along the stream are the streamside buffers. On the upper right is a finger buffer that follows a wades or soil depression, on the lower left is an arm-and-hand version that branches along wades or hollows where water can flow.

placed in normally dry ravines or wades. These provide added containment and capture where nutrient laden water is most likely to flow. In Figure 12.3, the buffer fingers extend uphill, perpendicular to the river and stream side buffer.

If nutrients are best retained close to where they originate, the riparian buffer becomes part of larger defensive scheme. Other vegetative additions throughout the landscape are the first safeguards. As part of larger scheme, waterside buffers need not be as fortified (require less width) but can be better formulated and managed.

In order to best reduce erosion and block the flow of water contaminants (essential nutrients included), a number of design options address the various situations encountered.[17] The design alternatives lie in the type of vegetation and management. Most buffers are composed of a mix of species. This generalizes nutrient capture and ensures that at least a few of the plants are active in the off-season. When flows are concentrated with one nutrient, it may be best to employ plants that seek more of this nutrient. A common example may be the planting of nitrogen-demanding species when water surges are high in this nutrient.

As buffers can become nutrient saturated, increased vegetative uptake helps keep the structure receptive. This implies some cropping, trees included, such that nutrients are taken away, but the floristic integrity is not compromised. Agroforests, with their continued harvests, are good in this regard.

Where well-watered, fertile planting sites are at a premium, the buffer may have to serve a productive role. This occurs with traditional waterside farming in dry lowland Peru and in the arid Sudan.[18] Here the types of crops and the agrotechnology employed (usually biodiverse with perennial elements) are important.

Waterbreaks

Bottomland inundation after heavy rains or during certain seasons may be common. Adding waterbreaks keeps soil in place or, better yet, captures the soil and nutrients being carried along in the flow. The classic layout, shown in an overview in Figure 12.4, has rows of permanent vegetation slowing the floodwaters.

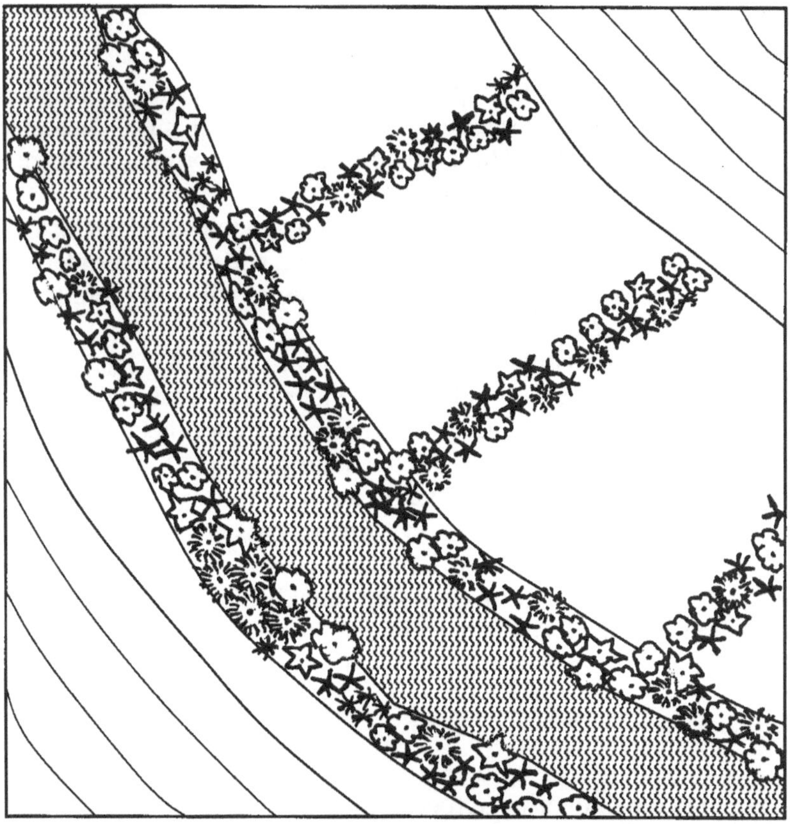

FIGURE 12.4. An overview of waterbreak design where, extending outward from and perpendicular to a streamside riparian buffer, the cross-field structures slow flood waters preventing soil scouring and depositing waterborne soils.

Wallace (1997) suggests a version in North America that has strips of taller vegetation (shrubs or higher) along steam banks or, for larger rivers, on the banks and spaced every 100 m along agricultural flatland, perpendicular to the water flow.

Not all subscribe to this design. Nabham and Sheridan (1977) describe an indigenous Mexican type that, instead of a wide strip, substitutes lines of short, woven-together willows. These serve fence-waterbreak-windbreak combination.[19]

Firebreaks

In some regions, fire is a problem and structures can be placed to control the spread. Most double as roads, but separate, dedicated structures are possible. These come in different forms: (1) cleared, (2) planted, and (3) fire transmitting.

The commonly thought of structure is a cleared strip. Such strips are expensive to maintain, subject to erosion, and offer no productive or facilitative value. Other options, if feasible, will add more to the ecology of the farm.

Two designs are associated with planted strips. Strips can be planted with a crop during periods of low fire danger (a rainy season) and harvested before the danger is greatest. For example, tuber crops that toss and bare the soil upon harvest are quite suitable. The other design incorporates perennial plants that do not burn (for example, cacti).

A fire-transmitting strip allows fire to pass, but slowly and at ground level. The purpose is to prevent ground fires from reaching into tree canopies or from spreading quickly. This is a function of vegetation-contained and/or a grazing regime that will trim the plants. These are found when combustable trees (e.g., pines) border crops.

Buffer Zones

When two neighboring agroecosystems are in ecological conflict, an in-between buffer zone may rectify this situation. A single plant-on-plant buffer species is presented in Chapter 2. More inclusive, the ecosystem version comes in many forms, in varying widths, and functions above- and/or belowground. In contrast to buffer species, buffer zones are not part of an ecosystem.

As with their single species counterparts, buffer vegetation can prevent branches of tall, woody plots from unfolding over and shading a field of shorter crops or by providing intermediate step where sunlight blockage is reduced (see Figure 2.8). Gains also occur in the other direction as, for tall, woody plots, horizontal light entering below the canopy can stimulate the growth of nutrient-robbing weeds (i.e., the edge effect).

Belowground, a buffer zone blocks nearby, nutrient-robbing roots from spreading into a cropping area (also in Figure 2.8). Buffer roots can also prevent underground creatures, such as nematodes, from an unwanted entry.

With well-chosen vegetation, and with a wider expanse to accomplish these tasks, these can be far more effective. A well-designed buffer zone minimizes interplot competition; the area usurped by the zone should be more than compensated for by yield increases in one or both of the abutting fields.

Live Fencing

Putting strips of vegetation around fields provides a barrier to the movement of animals and people. Clearly, a barrier can be a useful addition where hungry animals roam around eating crops. Fences come in two forms: (1) dead and (2) live (see Plate 12.2).

Standard wire fencing on inert posts is the dead version; it serves one purpose: to physically separate plots, and has no ecological function. Vegetative barriers, of which there are many types, are live versions (see Figure 12.5). A prime example is the traditional hedgerows in England.

The suggested DPCs for vegetative barriers are as follows:[20]

- dense branching;
- spines, rough branches, or other repellant properties;
- ease of establishment and management;
- nonspreading (no root sprouts);
- compatibility with crops;
- value as fodder (or grazing resistance, depending upon use);
- fire resistance; and
- good insect- and weed-countering properties.

There are many species that fit the bill. In arid zones, cacti work well. Other regions utilize other species, for exmple, *Eulychnia breviflora* from northern Chile and jumping cholla from the highlands of South America. Both have spines, as does the European living fence standard, the white thorn hedge.[21]

For added strength, hedge barriers can be covered with thorny vines. Roses provide this type of reinforcement. Trees that attract vicious ants also serve admirably.[22] Strength of a barrier can be aug-

(a)

(b)

PLATE 12.2. Two fence types: *(a)* wire on living posts and *(b)* a nonwire vegetative barrier.

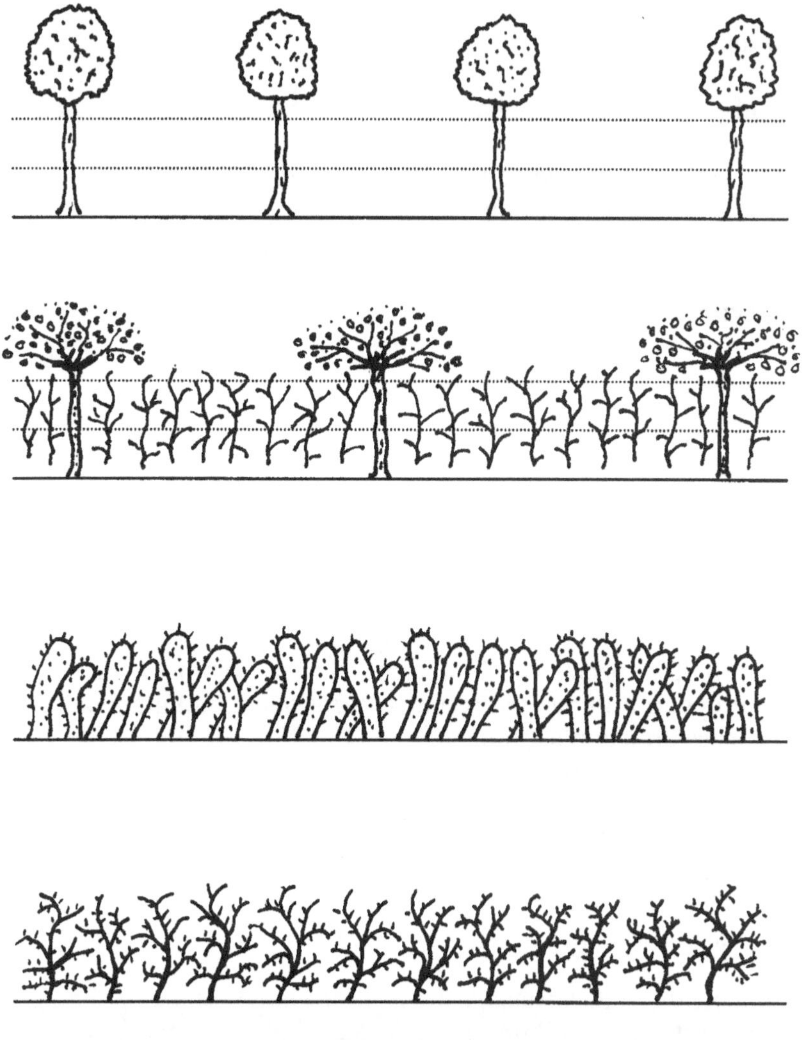

FIGURE 12.5. Four types of living fences from top to bottom: wire with living post, living post with the pruned branches interwoven into the wire, hedge type with cactus or some like plant, and shrub type.

mented if the branches are interwoven or, if of the same species, are pleached.[23]

Cut and Carry

The purpose of an auxiliary system may be to cut and carry, that is where vegetation is removed to another area to benefit a principal-mode system. Cut-and-carried mulch can boost productivity in the target system (by providing ecological functions such as weed control, water management, soil improvement, and/or by reducing the populations of herbivore insects).

Keeping a facilitative biomass source separate from productive plots is done for a number of reasons. With cut-and-carry strategy,

- a facilitatory species can be selected solely because it gathers nutrients, for example, a superior biomass plant, without regard to plant-plant competitiveness;
- the management, for example, planting or pruning schedule of the facilitative species, does not have to be reconciled with a primary species to reduce competition for light and water;
- the harvest of the biomass can be spread over a longer time frame with better labor scheduling;
- a buffer species can be planted between biomass source and crops to improve the resource partitioning and eliminate biomass-crop competition;
- biomass in dedicated locations can be directed to one or more locations and serve multiple tasks (e.g., green manure or a insect repellant); and
- incorporated in biomass plot can be a range of repellant species where, in addition to being a nutrient source, the organic material will also thwart herbivore insects.

If there is any disadvantage, transporting large volumes of biomass can be costly. As a soil improvement technique, the economics favor higher valued crops or intensely managed, multiple-output plots.[24] Cost reduction measures include keeping the biomass source near the recipient plots, employing labor-saving machinery, and/or having multipurpose biomass that expands the number of ecological functions and, in this way, reduces the overall cost.

Catchments

Catchments are areas set aside to gather rainfall for utilization in other parts of a farm landscape. These are commonplace in rainfall-deficient regions.[25] Some are simple, merely channeling runoff; others have complex pumping and storage capacity.

The area involved can be quite large, especially when compared with actual area under cultivation. Areas devoid of vegetation can be unequivocal in purpose, whereas areas covered with plants not so.

It is not clear as to what types of vegetation maximize water runoff (grassland, scrub, or forest).[26] There are advantages to slopes enveloped in lush growth. These offers productive output capacity. Whether it be low input grazing or some forestry activity, the only requirement is that water collection be at the top of the objective list.

Adding to the flexibility, water quality does not vary with the type of vegetation as long as human intervention is not highly intrusive. Perry et al. (2001) found no difference in water quality if the area was a tree plantation or a natural forest.

Chapter 13

Land-Modification Agrotechnologies

Not all land-use problems are addressed vegetatively. Under the heading of land-modification agroetechnologies are those practices where the land is physically altered, that is, holes are dug and earth moved to improve the agricultural possibilities.

Frequently, an expensive course of action, the land-modification agrotechnologies are often tools in overcoming hostile environments. These can promote the efficient use of water (on water-challenged sites) and/or to counter temperature extremes (often in dry, high-altitude environments). In near-desert (semiarid) regions, land changes are a reliable tool in coaxing yields from an otherwise reluctant landscape.[1]

Although powerful, these agrotechnologies do not always stand alone in countering climatic stresses. If insufficient for the assigned task, these can be vegetatively augmented, intensified with auxiliary systems, or enhanced through matching productive agrotechnology. Land-modification agrotechnologies differ from auxiliary systems in that these underlay and host principle-mode agrotechnologies.

ECOLOGICAL UNDERPINNINGS

Land-use modifications attempt to make the most of a site. In Chapter 3, the principles behind the barrier and ground cover are discussed. For the land-modification agrotechnologies, physical barriers, for example, contour mounds and ditches, provide much of the focus. As stated, the principle being exploited is that in-soil water

Introduction to Agroecology: Principles and Practices

doi:10.1300/5553_13

travels much slower than unimpeded surface flow and that properly constructed barriers encourage water to infiltrate into the ground, exposing water and waterborne nutrients to increased capture.

Land-modification plays a role in temperature modification. This can be to lower high temperatures or raise low temperatures, keeping the range right for optimal plant growth.

Many of the agrotechnologies that permit hillside cultivation without erosion, terraces, infiltration structures, etc., address temperature concerns. Hillsides shed frosts and can be slightly warmer than the bottoms of valley. This principle is also exploited with smaller structures such as mounds.

Water can also moderate temperature. Surrounding plants with surface water aids grow. Water evaporates and cools crops, depressing high daytime temperatures; pooled water collects daytime heat, raising low nighttime temperatures. Both benefit crops.

Another moderation mechanism, similar to that of surface water, lies with rocks or stones. Stones near small plants cool during the night, releasing heat and, during the day, the cooler stones moderate high temperatures.

A few of these technologies have habitat possibilities. This relationship applies to useful aquatic fauna where a pond may be the habitat and water channels a SIZ expanding corridor. In some cases, channels support a breeding population of insect-eating fish or frogs.

ECONOMIC UNDERPINNINGS

Most land-modification technologies are comparatively expensive to install and maintain. Therefore, these are feasible only where land-use intensity is high or where this is the only way to economically obtain viable harvests. The impetus is strongest when there is no other way to coax yields from the land.

Irrigation systems with catchments and/or channels seem universal. These address an immediate need and find favor in many drier regions. The gains are so great that societies cooperate to keep these in use. Where cross-landscape irrigation cannot be maintained (e.g., uphill water sources), one encounters plot-based modifications.

Far less universal are large cross-landscape structures that address climatic events. Frosts, heat, and flooding can be reduced and rainfall runoff increased through large-scale forests. However, preservation or replanting seems beyond the collective will of local societies. By default, these tasks are often assigned to each farmer. Being far less effective, disasters, as with large-scale deforestation and distant flooding, may occur with some regularity.

AGROTECHNOLOGIES

Despite what is often a formable economic barrier, land-use technologies abound. Those discussed here are as follows:

- Elevated earthen or other structures (mounds/raised beds)
- Collections or piles of rocks around specific plants (stone clusters)
- Small holes and/or biomass positioned to hold water for use by a single or a group of plants (absorption zones)
- Small channels to direct water to individual plants (micro-catchments)
- Vegetation or earthen contours placed to retard surface water flow and promote water infiltration (infiltration structures)
- Spaced ditches that catch, hold, and allow water to enter subsurface soil strata to feed bordering, downhill crops (cajetes)
- Constructs that provide flatten areas of hillsides or slopes allowing for less risky agriculture (terraces)
- Shallow flat area that can be filled with water to cultivate water loving plants (paddies)
- Small bodies of water that serve agricultural or other purposes (ponds)
- Ditches that convey water to different parts of a farm or rural landscape (water channels)

This listing does not exhaust the range of possibilities. In floating gardens crops are grown on soil-covered buoyant mats. As a land-modification method, this overcomes a lot of regional barriers, for example, poor soil (surmounted by transferring the nutrient-rich, lake bottom soil to the mats), low rainfall (the plant roots extend through the mats), and temperature extremes (the temperate moderation associated with a larger lake). These are more a curiosity in that, although entirely possible and a potential agrotechnology, few are found.[2]

Mounds/Raised Beds

Mounds in fields, flooded or otherwise, serve a number of purposes. One is to counter brief periods of low temperatures (see Figure 13.1). Three principles are being exploited: (1) the slight temperature difference due to elevation, (2) a mound above heat-retaining water, and/or (3) internal decaying vegetation that releases heat.

Although mounds may not seem a potent countermeasure, a slight change in temperate may prove decisive, especially in league with another countermeasure. With a 0.5 m mound, a two-degree temperature differential can exist between the top and the bottom. The more vulnerable and valuable species are planted uppermost.

Examples are found in the highland areas of New Guinea and Bolivia where cold or frost during the nights can be a growth hindrance.[3] Native Americans once employed 0.5 m high mounds for growing maize. This can counter waterlogging and/or protect early-season crops against sudden temperature dips.[4]

The other ecological gains from mounds should be noted. Due to the slightly drier soil at the top of the mound, root diseases may be less of a problem. With tuber crops (e.g., sweet potato in New Guinea and potato in Bolivia), the loose soil structure can encourage greater

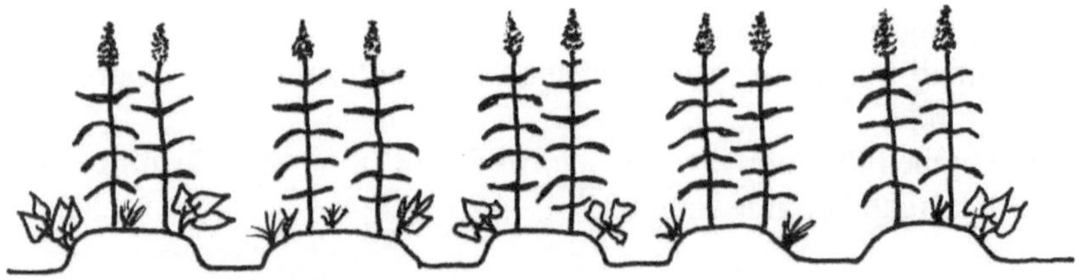

FIGURE 13.1. The land cross section for a system of mounds. The primary purpose can be to keep plants above waterlogged soils and/or up and away from freezing temperatures. As mentioned in the text, there are other uses as well.

root growth, and the lower bulk density of the soil facilitates harvests.[5]

Raised beds, often associated with private gardens or other intense land-use forms, can save labor. Although initial construction is a substantial undertaking, less bending by workers is labor reducing and a design-selling factor where labor inputs are continual.

Stone Clusters

Although rare, some societies do grow young trees or large annual plants between medium sized stones or within donut-shaped rock clusters. Generally, the rocks are from 5 to 25 cm in diameter.[6]

In addition to moderating nearby air and ground temperatures, clusters of rocks provide a more porous soil surface, retard soil drying around the plants, discourage some small animals, and attract windblown nutrients. Although seldom reported, crops do better in such situations. In Mexico, maize does better when planted in rocky as opposed to nonrocky plots.[7]

Absorption Zones

On water-challenged sites, such as those with brief, scattered, and intense periods of rainfall or those with short rainy seasons, water held near plants long after rains cease can prolong a growth period and ensure good yields. Absorption zones and small ditches dug slightly uphill of each plant catch and hold the scarce rainfall (see Figure 13.2). This is not possible with expansive crops such as wheat but can serve well with individual trees or with small clusters of mixed species.[8]

Under normal situations, the recipient plants are adjacent to and downhill from the dug cavity. In challenging climates, the plants may be located in the hole where some protection is afforded from drying winds.

If the collected moisture is in danger of evaporating or infiltrating away quickly, each hole can be packed with water-retaining biomass. Dead leaves serve this purpose and, in the Sahel of Africa, manure in holes holds rains while attracting dung-eating termites, further enhancing soil fertility.[9]

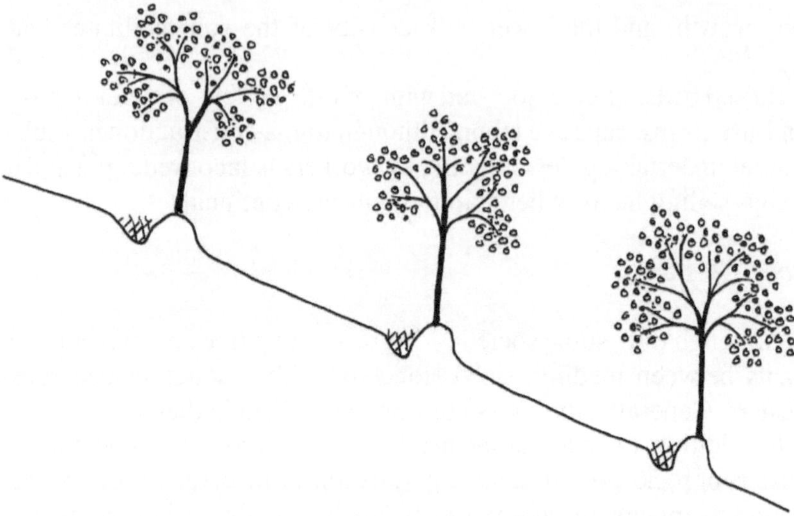

FIGURE 13.2. Absorption zones located above and near individual trees that catch and store infrequent rains though contained organic materials.

Microcatchments

Where rainfall is brief and infrequent, it helps to direct as much rain as possible to individual plants. Microcatchments are small-scale diversion channels, usually not extending more than one meter, that ensure that each plant, usually a tree, receives full benefit from a brief rainy period (see Figure 13.3). These are often found in conjunction with adsorption zones.

The combination of the microcatchments and adsorption zones have proved effective. In arid India, Gupta et al. (2000) noted, tree survival from microcatchment installation increased from 50 to 90 percent. This also increased the cost of planting by 20 to 30 percent.[10]

Infiltration Structures

Microcatchments direct water to individual plants, usually trees; however, this is not feasible with larger areas of small crops. To direct water to these crops, brief, infrequent, and strong rains must be cap-

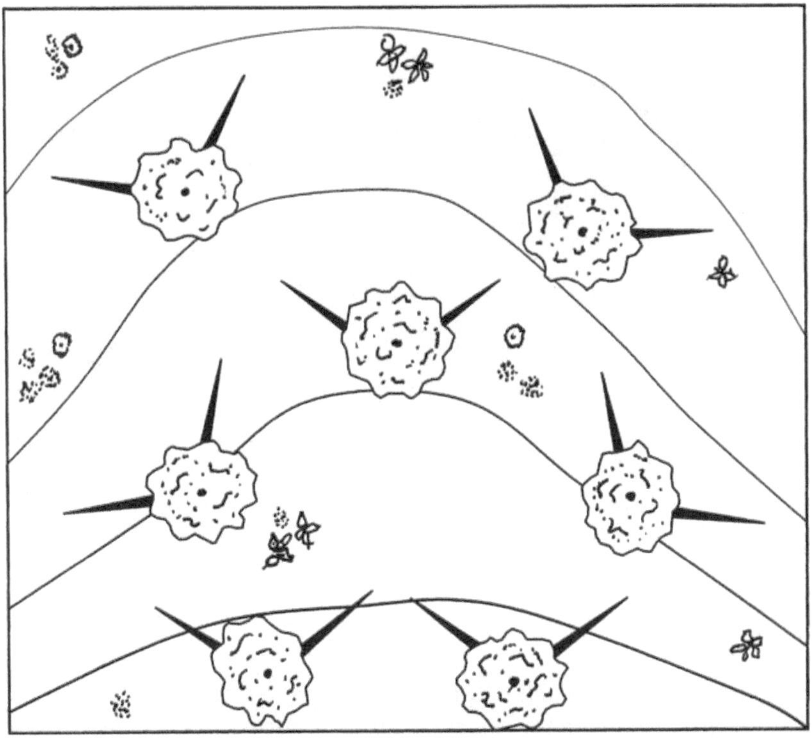

FIGURE 13.3. A system of microcatchments in overview where each plant has a V-shaped, uphill-facing, diversion ditch to channel water to each plant. The scale (size of the V-structure) is relative with plant size, rainfall, slope, and the area's erosion potential. The highest elevation is at the top.

tured and made available for an extended period across a wide area. Infiltration structures do this.

These structures (shallow ditches and/or bunds with or without vegetation) normally follow land contours (see Figure 13.4), but other layouts are possible. If conditions warrant, these can be positioned as illustrated in Figure 2.10.

To force water into the soil strata, the actual design depends on site and accompanying agroecosystems. There can be fewer, widely spaced, larger ditches, or series of smaller, closer-spaced structures.

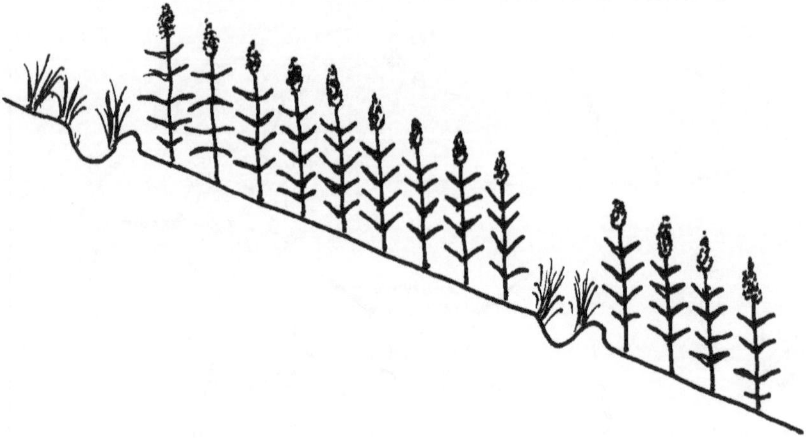

FIGURE 13.4. Two infiltration ditches in land cross section. Customarily these are contour structures with enough depth and spacing to capture water.

Published guidelines are a good starting point,[11] but long-term observation also serves well. The structures should contain, without spillage, higher than normal rainfall. When exact leveling is difficult, ditches are often segmented, preventing water from flowing lengthwise along the construct (see Plate 13.1). Other design preferences, such as the Andean practice of placing rocks or vegetation along the uphill side of a bund, may also increase effectiveness.[12]

Research shows that well-spaced, well-placed ridges are quite effectual at reducing erosion. Reductions are in the neighborhood of 40 to 60 percent over contour plowed fields without these structures and up to 90 percent when in conjunction with vegetative strips.[13]

Cajetes

Cajetes serve to contain and store brief periods of very high rainfall and feed this water to downhill plots. Here, individual, deep ditches, usually one meter square, are closely spaced or located at some distance. If one meter apart, placing dug dirt between structures helps channel the rainfall. If widely spaced, for example, at 2 or 3 m distance, these should be attended by diversion ditches to direct water

(a)

(b)

PLATE 13.1. Two hillside infiltration structures: *(a)* a segmented infiltration ditch and *(b)* individual cajetes spaced along a contour. Both of the photos are from Nicaragua.

flow into the hole.[14] Again the size and interhole distance is such that all rainfall is captured without spillage.

This mechanism directs water to crop fields located directly below the ditches not as direct irrigation but by way of belowground infiltration. There is a requirement that the soil strata be receptive to the mechanism, that is, the water reaches the crop (the arrow in Figure 13.5 shows the requisite flow).

In contrast to infiltration ditches, cajetes do not require precise leveling. This makes them ideally suited for irregular or rocky hillsides. Therefore, these are employed more with orchards and other systems suitable for topographically challenging, erosion-prone sites.

Terraces

Where steep hillside cultivation is a must and erosion a problem, terraces find application. Because of the expense of construction,

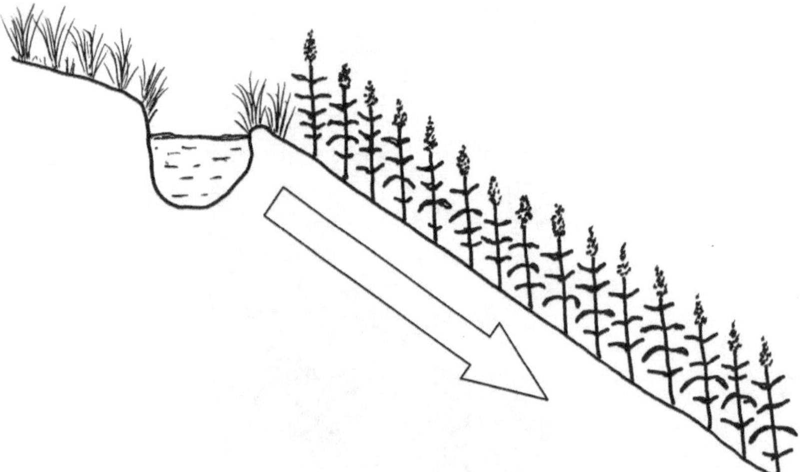

FIGURE 13.5. A cajete in land cross section where individual ditches, located uphill, store and release water from brief and intense rainfalls. This requires a favorable soil structure so that water can flow in the right direction (following the arrow). Vegetation located around the cajete filters waterborne soil, preventing the ditch from prematurely filling.

these are mostly found on sites too abrupt for strips. As an added benefit, these provide a flat place to work. Terraces come in many forms; some are constructed of local materials, usually stone, while others are earthen, supported by vegetative (see Plate 13.2).[15]

In regions where cold weather can negatively influence crop growth, stone terraces shed cold air while the stonework releases heat stored during warmer hours. These are utilized for heat retention in Pakistani mountains. In the Andes, the well-crafted structures assembled by the Incas and the more hasty constructs of their descendants serve a similar function.

The option exists to utilize trees both for added support and to augment the heat-retaining role. This is a somewhat tricker design as, if done poorly, the stone structure is weakened and less heat is absorbed. Mixed tree-stone terraces are found in the Andes.[16]

Where heat retention is not needed or where sufficient rocks are not at hand, purely vegetative structures can be employed. The vegetation can offer the more or less standard, antirisk, facilitative package. Two terrace options are shown in Figure 13.6: grass and hedge faced.

Terraces can be manually built over a short period. Earthen terraces, once constructed, are immediately covered with plants. In the tropics, vetiver and napier are candidate grasses. Among the DPCs, vetiver is nonspreading with deep vertical root system, while napier harbors predator insects. Woody perennials and cover crops also serve well.[17]

A unique variation is the buried hedge terrace. This starts with a contour hedge where a terrace is to be located. The hedge is trimmed and covered with soil. The now buried hedge prevents washout and, later, sprouts to strengthen the face of the structure.

FIGURE 13.6. Two terrace types: the one on the left is grass faced; the one on the right is hedge supported.

(a)

(b)

PLATE 13.2. Two land-leveling structures: *(a)* a rocky barrier and *(b)* a series of earthen terraces.

Terraces can be progressive, forming over time through natural forces thereby saving a lot of immediate effort. The progressive versions can be tree or grass based.[18] With this, rows or strips of grasses or low-cut trees, planted along contours, capture shifting soil and over time form a terrace.

This is not a rapid process. With a closely spaced pruned hedge, six years was required for natural erosion to shape terraces 35 cm high and 81 cm wide (with an interhedge spacing of 90 cm). The location was a 44 percent slope and adequate rainfall. It was also reported that crop yields (in this case maize) increased due to the soil augmenting hedge biomass.[19] Since these share characteristics with alley cropping, the user requirements of alley cropping must also be adhered to.

Paddies

Paddies are common and mostly associated with rice cultivation (see Figure 13.7). Rice *(Oryza sativa)* is not the only crop found in water-filled paddies; cranberries and wild rice *(Zizania aquatica)* are also well suited to a swamp environment.

As a modification for specific and general use, water-filled areas are often part of series of structures including catchment areas and channels to convey the water (as auxiliary systems, catchments are described in Chapter 12). Also possible are paddy-channel series where topography and the need to divide water between land users suggests a layout.[20]

Ponds

Small bodies of water can have an ecological and economic role. In many regions, these primarily serve as water storage for household use

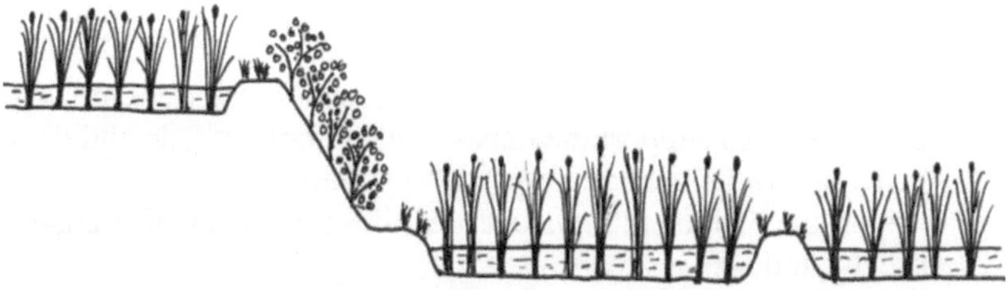

FIGURE 13.7. Some common paddy designs: *(left)* hillside established or *(right)* bund separated for flatland use.

and for irrigation. Along with these, ponds have various facilitative roles; these can store heat-lessening frosts and can be a source of nutrients for crops or water for domestic animals. Ponds can also be a principal-mode systems as with aquaforestry or as a stand-alone fish source.

Most of the gains come through multiple roles. Fish-filled ponds increase the nutrient content of the water, and later, if used for irrigation, these nutrients become available to crops. This is a case where waterborne nutrients are for good cause. A pond many also be a reservoir-habitat where water-loving, insect-eating insects can live and breed.[21]

Water Channels

The main purpose of water channels is to convey water across the farm landscape. Channels, especially those lined with concrete or other impermeable materials, can be regarded as a nonecological structure.

Plants help these keep their form. The key is to have tall vegetation only on the upper side of the structure (see Plate 13.3). The lower face is usually planted with short grasses or a cover crop. This expedites cleaning as the channel-clogging residues are shoveled downhill.[22]

In most cases, these are too valuable a resource not to exploit for other purpose. Secondary applications are as

- a direct irrigation source when trees or crops are raised along banks;
- a windbreak-channel combination: the windbreak trees are watered, the trees support the banks, and canopies shade and cool the stream (note the aquaforestry potential);
- a riparian/corridor that encourages the movement of predator insects; and
- the already mentioned habitat (pond)-corridor (channel) relationship where the channels radiate beneficial aquatic creatures, for example, fish that eat mosquito larva.

PLATE 13.3. A tree-supported water channel, now dry, that becomes active once a distance crop is in place and irrigation needed. Note the uphill placement of the trees.

Channels can be the basis of agriculture. The Aztecs farmed the banks of channels in a chinampa system. The key to extremely high productivity was the nutrient-laden sludge, dredged from the channel bottoms, deposited along the banks.[23]

Chapter 14

Social, Community, and Economic Agroecology

The first chapter assigns some lofty goals to agroecology. Beyond food and fiber needs, quality of life improvements are clearly possible and quite doable. The questions are how and in what form.

Before quality of life issues are addressed, it helps to have a better understanding on how societies interrelate with agriculture. This is a step back from the technical, instead, looking at the agricultural peripheral. In this line of development, human-imposed impediments in agriculture, along with social and community agroecology, are topics of relevance.

THE HUMAN EXPERIENCE

What is commonplace for one group may not apply to others. The nuclear family, a father, mother, and siblings, is the strived-for norm in many societies. Across the human experience, this is not universal. A family can be centered around a village or be extended through relatives or even multiple fathers.[1] The reasons may include risk, high in disease-prone areas where loss of one adult will have less of an impact on child survival and rearing. The same reasons may underlie extended families in West Africa.

If human variation exists at such an intimate level, differences certainly apply when people interact with terrestrial ecosystems. There may be a norm, but what happens in other cultures and climates demarcates the full range of agricultural thought. Again, risk avoidance may weigh heavily on what happens.

Introduction to Agroecology: Principles and Practices

doi:10.1300/5553_14

INDIGENOUS FARMING PRACTICES

One aspect of agroecology is in ascertaining the ecological and productive advantages of agriculture in diverse climates and in the far reaches of the human experience. Where the choice is between surviving or perishing, people tend to surmount local obstacles. The methods utilized, some quite unique, add to the body of agroecological knowledge.

Not all peoples, when confronted with challenge and change, adopt and thrive. History provides examples where, in not recognizing that change is required, people are forced to relocate or succumb.

A dramatic study is found with the Viking colonization of Greenland. The gathering strategy of the indigenous Inuit was locally better than introduced pastoral agriculture. Since relocation was not possible, the group died off. The failure on the part of the immigrants was in not recognizing the superiority of local practice.[2]

These lessons extend to the present. Remote, climatically accommodating, and/or long-lost practices, once understood, have been shown to coax higher yields with less effort or inputs than conventional agriculture. These often entail greater biodiversity, although unique approaches to land and crop management can be the decisive factor.[3]

The lessons are clear, as are the dangers. It is folly to replace perfectly good, ecologically beneficial or ecologically harmonious indigenous practices with less productive, input-demanding, less climatically forgiving, alien systems. This can be a cultural perspective as when well-intentioned foreign experts dispense advice that goes against agroecological logic and local traditions.

Some have stated that slash-and-burn agriculture, especially in tropical forests, is destructive and inefficient. There is some truth in this, but with numerous caveats. Slash-and-burn agriculture is ecologically efficient as this duplicates a forest cycle where natural gaps fill with nonwoody vegetation. When confined to areas of low-population density, this practice may promote, not destroy, natural biodiversity.

HUMAN IMPEDIMENTS AND ENCOURAGEMENTS

Agriculture, instead of being a technical endeavor, must contend with a range of nuances. Seldom overtly noted are the legal issues, cultural norms, and a host of societal and religious admonitions that steer agriculture.[4] Some are a formidable barriers to change, while others can be overcome without much effort. Many fall under the heading of cultural diversity.

Societal Intrusions

Few societies permit or encourage the full exercise of land rights, even if a person has unrestricted legal possession. This list of intrusions is quite long. At the very least, there is the pressure to conform. More meddling are land-zoning restrictions. Others still must give way to government support or other programs that encourage the raising of specific crops in a specified manner.

In further limiting what can be done, land ownership is not the same concept in all regions. Some have communal views where, as in part of Peru, land is apportioned through the village structure. In parts of West Africa, if one plants a tree, this person can claim the fruit, even if this tree is on someone else's property.

Ownership can be a barrier where the land is not farmed by the owner.[5] Sharecroppers rent land on a seasonal basis with a percentage of the harvest as payment. An absentee landowner may be spatially divorced and not cognizant of or interested in faraway agroecological opportunities. The renter is involved but temporally divorced from and equally uninterested in long-term ecology, especially when this interferes with the immediate gains. This can exclude soil and site-improving rotations, similarly beneficial tree-based systems, or climatically positive land modifications such as windbreaks.

Local interests take away, but also add opportunities. With intercropping comes the possibility for multiparticipant agroecosystems. This is where the landowner and a partner(s) simultaneously use the land, each focusing on a different crop. This is not uncommon with simple taungyas as described in Chapter 11: the landowner raises the long-term crop, usually tree crops or forest trees, while the partaking partner grows seasonal species.[6]

If land is scarce, the owner may require the partner to both prepare and weed the land. A more equitable distribution has the owner doing the preparation; the partner contributes the weeding.

Although fraught with potential conflicts, mutual benefits come in the form of the specialized cropping knowledge each participant brings to bear. The benefits magnify with shared chores and expenses (land preparation, weeding, nutrient application, etc.), more so if the crops are complementary and facilitative. Conflicts are minimized if the intercrop is fairly simple, the benefits are apparent, and the agreement or contract well thought-out.[7] As complexity grows and/or the crops become competitive, so does the potential for discord. As a result, complex, multispecies intercrops, as with agroforests, are usually undertaken by one person or one entity.

Foods

Humankind does not accept all of what nature offers. Individuals and cultures have dietary preferences. Rice is the overwhelming favorite as a staple food in South Asia and parts of West Africa. Wheat, as bread, serves well in a large swath of the world, including North Africa, Europe, North America, and southern South America.

It is ideal when a staple food is grown where consumed, more problematic when the underlying plant does not thrive in the consuming region. Bread is the staple in some humid tropical regions although outside the efficient climatic range of wheat.

Other products, direct or indirect, can be shunned or accepted by a society. Insects are looked upon as a delicacy in many regions while others view the eating of bugs with revulsion. The population of North America generally shuns dogs and horses as consumables, instead valuing these as pets or work animals. These are cultural or value judgments or, as with vegetarians, rooted in beliefs.

An indirect influence are food properties. Sticky or glutinous rice (i.e., which clings together upon cooking) may be favored by societies that eat with chopsticks. When the spoon is the accepted eating utensil, loose rice may be preferred (e.g., in traditional West African societies). This is a varietal-cultural manifestation.

Another facet is how cultures view eating. Some, out of necessity or choice, place a utilitarian view on cooking and partaking. Others spend time and effort to ensure eating is an enjoyable experience.[8]

These views are reflected in the plants raised. Where a strong culinary effort is made, traditional or heritage varieties may be preferred over less tasty commercial offerings. Where high cuisine rules, spices would be far more common and multiple varieties of fruits and vegetables prized. It can be argued that this promotes a more agrodiverse farm landscape, favoring small farms and small-scale producers with specialized outputs.

Recognition of many of these is contingent upon close examination of a society and culture. Cultural norms may be so embedded as to be as potent as a religious edict. The enforcement mechanism—society at large—can be omnipresent and persuasive. Deviation from some norms is highly discouraged.

Religion

Religion enters the picture by placing taboos on foods or even means of production. The Koran and the Talmud both have prohibitions on eating pork. The underlying logic many have been to control disease spread (e.g., trichinosis) and/or to discourage the raising of pigs where pigs and people vie for the same foods.

These are just a few of the many examples where foodstuffs are preferred or prohibited. Other cultures place limits on the production methods. For example, the Talmud warns against picking the fruit off trees for the first seven years or against continual planting without a fallow period. This prohibition may have been rooted in the agroecological thought of the time.

Advice, Adages, and Admonitions

Many groups, especially those with a long unbroken history of agricultural activity, have anecdotes that serve as practical guides. These are a means to pass down knowledge and, although these may be embellished with nonagricultural details or delivered with almost mythical veracity, these can be based on ecological and biological truths.

These vary in form and tone. An early quote was

> I was told by an experienced farmer, that if you girdle, or cut brush in the months of May, June or July, in the old of the moon,

> that day the sign removed out of the foot into the head, if that day be cloudy, it will kill all before it: . . .

This refers, in a convoluted way, to the best time to cut brush so that it dies and does not resprout. An application lies with temporary vine-over-tree systems.

Another is brief and more rhythmic

> Sow flax in the mire
> Sow barley in the fire

This addresses the best planting conditions under the belief that manure warms (fires) the land.[9] The lack of scientific proof does not always disqualify handed-down traditional advice. One must remember that scientific studies are true without any assurance of being functional. Agricultural adages can be functional but lack the assurance of full effectiveness.[10]

QUALITY OF LIFE

The idea that, as a side benefit of agriculture, the human experience can be upgraded is well entrenched. Country homes allow some city dwellers to enjoy the rural experience. Although much can be included under this broad heading, a few topics can be singled out.

Adding to the difficulty of incorporating quality of life are the perceptions, much of which is not qualifiable using monetary criteria; for example, yields can be expressed in monetary terms, but a beautiful landscape cannot.

Water

Water runoff is a source of drinking water for much of the world's population and much of this comes off agricultural or forestry lands. It goes without saying that there are clear linkages between human health and water quality.[11]

If the water falls on farm fields, ensuring clear, clean, contamination-free runoff is a responsibility of agroecology. The methods for accomplishing this are covered in previous chapters.

The importance of maintaining high-quality runoff or of stocking groundwater through nonpolluting agriculture should not be understated. The City of New York finds that controlling the source substantially reduces the cost of water treatment.

Aesthetics

For a few landscapes, especially those with tall, snow-capped mountains, lakes, and other dramatic features as a backdrop, the view is spectacular. Other landscapes do not rise to this level, but are still aesthetically pleasing.

Within an agricultural context, considerable latitude exists to plant flowering species or to place flowing crops to heighten aesthetics. The diversity of decorative species found around farm dwellings is indicative of the importance that many place on nice surroundings.[12]

To heighten the visual experience, crops and gardens can be employed in an aesthetic role, i.e., the edible landscape. This is a well-explored topic where vegetable and/or herb gardens double as but do not replace flower gardens (see Plate 14.1). The key is placement so that the flowering components are complementary in time and space.[13]

The garden is only one element in the larger landscape. Commercial farms can be decorative and productive and, with a little forethought, arranged to maximize the visual impact, remembering that flowers and flower crops can add to the overall ecology. Farms of this type are far from unique and cross both cultures and time periods.[14]

Beauty can be incorporated on a larger scale. The shoulders of Texas public highways, seeded with abundant wildflowers, provide a welcome visual addition. Although it is not advisable to plant fruiting species near roadways, flowering and fruiting trees can adorn other public areas.[15]

Lacking acknowledged scenery and colorful crops, a fragmented landscape, one with plenty of visual elief, well-placed trees, and flowers are a good start. For greater overall effect, unique land-use patterns, field positioning, spatial layouts, plot content, and boundary patterns can be brought together into a visually pleasing scene.[16]

(a)

(b)

PLATE 14.1. Two decorative and edible gardens: *(a)* mixed herbs and *(b)* intermingled flowers and vegetables.

ECONOMIC GAINS

As agriculture is a livelihood, the economic outcome is of paramount importance. In agroecology, the potential gains are often expressed through the LER. This avoids the clouding effect of monetary units, putting the focus on plant-plant interactions. Still, if agroecology is to succeed, it must progress on many fronts. Especially important is profit and loss.

Without indications, such as the LER, economic studies are often not clear on how much agroecology has been brought to bear in any one situation. Further confusion results from an imprecision of terms, for example, organic versus reduced-input agriculture, and how well a specific crop situation is agroecologically addressed. The result is an uneven economic picture.

Revenue orientation, especially agrobiodiversity expressed through multiple outputs, frequently results in very efficient resource and land use. This can, but does not always, translate into a profitable relationship, but oftentimes secondary considerations are more of a factor. Clearly, where land is relatively scarce, as near cities or other high population areas, there is impetus for greater per area outputs. Also, many land-constrained gardeners benefit from intense intercropping.[17]

The same holds true with revenue-oriented taungyas. These have a long history and studies continue to show positive economic gains.[18] The drawback is that taungyas address a restricted, early plantation situation and are not for everyone.

Better reported is the full farm impact of cost-oriented, agroecological changeover. This is where output intensity purposely drops in order to implement or add on-farm, proagroecology attributes. There is a greater understanding of what can happen. In large-scale plantings of commercial crops, yields are approximately 10 to 40 percent less than those found with conventional farming. On the plus side, costs go down substantially, increasing the per area gross margins. Profitability ranges from slightly lower (–4 percent) to much higher (+40 percent) as compared with a conventional approach.[19]

Chapter 15

Agricultural Alternatives

Except for the brief look at social and community agroecology, the previous chapters have taken a mechanistic view, that of agrobiodiversity implemented through the agrotechnologies. Under the agrotechnological approach, one selects the most promising of the agrotechnologies (one or more) from the list provided. These are modified as needed.

This is not the only view—some perceive agroecology as alternative agriculture. Just as the productive agrotechnologies are a composite of ecomechanisms, the same holds true with the various forms of alternative agriculture.

In contrast to the technically eloquent agrotechnologies, the alternatives tend to be broad statements of purpose, rather than concise descriptions. As composites, they are more accurately depicted. This chapter offers a speculative look at this breakdown.

THE BASE

Throughout this text, a de facto base or core system is put forth. This is a genetically pure monoculture on a flat site, without rotational planning, and without nearby ecologically bolstering systems.[1] In order to address end goals, overcome threats, and so forth, development approaches take this simplest of systems in any number of directions (i.e., down different development paths).

Introduction to Agroecology: Principles and Practices

doi:10.1300/5553_15

CORE APPROACHES

A core approach is the first step away from the de facto base. Undiluted, each is capable, on its own, of making a large contribution toward yield improvements or in reducing forms of risk. Allied with other approaches, or as is, these can be pushed along as development paths. Alone or in combination, each is subject to considerable branching.[2] The unallied approaches are as follows:

Genetic: Genetic improvement underlies much of commercial, mainstream agriculture where a few plant varieties, resistant to adversities (e.g., climatic variation, disease, and insect attack), are further armored against whatever may be encountered. Conventional agronomy partially follows this development path.

Varietal: There are numerous varieties of many crop species. Rather than relying on a single genetic base, the genetic strengths of the different varieties are spatially maneuvered and optimally located within individual agroecosystems to accomplish agroecological objectives.

Microbial: A mostly unexplored side of agriculture, the microbial alternative proposes utilizing the vast array of microbes (bacteria, fungi, etc.) as specialized and general purpose agroecological tools.

Agrobiodiversity: This is a large heading in which different agricultural species are raised in close proximity for mutual benefit. The different forms of productive intercropping (seasonal and agroforestry) fall under this category.

Biodiversity: Rather than employing a mix of agricultural species associated with agrobiodiversity, nonproductive plants support agricultural production through favorable interspecies mixes. The facilitative agrotechnologies (intercropping and agroforestry) are based around facilitative biodiversity.

Rotational: Soil nutrients, insect populations, and other productive factors are affected by sequencing crops. This approach puts the ecological emphasis on cross-seasonal, cross-species dynamics.

Landscape associations: Instead of relying on ecologically self-contained plots, objectives can be addressed through inter-plot associations. To maximize ecological cross effects, the hallmarks of a strong approach are small plots and/or narrow strips. Auxiliary systems help in this.

Location: Instead of varietal matching to make crops compatible with a site (a crop-first approach), crops can be placed where they grow best (a plot or site-first approach). Topography, soils, microclimate, etc., are utilized in this quest.

Physical land change: Local obstacles to agricultural production can be overcome by modifying the land form. Mainly limited to areas where agriculture is marginal or where there is difficulty in raising key crops, this approach can change an unproductive or chancy landscape to one that is reliably fruitful.

Ex-farm inputs: Some feel the key to good agriculture is through ex-farm, often ex-nature, inputs. Farm chemicals such as insecticides, fungicides, herbicides, and fertilizers represent a major development path in mainstream agriculture.

Environmental setting: Implicit in agroecosystems is the environmental setting brought about by the management inputs. A few management practices clearly distinguish alternatives, for example, plowing method (till or no-till) and/or nutrient form (composting or manures), and, in doing so, set the agroecological stage. This can occur at the micro and/or macro level.

These represent a first step in an intended direction with immediate subdivisions. For ex-farm chemicals, the first divergence is into insecticides, herbicides, fertilizers, and, as a lesser offshoot, pheromones. For insecticides, the choice then becomes the synthetic insecticide or the home remedy path.

Some subdivisions are known, while others are hidden. Environmental setting seems especially rich with unidentified path options. For the purpose of discussion, the agroecological argument for each is made on the most prominent thrust or, lacking direction, where each could be heading. No attempt is made here to enumerate subdivisions.

Genetic

A genetic improvement approach, often implemented solely through the monoculture, places the onus on individual species where strongly positioned DPCs fight off all reasonable threats (i.e., climatic variation, diseases, and insect attack). The ideal super plant is drought resistant, tolerates flooding, counters all plant and insects attacks, and is distasteful or toxic for a range of harmful and unwanted fauna. Most of the current work done to advance the super plants comes through genetic engineering.

The positive gains and some negative consequences are discussed under domestication. As with too much of anything, things can go awry. A single super species, the antithesis of biodiversity, is the equivalent of putting all eggs into one basket. The basket may be strong but, if breakdown occurs, the risk is correspondingly large. The risk can be direct or indirect.

In an illustrative case from the Congo, an introduced, disease-resistant variety of cassava flourished. In displacing alternative, more nutritious foods in the commercial quest, there was no longer any impetus for site-improving and disease-reducing countermeasures, for example, rotations and intercropping. In this breakdown, the land suffered.[3]

The reliance upon single varieties can have other, unintended consequences, as a link has been established between too much of a staple crop and poor diet. This argument has been made for maize in North America and certainly applies to poor countries when the productive focus narrows.[4] The problem can aggravate when crops are destined for industrial purpose with little motive for greater genetic diversity.[5]

Genetic engineering enters the sphere of diet through the super food. This has a single species producing a highly nutritious output that can, in itself, provide for or go far in supplying a balanced diet without reliance on other foods. The perfect super food is more an ideal than it is a current reality.

The folly in this is that nonstaple foods add both known and unidentified compounds that benefit human health in many ways; the one or two varieties, as found in commercial markets, sideline taste, presentation, and other quality of life issues. In contrast, farmers, as consumers of their own produce, do put plant disease and pest resistance comparatively low on the list of end-use requirements.[6] Factors such as taste are regarded as being more important.

From a broader perspective, seemingly superior, improved varieties can reduce or cause the extinction of much existing and useful genetic material. As techniques improve, as through genetic engineering, there is danger in attempting and achieving too much. There are risks, mostly undefined, in unleashing modified plants without long-term guidance. To be fair, there are gains and these should not be over- or understated.[7]

Varietal

Rather than confining agriculture to a few super-plant varieties, there are ecological and environmental gains from exploiting and expanding the genetic diversity of common and not-so-common agricultural species. For most commercial species and large segments of non-commercial species, there exists or once had existed tens, hundreds, and, in some cases, thousands of varieties.[8]

This advances and implements the concept of heritage varieties. The 600 identified varieties of wheat are indicative of what one can find.[9] What is exploited is an intervarietal range of field-usable DPCs and prized end-use properties (color, taste, etc.).

Although less utilized than other development paths, the matching of varieties to locations is a tried and true technique with a long use history. Early farmers exploited the existing genetic base, selecting and interplanting multiple varieties to fit sites and counter threats.[10]

In a varietal approach, yield gains occur when the resource needs of each variety are matched against the resource profile of a site (as in Figure 2.1, species *b*). This strategy is found in Mexico where maize varieties are placed such that each conforms well to the soil characteristics of a site.[11]

Although the base for many crops is wide, more may be needed. Existing varieties offer the genetic stock from which to address further needs and/or expand cultivation to marginal climates and conditions. This is an ongoing process that, through better selection of the underlying stock, can be accelerated.[12]

Microbial

In agriculture, microbes are mostly unseen but can contribute to productive objectives. As a form of governance and a contributor to ecosystem gains, microorganisms are briefly presented under ecosystem governance (Chapter 3). Carried further, it is possible to expand microbiodiversity to greater utility.[13]

There are major unanswered questions and a number of possible directions from which to proceed. Microbial dynamics could be exploited through (1) inoculation, (2) formulating the right conditions to support the desired microecosystem, or (3) employing microorganisms to prop up a range of slightly larger but still beneficial organisms (the bottom up approach).

Some forms of inoculation are well recognized, for example, species-associated nitrogen-fixing mycorrhizae. There is the option of adding a range of soil-enhancing organisms. In addition to seeking nitrogen, some promote the more efficient utilization of in-soil or added nutrients.[14] This includes breaking down chemically inert sources or extracting for use nutrients physically entrapped in rocks.

Less known are those organisms that help ecosystems and plants hold water, make component species less palatable for herbivore insects, delay or prevent disease spread, reduce the effects of temperature, and undertake other facilitative chores.[15] Inoculation can be either through plants or soil based.

The in-plant version has shown remarkable ability to increase the resistance of plants to a range of climatic threats. These mechanisms are endophytes—in-plant living fungi. Pennisi (2003) mentions proven gains where watermelons and tomatoes can withstand root temperatures of 10° C, higher than possible without endophytes. Wheat inoculated with these cross-species fungi can resist drought better, that is, it can survive ten additional days without water. Fungi can even reduce the need for sunlight.[16]

Whereas inoculation is with species-compatible fungi, bacteria, and other microorganisms, it is possible to formulate a microecosystem of free-ranging, in-soil organisms. These, in sufficient numbers and diversity, can facilitate growth. As with in-plant inoculation, the scientific frontier may reach a point where there is a planned role for each addition.[17]

The in-soil microbes can be used alone or be part of a bottom-up ecosystem. This can be where the microcomponents do not directly facilitate plant growth but support larger organisms that do. Worms, ants, and other small creatures are the primary vehicles of facilitation encouraged through a healthy, well-balanced ecosystem.[18]

Agrobiodiversity

Proceeding down the biodiversity path is an attempt to duplicate what nature does: utilize a lot of plant species. As a result, this is a key agroecological direction off the de facto base. Agrobiodiversity,[19] where each of the coexisting species offers a useful output, breaks down into the following:

- *Archetype agrobiodiversity,* where each species has a preplanned role in the productive outcome (e.g., as with seasonal, fully productive intercrop)
- *Expanded agrobiodiversity,* which represents the idea that some minor niche substitution can take place without changing the ecological character or economic emphasis of the planned system (e.g., pastures with multiple grass species)
- *Agro-enrichment,* where a land user, sensing an unused niche or other opening, inserts more crop species or, having found something useful growing on a site, plants around the existing vegetation; this might involve adding a short-duration species (e.g., early-maturing radish or lettuce in a planned tomato patch)
- *Casual agrobiodiversity,* which has plants interfering with the overall outcome but, because the intrusion offers or will offer a valued output, it remains (e.g., a fruit tree that remains from a previous land use or one that spontaneously spouts without any prompting)
- *Supplementary agrobiodiversity,* where a productive species can be added without subtracting from the overall ecosystem (e.g., a low-density vine on a few trees).

Biodiversity

A step off the monocultural base system can involve adding nonproductive biodiversity. These are nonyielding species that have or assume a facilitative role in the ecological functioning of an ecosystem, or are plants that are just useful to have around. The breakdown is into the following:[20]

- *Facilitative biodiversity,* those preplanned additions with a defined facilitative role (e.g., a cover crop)
- *Expanded facilitative biodiversity,* where one facilitative species is replaced by many (for example, multiple cover crop species)
- *Casual biodiversity,* based on those species, most often unintended, that colonize a site but, because these do not interfere in any substantial way with crop yields, are allowed to remain (e.g., a shade tree that is nice to have)

Confusing to the casual observer is the overlap between biodiversity and agrobiodiversity. This includes the nonharvest option where fruit, nut, or other trees are utilized in a facilitative role but where only a small percentage of the output is collected or harvested, while the remainder biodegrades, adding to the seasonal nutrient carryover.

Rotational

Rather that focusing on static agroecology, rotational or cross-seasonal ecodynamics can be the prime mission. Many threats, climatic or other, can be countered though a well-planned succession of different species. This can be choreographed with temporally complementary species or formulated around temporal DPC. The latter takes into account the condition each species leaves the soil in and what an upcoming plant requires from the site.

Landscape Associations

Instead of relying upon interplot associations, it is possible to put the agroecological emphasis on cross-plot spillover dynamics. The landscape alternative, although little explored, has as much potential as other options. The complexity of the approach is partially demonstrated in Chapter 17. Although not always needed, much more can be accomplished utilizing small plots or narrow strips to maximize cross-ecosystem relationships. There are often strong cultural elements that underlie individual landscapes.

Location

With a varietal approach, one first selects the crop and then finds the best variety for a preselected site. Location takes the opposite tack, matching the site to the crop; descriptively, the crop goes in search of the best location.

A hillside plot, due to contained nutrients, erosion risk, or other reasons, may not be the best location for a maize crop. Rather than force the issue and plant the best available variety in the most accommodating way, something else is planted. For example, an orchard may be far more suitable on a hillside than maize. The maize goes elsewhere.

This is an unappreciated approach with little research backing. The potential is not only in comporting crops with the correct soil nutrient and moisture content but extends to ambient temperature, site preferences of pest and/or predator insects, length of the growing season, sunlight intensity, and other factors.

This can be a microsite in which plants are placed within an agroecosystem where they grow best. The criteria may be local microtopography; such placements are often associated with mounds and other physical structures (e.g., one crop on the top of a mound, another among the sides, a third between).[21]

This plot-first approach may seem a subtle variation of a crop-first, varietal approach. However, on the larger scale, the outcome is profound. Farms in parts of East Africa and in parts of Peru are noncontinuous, with fields scattered across the landscape.[22] Much of this involves matching crops with the site. Not documented is how much advantage is gained. The base comes in diffusing risk, that is, not all crops on a scattered farm are exposed to the same threats or threat levels. This is further reduced if the risk potential on the sites chosen is less than on those not chosen.

In exploiting land features, the resulting farms tend to have smaller plots, each containing something different. Subdivided fields and the placing of crops along contour strips, provided that the best site-crop match is made, can be locational expression. Plate 15.1 is of a site-first approach.

Physical Land Change

In climatically challenged agricultural regions, land modifications can and do overcome obstacles. Terraces, mounds, canals, and infiltration ditches, all help to increase yields on otherwise reluctant sites. In different ways, the intent is to either concentrate or diffuse nature. Concentrating gathers essential resources and directs these to one

PLATE 15.1. A location approach where a small internal section of a cabbage field, an area unsuitable for cabbage, has been planted with beans.

plant or agroecosystem (e.g., irrigation). Diffusion scatters or subverts the risk vectors so they become ineffective and less damaging (e.g., directing freezing air away from crops).

Ex-Farm Inputs

The line of development holds that chemicals are the key to more productive, risk-free agriculture. Although with a checkered history, as when introduced chemicals have proven overly toxic, some downright dangerous, many still seek synthetic compounds (herbicides, fungicides, fertilizers, insecticides, etc.) to advance farm goals. Also included are the ex-nature, benign home remedies.[23]

Without paying much attention to home remedies, conventional agriculture has, for decades, embraced the chemical path. Only recently has a shift taken place, with more emphasis on genetic engineering. This is not always a total change in course but often a fusion. The development of herbicide-resistant crop strains combines ex-farm inputs with genetic modification.

As can be the case, a fusion of paths may not always provoke the required results. This has been observed with herbicide-resistant soybeans. As the targeted weeds grow more resistant, increasing amounts of herbicide are needed. This multiplies the cost, both to farmers and to the environment.[24]

Environmental Setting

How a system is managed affects the outcome. It is possible to proceed from the de facto base by focusing on the management inputs and the changes wrought on the internal cropping environment. This is a large category of approaches with little accompanying understanding of the options and possibilities.

It has been noted that animal manure, rather than chemical fertilizers, negatively affected aphid populations.[25] Rather than being the mere pouring on of plant nutrients, nutrients in an ecologically agreeable form stimulate an environment upheaval. The basic premise is that by setting the stage for all forms of microbiodiversity, more happens. This may be the tip of the preverbal iceberg.

Speculating under this premise, manures and composting resupply plots with nutrients and, equally important, with microorganisms. As a bottom-up approach, one explanation is that a miniecosystem develops for a short period, longer if the cropping system is more receptive. If the mini- or subsystem is potent enough and the effects persistent enough, larger organisms, such as harmful insects, may be inauspiciously affected. Ideally, the even larger and more mobile predators may escape unscathed.[26]

Another interpretation is that in providing an environmentally compatible nutrient soup rich in elements and compounds, interactions occur. These will be beyond current research, not yet recognized for their importance to plants or to a bottom-up ecosystem.

Chemical fertilizers, fungicides, insecticides, or herbicides may either shock or unbalance the system. As ecological theory suggests, out of balance ecosystems favor lesser, not greater, biodiversity. This applies on the microlevel.[27]

Not all need be microbial. Plowing and mowing are part of the environmental setting. Rotary-type plows chop up earthworms while high-speed rotary mowers grind up predator insects. The loss of worms and insect predators does not bring about favorable agrodynamics. Less intrusive methods are far better.

COMBINATIONS

In wanting to replace costly inputs, in desiring risk-free agriculture, or in promoting other ideals, the core approaches come into play. One, if well implemented, can yield much, two may be better, and three, functioning in harmony, better yet. Cross harmony derives from each being

- capable of overcoming one or more threats and/or maintaining certain aspects of cropping viability and
- in doing so, any one approach can be enacted in conjunction with another (nonexclusivity) to further site goals.

With regard to common threats or obstacles to growth, some are ecologically more compelling, while some are easier to implement.

Biodiversity variants address threats while ensuring many aspects of surrounding natural ecosystems. Physical land change does less to advance natural sustainability but does help in overcoming climatic stresses and in securing otherwise elusive crop yields. The easiest to implement are the ex-farm inputs which come without a sustainability pretext.

COMPOSITE APPROACHES (ALTERNATIVE AGRICULTURE)

If there was one superior agroecosystem type, all would subscribe. Despite attempts, none is yet a technically stand-alone system. To encourage ample harvests without incurring undue risk, each requires a second, third, or even more approaches. Multiple approaches, drawing from those listed, can be employed to great benefit.

- Genetic improvement
- Varietal
- Microbial
- Agrobiodiversity
- Biodiversity (facilitative associations)
- Rotational
- Landscape associations
- Location
- Physical land modifications
- Ex-farm inputs
- Environmental setting

In combination, these represent a trove of possibilities. Take hedgerow alley cropping, which is a pattern-based expression of facilitative agroforestry (biodiversity). Mostly ignored in the standard design are other developmental approaches. To name a few, there are varietal questions along with eventualities for increased biodiversity (within the hedge) or agrobiodiversity (within the alley). Also possible are rotational considerations (within the alley), microbial inclusions, or the potentially beneficial landscape associations.

Developing systems around the above list, without insight or direction (as provided through the standard agrotechnologies), is an onerous endeavor.[28] Also, in contrast to the technically expressive agrotechnologies, alternatives are cloaked within broader philosophical outlooks. This adds to the confusion.

Inclusive under the alternative heading are composite expressions such as organic gardening, low-input agriculture, biological farming, regenerative agriculture, permaculture, mimicry, and productive harmony.[29] This section will attempt to put a technical face on these composites.

Plot Centered

Many of the alternatives center on plots. This is understandable, as the bulk of the practitioners are gardeners with comparatively small areas. The techniques, often lumped under an organic label, follow agrobionomic principles and incorporate a range of insect-control methods. The difference, a heavy reliance on labor, makes many of the techniques less interesting in a commercial setting. Still, commercial variations of these alternatives can and do exist.[30]

Organic Gardening

Organic agriculture has appeal with a set crowd. In shunning the application of synthetic inputs, practitioners and consumers hope for a diet more as nature intended and food cultivation with fewer natural conflicts. Needless to say, there are multiple variations of this theme.

Within this group, man-made chemicals (i.e., fertilizers, insecticides, etc.) are completely out of the picture. Benign home remedies are generally sanctioned, but most would prefer organically obtained nutrients: green biomass, compost, animal manure, and nature-harmonious insects controls.

Although some advocate the use of heritage varieties, commercial varieties may be shunned because of taste or other quality considerations. Genetically modified (GM) foods fall outside the scope of organic agriculture.

Farming, and obtaining the premium price for outputs, adheres to set standards. These are sponsored and supported across North America, Europe, and in a few other developed countries through product labeling and a sales network.

Low-Input Agriculture

The aim of low-input agriculture is to optimize on-farm resources, applying as little as possible of ex-farm chemicals. More than acceptable is the on-plot, simultaneous (crops plus facilitative species) and sequential (rotational) nutrient accumulation. The interplot and intrafarm transfer of nutrients, repellant biomass, and other useful byproducts is within the sphere, but may be expensive in terms of labor expended in the transfer. Seldom mentioned, but also within the sphere, are the facilitative versions of agroforestry.

Biological Farming

The biological alternatives, as with many under these headings, are directed toward cutting out chemical inputs, in this case, reductions in or the elimination of harsh ex-farm chemicals. There may be some leniency in using chemical fertilizers.

Regenerative Agriculture

With regenerative agriculture, the emphasis is on in-place or plot renewability. Because of the tasks required, rotations are a large part of the process. Simultaneous cycling, where the nutrients removed through crop harvest equal those input during that same season, is a part of this. This can involve facilitative and productive intercropping. Agroforestry is usually not mentioned as being inclusive in regenerative agriculture.

Landscapewide

Going beyond individual plots (and the manual interplot transfer of resources), there are landscape-based alternatives. These take advantage of wider dynamics and the direct or indirect interactions to achieve environmental goals. One difference is in how the alternatives view (and use) wildlife.

Permaculture

Stating that permaculture is an expression of organic farming may be an oversimplification. It is broader than this, but not all the elements of landscape agroecology are to be found.

Permaculture carries many of the same underpinning as organic or regenerative agriculture, but the ecological core revolves more on the landscape.[31] Location and the integration of productive intercropping, productive agroforestry, facilitative intercropping, facilitative agroforestry, auxiliary systems, and rotations are all part of this. There seems to be a passive view on wildlife and room for an active crop-fauna interface.

Mimicry

A step further in duplicating what nature had intended is natural mimicry. The cornerstone of this approach is to formulate and position agroecosystems so that these are ecologically, and often visually, analogous to natural landscapes.[32]

Through mimicry, an attempt is made to offer an agricultural surrogate for what nature may have placed on a site. If achieved, native plants and animals, but mainly birds, will be better accommodated than with insensitive, more intrusive farm landscapes. Among the agriculture facsimiles of nature are where rice paddies replace swamps, woodlots and agroforests replace natural forests, parklands replace open savannas, and so on.[33]

There is a successional, often destructive, facet. Over the long-term history of any region, natural systems are seldom static. A large body of evidence supports the idea that landscapes, even those without human intervention, are quite dynamic. High winds, fire, and heavy grazing are among those forces that shape nature. Ecosystems and the wildlife they contain were once resigned to this (i.e., they have evolved) and may still depend on a high level of disturbance. This transcends what is visually apparent.

Rather than persisting static systems, mimicry may include natural successions with a planned progression of plants. It is possible to emulate natural destruction including a bareground phase at prescribed intervals.

Unabated fires, floods, strong winds, droughts, and other periodic stresses can, in untouched natural ecosystems, produce a less than uniform, patch-type landscape. These, in league with now-extinct mega-fauna and substantial populations of other plant and tree eaters, can return a landscape similar to that now found in the savanna regions of Africa.[34] Evidence exists that the landscapes of North America, Asia, and other regions of the world were influenced in this way.[35]

Harvesting is a form of destruction which need not be out of sync with what nature does. The now-missing savanna (patched) pattern can be fronted by parklands with grazing or parklands where crops substitute for grasses. Forest blocks, windbreaks, and other tall features are part of, emulate, and add ecological credence to a mixed farm landscape. If formulated for the task, even features such as orchards can exhibit favorable grassland ecology.

Other regions are shaped in other ways. A gap landscape is current wherever untouched humid tropical high forests exist. These are formed where very large trees, through internal decay, winds, and rain, crash to the ground, opening gaps in the closed canopy.

Traditional slash-and-burn farmers work within self-imposed gaps, clearing small areas, burning the opening, and planting crops. These are abandoned after one or two seasons, and another gap is made for a subsequent planting. Slash-and-burn agriculture, if not overwhelming, follows from, and can be ecologically similar to, this natural gap pattern.

Evolutionary Convergence

Mimicry has another side, one where the nature and agriculture converge and long-evolved agricultural landscapes exhibit, through mimicry, a high degree of harmony with nature. Attenborough et al. (2000) illustrate a Japanese landscape where mixed-element agriculture offers both habitat and enduring productivity. The key seems to not be vested in one output or one ecosystem type, but to utilize the different ecosystems (or agrotechnologies) to produce a variety of salable products. Mimicry comes through the agroecosystem-natural ecosystem dichotomy, that is, with swamps (paddies), ponds (aqua-agriculture), grasslands (grazing and forage), forests (woodlots), and streams (irrigation channels).

The temporal sphere is equally rich with dynamics, reaching back hundreds or even thousands of years. Given the time frame, flora and fauna, both agricultural and natural, have achieved a level of cross harmony and become selected and/or domesticated within the broader multipurpose landscape.

Productive Harmony

Whereas permaculture and mimicry are passive with regard to the nature-human interface, productive harmony seeks direct engagement. The relationship between agriculture and natural flora and fauna offers a number of possibilities:

1. *Indulgence,* where native flora and fauna survive in the face of agriculture, is often the intent behind organic agriculture.
2. *Prosperity,* where nonproductive plants and animals thrive in league with agriculture, is a good vehicle for encouraging biodiversity (rather than agrobiodiversity). Wildlife harvest losses are accepted if overall gains accrue, for example, through semihusbandry and/or to control harmful insects, rodents, and so on.
3. *Productive harmony,* where fauna directly contribute to a positive productive outcome. Productive harmony is a step back from allowing natural flora and fauna to thrive. It is, however, a step forward in the productive exploitation of local wildlife.

Instead of offering what nature intended, miniecosystems are put in place. Fauna that benefit agriculture are encouraged, while those less favorable are discouraged. This can be with birds, bats, predator insects, and whatever else can be agriculturally exploited. This is a balancing act as it can be difficult to determine when lines have been crossed, that is, when ecosystem harbors good rather that detrimental fauna. Wildlife control, hunting included, may be employed to tip the balance in favor of sought-after wild species.

Chapter 16

Agroecosystem Design

Design theory, formulating the best agroecosystems for any one use, is a critical and evolving topic. Chapters 7 through 10 offer, as standards, temporally set agrotechnologies. Not all fit; some must be redesigned to be ecologically and/or economically suitable. This is the design problem, although with an entry point.

Farms are seldom static, and active agroecosystems do undergo a gradual evolution. This can be visually pronounced if a system is biocomplex with more variables from which a redesign can unfold. Generally, however, land users are loath to abandon what they know unless prodded by inside failures or outside economics. Because of this, a major reformulation of the existing or the adoption of new agrotechnologies is often best accomplished under the guise of change.

Gradual change, through increasing populations, can shift the emphasis of regional producers. Movements in farm policy, price supports, or shifting international trade realizes a quicker transformation. At the extreme, situations have been documented where the shift was sudden, as when the loss of a railroad cut access to markets, switching local farmers from commercial to subsistence agriculture.[1]

Whatever happens, farms and agroecosystems must adapt. With biocomplexity and/or with long-term systems, this can be problematic.[2]

The main hindrance with agro-biocomplex systems is the abundance of variables. However, if well presented, these same variables make redesign manageable. This chapter will attempt to put a practical face on a difficult topic: the design or redesign of economically viable, reduced-input, threat-free, sustainable agriculture.[3]

Introduction to Agroecology: Principles and Practices

doi:10.1300/5553_16

OPTIMIZATION

Agroecology is a multiple-objective optimization problem where land users, full- or part-time farmers, attempt to maximize the landscape to achieve their varying goals. In subsistence farming, goals are acceptable output levels for staple and other crops with minimal inputs and without much risk. Commercial farmers seek a similar outcome, but with enough profitability to purchase what is needed. Thrown into the mix, as objectives, are sustainability, quality-of-life, environmental, and other issues.

Well-designed plots, well-placed across an area, are the chief tools. Achieving the right mix is an extremely complex problem.

According to optimization theory, as complexity grows, so does the number of optimal and equal solutions. These solutions need not be visually or ecologically similar, but they can be quite diverse. The lesson is that there is no right way to do things. Each person and group strives along different development paths (often one or more), employing ever-shifting objective criteria in their attempt at their perfect landscape mix.

VARIABLES

In the previous chapter, the core and development paths, singly or in composite, are different routes to what may be the same set of objectives. The standard agrotechnologies do much the same, that is, offer a starting point, but these are more advanced and therefore more in tune with preestablished objectives. From these advanced points, design variables (1) add productivity, (2) trim costs, and/or (3) reduce risk threats.[4]

In regard to crop requirements, there are two opposing notions: (1) that people provide the inputs or (2) that these come through natural (nature-furnished) means. Since most of agriculture is a mix, this is a continuum where, on one side, all the inputs are provided ex-nature, and on the other, nature affords all.

The extremes are not well represented. Hydroponics are a prime example of nature-free agriculture.[5] On the other extreme, agroforests are as close as one gets to nature-indulged, input-free agriculture.[6]

In practice, conventional farmers tend to reside in the center, providing a lot of what crops need, for example, fertilizers, weeding, insect control, and so on, allowing nature to furnish rainfall and some minerals. This is more revenue-oriented agriculture.

Although many organic gardeners use a lot of inputs and achieve impressive productivity, those championing commercial versions of alternative agriculture generally seek a contrary, off-center position: nature delivering the bulk of crop needs, people proffering less.

With this pronature stance, through biodiversity, a nature-friendly environmental setting, rotations, and other paths, there is a tendency toward cost orientation. The economic studies cited in Chapter 14 encapsulate this view. It should be noted that many of the agrotechnologies are economically neutral and can proceed in either direction.

Productivity

As yields increase, upper and absolute limits begin to take hold. It is possible to keep soil moist and well stocked with a range of nutrients, but sunlight is problematic. Economics also provides a disincentive: forcing every bit of productivity from the land may mean that yields above normal are, on a per-unit basis, comparatively expensive.[7] Still, if land is in short supply and market prices robust, high yields are very enticing.

Agroecology does address this situation. Agrobiodiversity does coax a lot from the land with the proviso that the high LER values translate into high RVT values which, in turn, mean high profitability. The route to this is through seasonal or agroforestry intercropping where, by apportioning resources efficiently, resource limits imposed upon a single crop are overcome. Cost savings, where the weeding and fertilizing of one crop benefits a second or third, can further ensure a favorable financial outcome.

The danger lies in going in two economic directions at the same time. If revenue increases are the goal, one should not invoke yield-declining, cost-cutting options. Cost reductions should only be a welcome side effect of a productivity-increasing measure. As a universal economic precept, the point should never be passed where the cost of the input is greater than the additional revenue received.

Cost Containment

Often the incentive is to eschew high-input, revenue-oriented agriculture and to adopt less costly practices. Growing crops with less has the added advantage of assured profitability, even when selling prices are low.

Studies (as cited in Chapter 14) show that with planned alternatives and lesser amounts of outside inputs, yields decline, but so do the costs. Through rotations (to improve soil mineral content and reduce weeds), plot size reductions, and auxiliary habitats and corridors (to reduce harmful insects), and better timing (helps reduce weeds), reductions in yields are balanced by equal or greater cost reductions. Profitability remains more or less constant.

As with revenue orientation, the economic rule of agroecology is not to mix revenue and cost orientation. If cost cutting is the road chosen, costly yield improvements are not considered. Revenue (yield) enhancements are applied only if free from costs (e.g., as with a shift in plant variety) or cost reducing (e.g., a variety that also reduces anti-insect countermeasures). In a rephrasing of a universal economic precept, losses in revenue should never be greater than a cost savings.

Threats

It is comparatively easy to identify and implement specific threat counters, and many are tagged in the proceeding chapters. The task lies in achieving farm goals, including countering multiple threats, with comparatively few design variables, that is, fields and farms in cross-ecological harmony. As this is site specific, the best that can be done here is to put a friendlier face on a difficult topic.

In Chapter 3, the frequency (number of incidences), magnitude, and, in some cases, the duration of threats is discussed. This can be expanded.

Each approach employed should add something to an ecosystem. As composites, these can write or be cumulative against a single threat, or each approach employed can address an individual threat to a single crop.

Take the case of frost. This can be counted with the following:

- A single cold-resistant species (genetic improvement)
- Different varieties (varietal selection)
- Inoculation (microbial)
- An ecosystem that offers perennial protection (agroforestry)
- Landscape affiliations, that is, tree strips (facilitative landscape affiliations)
- Topographical, for example, hillside, positioning can help (placement)
- Land modifications (mounds, rock heat sinks, and/or flooding)

This does not mean that if all are employed in tandem, cold-sensitive tomatoes can survive an Antarctic winter.

Through well-chosen, cumulative alternatives, slightly lower temperatures can be countered (threats of greater magnitude can be endured) or the amount of time freezing air is present can increase without harm (the effective duration of the remedy). These work only to a certain point. In employing too many counters, the cumulative effects and the amount of added risk reduction may, in large part, be overkill.

For example, one insect control countermeasure (e.g., biodiversity) may reduce incidence by 70 percent, a second (e.g., landscape associations) by another 20 percent, with another added mechanisms virtually eliminating the problem. A land user may be content with a 90 percent reduction in risk and consider the problem resolved.

This is an ecologically legitimate line of reasoning. A 100 percent reduction may be nice, but this can be expensive to achieve. In some situations, overkill can prove disadvantageous. Besides being potentially costly, bugs, if entirely removed, can result in a dearth of predators and a quick reinfestation with herbivore types. Better is to maintain an unnoticeable or tolerable base population.

This argument crosses most of the adversities. Controlling weeds at the 100 percent level may seem desirable, but allowing some to survive, especially those with facilitative potential, may prove preferable.[8]

Cross Effects

Rather than relying on a one-on-one pairing of changes or inputs against productive needs and threats, a more efficient course is to exploit the protective and productive overlaps. In application, the first addition many confer some insect protection and a lot of frost protection while the second offers good insect protection while helping to ward off frosts in some small way. Together, these may confer 90 percent protection on both counts.

To cite specific examples, sunflower and Jerusalem artichoke are mentioned as seasonal windbreak species. These also provide output and have decoy potential for birds. The lining of microcatchment holes with manure helps maintain on-site moisture. This also fertilizes plants and may add aphid control. On the larger scale, a windbreak serves admirably as habitat-travel corridor, waterbreak, and riparian buffer. In short, the summing of many cross effects, those conferred by numerous internal and external agroecological additions, can lead to a higher agroecological plateau.

Implementation

In addressing need, a good starting point is the form of alternative agriculture to which locals currently subscribe. Presumedly, land users are predisposed to continue along well-worn paths and, through practice, have acquired enough understanding to advance the chosen approach further.

As seen, it clearly helps on all counts if a second, or even third, approach can be brought to bear. Users may know one reason for why something is done but be unaware of the cross effects.

It is the gradual slackening of design characteristics and elimination of cross effects that deimplements agroecology. What often happens is a slow, gradually accelerating, loss of ecological function. Few notice when change increases major bug infestations from 2 to 3 percent (1 in 40 seasons) to 5 percent (1 in 20 seasons).

As time passes, land users become accustomed to the 5 percent risk factor. A subsequent ecosystem change that increases the risk to 6 to 7 percent (1 in 15 seasons) may merit mention but little action. At this point, enough of the earlier, less risky practice may be retrievable. These would be a basis for ecological progress.

As agroecological deimplementations retrogress, more function is lost and a crash is eminent (the risk is becoming intolerable at say one in three or one in four seasons). At this stage, instead of restoring the ecosystem that was gradually lost (and maybe forgotten), land users often seek more immediate and simpler ex-farm, chemical-based solutions.

This is where local understanding is required. If user knowledge still exists or a practice is still locally accepted, it may be possible to make a U-turn.

If the once-encountered, pro-ecological features have vanished and are no longer in user consciousness, the process of simplification can continue but take another, slightly different path. The development paths that embrace agroecology lead away from large monocultures and costly outside inputs. The solutions can lie in strips, rotations, habitat-travel corridors, windbreaks, and other in- and ex-plot features that add to and ecologically enhance monocropping.

AGROTECHNOLOGIES

These de- and reimplementation scenarios depend upon following an appropriate development path. Alternatively, one might prefer the comfort of well-guided jumps instead of a progression of small steps. The agrotechnologies are a framework, not only for utilizing conventional knowledge of crop production but also for implementing agricultural alternatives.

Chapters 7 to 10 discuss the productive agrotechnologies in order of their inherent biocomplexity. Others contribute in other ways/dimensions. The full list of agrotechnologies is as follows:[9]

- Productive agrotechnologies
  - Monocultures
    1. Pure
    2. Varietal/genus
    3. Perceived and managed
  - Facilitative intercropping
    1. Simple mixes
    2. Strip cropping
    3. Boundary
    4. Cover crops
    5. Clusters
    6. Complex agroecosystems

— Productive intercropping
  1. Simple mixes
  2. Strip cropping
  3. Boundary
  4. Complex agroecosystems
— Facilitative agroforestry
  1. Isolated tree
  2. Parkland
  3. Protective barrier
  4. Alley cropping (hedgerow)
  5. Strip cropping
  6. Crop over tree
  7. Physical support systems
  8. Shade systems (heavy and light)
— Productive agroforestry
  1. Agroforestry intercropping
  2. Alley cropping (treerow)
  3. Strip cropping
  4. Shade systems (heavy and light)
  5. Clusters
  6. Agroforests
— Feed systems
  1. Silvopastoral
     a. Pastures
     b. Pastures with trees
     c. Forage trees with pasture
     d. Forage trees
  2. Aquaforestry
  3. Entomo-agriculture
  4. Avian-agriculture

- Auxiliary agrotechnologies
  — Reservoirs/habitats
  — Corridors
  — Windbreaks
  — Waterbreaks
  — Firebreaks
  — Riparian buffers
  — Buffer zones

  - Living fences
  - Infiltration structures
  - Cajetes
  - Cut-and-carry
  - Catchments
  - Ponds
  - Water channels
- Temporal agrotechnologies
  - Single rotations
  - Series rotations
  - Overlapping cycles
  - Taungyas
    1. Simple
    2. Extended
    3. Multistage
    4. End stage
  - Fallow sequenced
  - Continual
- Land-modification agrotechnologies
  - Absorption zones
  - Microcatchments
  - Infiltration structures
  - Cajetes
  - Terraces
  - Paddies
  - Water channels
  - Mounds/beds
  - Stone clusters

Selection Criteria

With agrotechnologies, the DAPs help match each with the ideal application. Although current DAP lists are spotty at best, if amplified, they can go far in promoting this version of, or approach to, agroecology.

In practice, eliminating factors helps to narrow the list.[10] Three such criteria are given in the following, along with those productive agrotechnologies that fall under these headings.

- Machine (harvestable) agroecosystems:
  — monocultures (all types)
  — alley cropping (tree and hedgerow)
  — parklands
  — strip cropping
  — alley cropping (tree and hedgerow)
  — barriers
  — shade systems (light and heavy)
  — facilitative intercrops (with cover crops)
- Forestry augmented (able to supply marketable poles and logs):
  — isolated tree
  — parkland
  — alley cropping (treerow)
  — physical support (tree over vine)
  — shade systems (heavy and light)
  — agroforests
- Maize compatible[11] (a similar list can be derived for any crop):
  — monocultures (all types)
  — strip cropping
  — alley cropping
  — parkland
  — barriers
  — clusters (depends on the cluster type and internal location)

The lists of candidate agrotechnologies are further reduced through other criteria. Strength in combating erosion can separate out systems from the full list or further pare a subgroup (as above). Revenue or cost orientation are often factors in making the final selection. Without going into greater detail, it is easy to see how lists are trimmed.

Redesign

Instead of selecting anew from the list of productive agrotechnologies, many may be content to redesign what they already have. Internally, agrotechnologies often start by modifying or changing the species mix or pattern arrangement. This goes far in fine-tuning systems.

A greater boost may be needed. The major grouping, productive, auxiliary, temporal, and land modification, are not exclusive and, by combining the groupings, these encompass considerable variation. Under this scenario, the nearby positioning of an auxiliary system can correct an ecological imbalance, threat, or shortcoming in a productive ecosystem. A change in rotational pattern is another agrotechnology redesign tool.

Development paths (core approaches) are also employed in redesigns. Approaches and agrotechnologies can be viewed as two sides of the same coin and agroecology can be implemented from either direction with equal effectiveness. This view is one of a continuum.

agrotechnologies <——————————> composite approaches

The more technical productive agrotechnologies versus the philosophically laden composites, dropping one to evoke another.

One may wish to start with an agrotechnology and proceed down one or more development paths. Figure 16.1 shows this more involved relationship and how, when a farmer subscribes to one approach, either an agrotechnology or a composite approach, this means taking

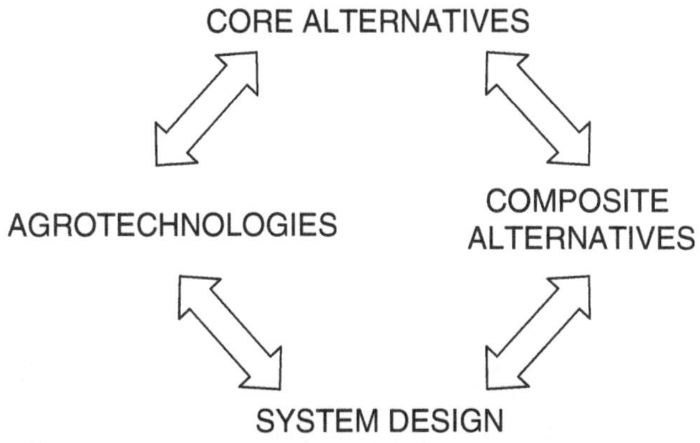

FIGURE 16.1. The relationship between the composite approaches and agrotechnologies. This can be one of a continuum, where as one weakens, the other strengthens, or core alternatives that contain elements of both. Both the agrotechnologies and composite alternatives contribute in system design.

lessons or borrowing techniques from the other. For example, a commercial farm may start as a series of agrotechnologies and, in wishing to market chemical-free produce, apply appropriate approaches to bring the farm into organic compliance.

As a clarifying point, the productive agrotechnologies and forms of alternative agriculture contain the same underlying elements, but they only emphasize different development paths in reaching their present form. The composites (forms of alternative agriculture) are less encumbered but also less technically descriptive and less defined in the goals and threats addressed.

Agrotechnological Variation

How the agrotechnology is viewed has a lot to do with implementation. The agrotechnology can be a faithful depository of agricultural knowledge when each agrotechnology has a complete design package. The proceeding chapters present the agrotechnologies in this knowledge-layout rigid form. For this, each is visually and ecologically comparable. This is not the only view.

Divisional Agrotechnologies

Rather than agrotechnologies as unalterable, rigid systems imported into a farm landscape as is, these can vary in form, accepting local advice and new developments to produce variations on the standard design. With this divisional view, the variations are along set lines.

Change commonly comes in species form. One can substitute for the primary or for a secondary, facilitative species. A single change might occur when, in hedgerow alley cropping, one hedge species is substituted for another.

More profound, but in keeping with attributes and design package of the original system, maize alley cropping can differ from wheat alley cropping in both the primary and secondary species. Paired changes ensure tree-crop complementarity and favorable yields while keeping the underlying purpose.

Formulation Agrotechnologies

Further along this design track is formulation. This view encourages much more variation off the standard theme. The agrotechnologies still have a central or defining characteristic but are formulated for a specific purpose and will vary considerably across the landscape (see Plate 16.1). In some cases, only close examination will reveal their common base.

Some systems rely upon one or two strong effects to confer advantage, for example, nutrient facilitation couples with good predator-prey dynamics. Preserving these keeps the agrotechnology ecologically intact even if the changes seem severe.

The opposite is where a system relies upon a multitude of agroecological interactions to confer the net benefit. Although the loss of one or two may seem unimportant, the variation does suffer, especially when those lost mitigate risk or champion other, less apparent, issues.

PLATE 16.1. A traditional East African farm. These can be free-form, lacking in agrotechnological definition but well laden with agroecology-promoting features.

Such a situation can occur with a biodiverse, heavy-shade system, one that operates under the parameters of ecosystem governance. Much or little can be lost in changing to an overstory containing fewer species. Such a changeover can be perceptibly negligible but, in actuality, ecologically significant. This is where an understanding of the internal dynamics ensures that the economic and ecological character of the original design is not compromised in a major changeover.

Chapter 17

Agricultural Landscapes

The previous two chapters sojourn into the conceptional core of agroecology. This chapter goes back on course, tackling the broadest individual expression of agroecology: the farm landscape. This can be either as a collection of agrotechnologies or, in encompassing non-agrotechnological thought, some other ecologically agreeable form.

A series of ecologically independent plots carries the assumption that each is ecologically self-contained or is provided, on an individual basis, with any needed outside inputs. In agroecology, self-containment is nice, but it is not the only alternative.

Landscape associations can and do underwrite agroecology. In intent, auxiliary systems are best if placed next to target productive systems, and so it goes. Plots and farms are best operated with some or a high degree of cross-landscape ecological harmony. There is large and growing body of theory under this heading.[1]

FACTORS

A number of factors come together to drive or derive a farm landscape. Agricultural intensity, often expressed as per area population, is part of this mix, as is economic intensity, that is, how farm resources are allocated between plots or areas. Inclusive in this are the cropping systems and texture of landscape. This can produce a uniform landscape, as found with a series of similar systems, or one that is visually dramatic, interspersed with plots of trees and plots of annual crops.

A landscape is also influenced by cultural factors. Most apparent are whether geometric plots (square, rectangular, etc.) are adhered to

Introduction to Agroecology: Principles and Practices

doi:10.1300/5553_17

or whether plot layouts are dominated by topography or soil characteristics. It is also possible to have landscapes devoid of clearly defined plots.

The artificial boundaries that cross landscapes limit what can be accomplished. These can be legal (as when farms defy topographical acquiescence) or cultural (academically derived as when agriculture is spatially segregated from forestry). Although only touching on a complex topic, this suggests what might be encountered on familiar or foreign landscapes.

LEGAL BOUNDARIES

Land ownership patterns can be ecological, but, as few legal boundaries conform to ecologically distinct areas, this is more the exception than the rule. Social conventions dictate what can be done; described here are the limitations that result from less than optimal spatial layouts and farm placements.

Most often, ownership patterns are geometric, rather than topographical. Valley bottoms can be under one management, while hillsides can be under another. Breaking up the land in this way should, but does not, promote agrotechnological diversity. To produce a staple crop in an environmentally sound manner, the farming bottomlands should focus on one practice (agrotechnology), and those on high-risk hillsides should utilize another. As this seldom happens, hillside land degradation, caused by the adoption of unsuitable bottomland practice, can and often does follow.

Land ownership formats can be in response to past traditions and practices. As a form of alley cropping, long narrow plots bordered by hedges were once the norm in Medieval Europe. In addition to the more or less standard ecological gains, these favor draft animal efficiency as, when plowing, they have to turn a fewer number of times.

As the situation changed—rotations ecologically replacing alley cropping—the land was reordered. This change may have been spurred by widespread cattle possession and the need for less constricted pastures.

Patterns have endured simply because of the difficulty in reordering land ownership or have persisted because they continue to confer ecological gains.[2] With the advent of agroecology, narrow fields, bordered by hedges or grass strips, may turn out to be the best option, and with this design, the maneuvering of tractors is also efficient (as was the case with draft animals).

Ownership formats also hinder the placement of cross-landscape structures. With little incentive for a coordinated landscape, farmers let slip opportunities for corridors, windbreaks, waterbreaks, firebreaks, riparian buffers, and other proagricultural features.

The exception lies in irrigation projects where the benefits are immediate and observable. In contrast, the ecological gains from other auxiliary structures are usually long term, sporadic, and seldom visually apparent.

Other negative aspects come to the fore: a poorly managed farm can be a habitat for insect pests and plant diseases; a poorly coordinated landscape does much the same and can restock well-managed areas with unwanted organisms.

Along the same lines is when one farmer decides to employ toxic sprays. Environmentally aware neighbors, through spray drift, can lose the organic farming option.

OVERALL LAND-USE INTENSITY

Through observation, the degree of land-use intensity is more than apparent. Observation aside, this can be troublesome to formally measure.[3]

In high-intensity landscapes, most of the area is under cultivation. This is visually obvious when marginal, erosion-prone hillsides are utilized to produce staple crops.

In contrast, a landscape may be where large areas are utilized but in low-intensity form. This is the case when cattle freely roam across large tracts.

There are variations off this intensity continuum. The land can be minimally impacted with small areas of intense cultivation and large areas with little or no agriculture. Slash-and-burn agriculture (areas of intense cultivation within a mostly untouched forest) epitomizes this high-low dichotomy. Figure 17.1 shows mixed intensity where a few plots (left) are high input but most of area, in a few large plots, is low input (right).

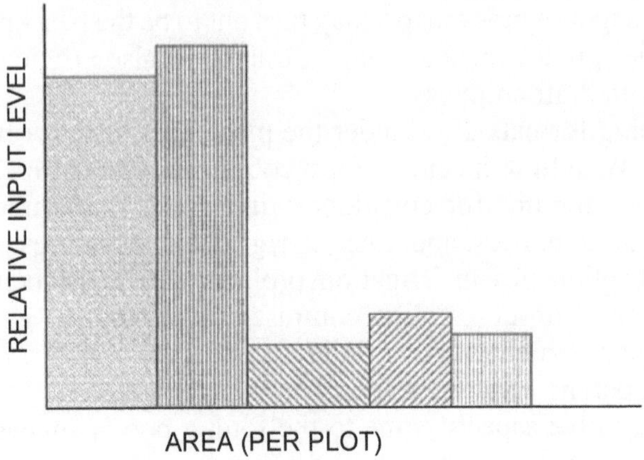

FIGURE 17.1. Plot intensity across a five-plot farm landscape. Two plots, those on the right, are high in intensity, while the three on the right are of low intensity.

Figure 17.2 illustrates this continuum and the high-low dichotomy. All plots can be intensely managed (Figure 17.2, lower left), all can be low input (lower right), or this can be mix of high- and low-input plots (upper).

ECONOMIC FACTORS

A number of agricultural economic theories bear on cropping intensity. If strictly followed, these can dictate the landscape form.

Economic uniformity stems from marginal gains.[4] As the marginal gains diminish on more intense plots, resources and production are transferred to areas with greater marginal gain potential (i.e., where yields increase dramatically with relatively small levels of inputs). The idea being that, across plots, inputs are applied such that eventually the ratio of the value of inputs (e.g., labor) gives the same ratio in the output value (the harvest).[5]

This concept is easiest to implement when the land is topographically unchanging and/or when cross-landscape ecological influences are not being contemplated. This results in more uniform landscape. This is illustrated in the lower high-low continuum in Figure 17.2.

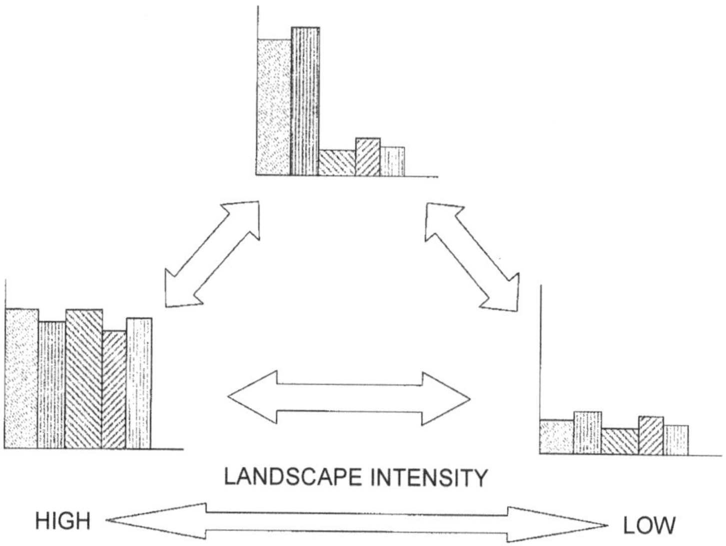

FIGURE 17.2. The intensity options: on the lower left is farm totally high intensity; the lower right has all plots being low intensity; the upper is of a mixed farm. The graphs are labeled as in Figure 17.1.

Theory does not always trump practice. Inherent, and seemingly extraneous, factors intercede to support one landscape type or another. These include the following:

- Agricultural alternative espoused (e.g., more set aside natural areas foster nonuniformity whereas plots containing internal agrobiodiversity tend to be uniform)
- Topography can force activity into small areas
- Per unit harvest costs (these are lower when per area yields are higher and, when resources are lacking, this comes from putting a greater share of the resources into a few plots)
- Relative distance from the farm center (more actual resources are applied to closer plots but, if travel time is considered, the labor resources expended may be more or less equal)
- Farm infrastructure (roads, labor, tractors) may not be fully up to the task of high-intensity cultivation without more investment

- Environmental risk from overall higher intensity strategy is high, for example, increased erosion danger in converting intensely managed, high-risk hillsides to seasonal agriculture
- Gains that occur with interspersed, less intense areas, for example, as with habitat reservoirs, are difficult to fiscally qualify, but can still be economically significant, adding, through cost controls, to the overall financial picture

PLOT TYPES

In many societies, the plot is the basis of agriculture where many plots combine to form the farm landscape. To most, this is unchallenged, but cultures differ and so does plot definition.

Two plot forms are evident (1) geometric and (2) terrain based. There can be some overlap between the two types.

Geometrically Dominant

Square, rectangular, or circular plots characterize a geometrically dominant landscape. Common in flat regions, usage extends to hilly and even slope-challenged mountain regions. As farm machinery or draft animals can operate more efficiently in a straight line, there is some facility in this. Erosion dangers can magnify when plot geometry ignores the land contours. As has been mentioned within this text, this can be compensated for with the correct agroecosystem (terraces included). A geometric landscape is illustrated in Figure 17.3 (right).

Terrain Dominant

In high-relief, challenging mountain areas, users may opt for plots that conform to the slope, that is, field outlines are set by the degree of flatness. This approach has wider implications where land relief is a factor.

Plot boundaries can also be set by soil characteristics, for example, soil moisture, density, nutrient content, and/or pH. When land-use intensity is high, this produces a landscape with oddly shaped, inter-

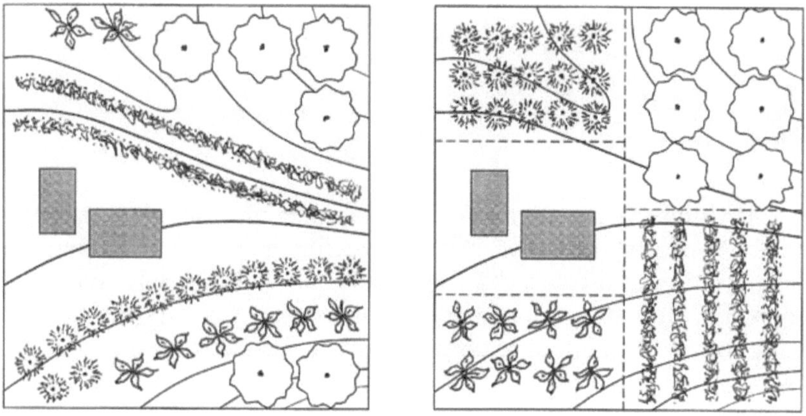

FIGURE 17.3. A geometrically dominant landscape *(right)* where geometric-shaped plots determine the layout versus a terrain-dominant landscape *(left)* where a topography-crop match may be the underlying motive.

locking fields, but does encourage a closer match between crop types or varieties and attributes of the soil and subsite. A contour-conforming, nongeometric layout is shown in Figure 17.3 (left).

INTERECOSYSTEM COMPLEMENTARITY

An agroecologically sound landscape is cross-hatched with a web of interactions. Auxiliary systems can do much to promote these. Principal-mode systems, through a comprehensive listing of DAPs, can confer to neighboring plots many of the same ecological gains as obtainable with neighboring auxiliary systems.

The landscape, through immediate and distant plot-plot associations, can maintain, and even amplify, certain positive interactions. Plot-plot or interecosystem complementarity underlies a well-designed landscape.

Whereas certain DPCs have an accompanying SIZ and help determine the best plot formulation, some of the DAPs of ecosystems have accompanying SIZs. This is part of landscape formulation.

On this larger scale, more variation exists. The influences are (1) the simple SIZ, (2) SIZ extensions and augmentations, and (3) multiple SIZs.

Simple SIZs

The simple SIZ is where one influence encompasses all or part of a neighboring plot. Predator insects spreading into and consuming the herbivore insects in a nearby crop plot is one such example. With this approach, plots are comparatively small as nearby positive influences must convey with enough force to guarantee effectiveness.

SIZ Extensions and Augmentations

With some planning, the SIZ influences are upheld or amplified by internal design of target plot. Predator birds or insects may require a habitat but, if the crop plots are favorably formulated, their spread and their stay in the crop zone can be greatly enhanced. Crop plots so formulated can be larger than when one relies upon a simple, juxtaposed SIZ.

There are all kinds of DAPs exploitable in this manner. Parkland systems can provide a favorable hunting ground but little habitat for rodent- or insect-hunting birds. Parkland systems can augment but not initiate a windbreak effect. The previously mentioned practice of placing T-shaped poles in fields is part of this, keeping beneficial birds on site.

Some systems can be favorable to certain effects. An understory of a flowering cover crop may go far as a habitat for predator insects. The system may fall short of being fully self-contained in predator insects but require only a nearby ecosystem for restocking purpose. These are but a few of the mostly undocumented expansive or augmentative impacts of various agrotechnologies.

Multiple SIZs

Farmers do not live in single-dimensional world; instead, multiple concerns and threats are the norm. One way to deal with these is through agrotechnologies that address some of the issues. Often a system is vulnerable and a landscape SIZ, though a well-endowed neighboring system, can instill enough ecological strength to counter the vulnerability. Better yet, a system may be a redesign where a susceptible ecosystem is internally strengthened and this strength is further solidified by the design and the spillover ecology from the neighboring system.

This multiple protection format is illustrated in Figure 17.4 with three wind-protected areas. All three, within the shelterbelt-windbreak system, are further protected by an internal plot feature. On the left is a series of smaller, possibly seasonal, windbreaks; the middle plot is wind shielded by parkland trees; and, on the right, the plot center is planted with a wind-resistant crop.

In this complex world, little formal guidance is in place. One must rank concerns and threats and selected counters.[6] More than one

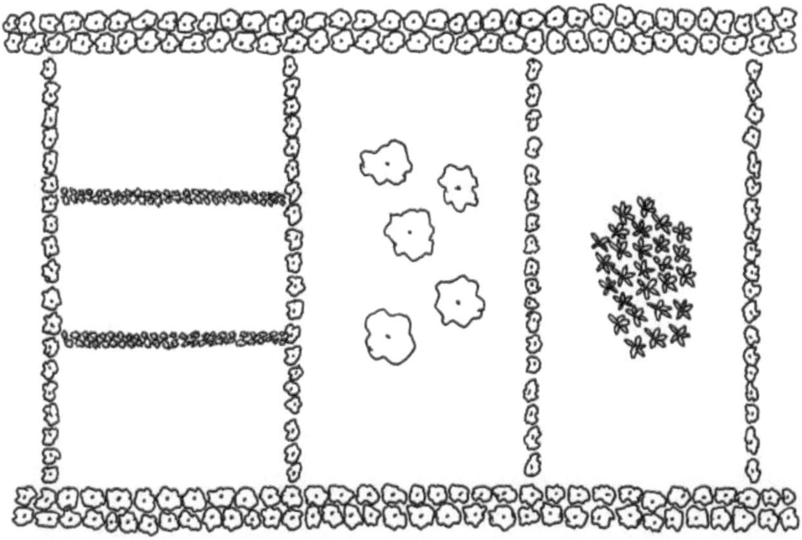

FIGURE 17.4. The simple integration of shelterbelts and windbreaks and the internal plot. On the left, the plot is subdivided by seasonal windbreaks, the center plot is wind reinforced through parkland trees, and, on the right, the exposed center is planted with a less wind-susceptible species.

thread or concern is often the norm; this requires a high degree of sophistication on the part of the land user.

BASIC LANDSCAPE FORMS

The uniformity of a landscape can be due to the economic apportionment of farm resources or based on differences in type and structure of ecosystems. Often visually indicated, some farms are a series of ecologically like ecosystems, while others express profound agroecological differences across the landscape.

Part of this goes to the idea of academic expression, that is, where some cultures separate agriculture and forestry, others mix systems as through agroforestry. The outcome may be referred to as texture, where the landscape is not vegetatively uniform but is patterned in the intermingling of diverse ecosystems.

Uniform Cropping

The most recognized landscape forms contain uniform cropping. This is found in large agricultural areas and epitomized by endless fields of monocultural grains, that is, a series of abutting monocultural agrotechnologies. Greater texture and ecological cross-effect comes when different seasonal or perennial crops, located in individual plots, are interspersed across the landscape.

Better yet is when the landscape is maintained by large, ecologically self-contained plots supported by internal agrobiodiversity. Figure 17.5 (lower left) is of a farm with agrobiodiverse farm fields that, in being self-contained, require less support from nearby areas. These can cover a wider expanse without ecological reinforcement.

Fragmented

With some crops, internal agrobiodiversity is less of an option. For these, monocultural strips, supported by nearby crops or corridors, complete the agroecological picture. Figure 17.5 (upper right) has this layout type.

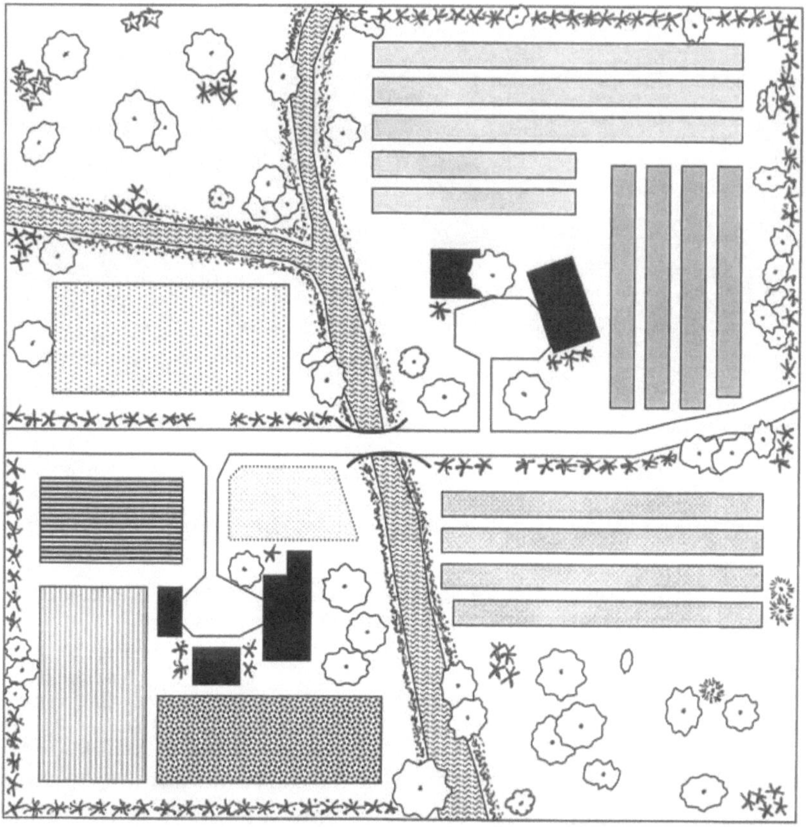

FIGURE 17.5. Comparison of monocultural versus agrobiodiversity-based farm designs. The bottom left farm has ecologically self-contained, mixed species plots that allow for larger, more uniform plots. The top right farm is monocultural and more risk prone and is ecologically guarded, and fragmented, by auxiliary systems and extending strips and corridors.

Further strengthening movement corridors and cross-field strips are productively inert habitats. Orchards, stands of tree crops, woodlots, or forest fragments amid crop plots produce a similar outcome. This results in a fragmented landscape (see Plate 17.1).

When forest fragments are involved, change is most often responsible. Forest fragments often result when farmers clear a forested landscape. At an advanced stage in the deforestation process, residual

PLATE 17.1. A fragmented landscape with small plots, the three juxtaposed systems are a pine plantation, an apple orchard, and a hayfield.

forest remnants stand in adjoining cropping areas. Reforestation does much the same when parcels revert to forest or are put into less intense tree plantations. The remaining more intense plots create the mixed fragmented landscape form.[7]

These more ecologically endowed stands are an on-farm asset through simple SIZs without expansion or augmentation. To be fully effective, comparatively small plot areas and connected forest or other wild areas best promote this landscape form.

Agroforestry

The alternative to comingling tree and crop plots is through agroforestry. The bulk of the agrotechnologies presented in earlier chapters are tree based and together, through DAPs, these can be utilized to great effect. Mixing different agroforestry technologies can produce a very nonuniform texture.

Keeping within one agroforestry agrotechnology, provided that the tree-crop mix is ecologically potent and to a large degree ecologically self-contained, can produce a uniform, less textually diverse landscape. Large cocoa or coffee farms found in South America are very unchanging. This is because the shade systems are ecologically strong, are self-contained, and require no interplot influences.

LANDSCAPE INTEGRATION

Clearly, there is much to be gained with cross-plot effects. This covers all aspects of agroecology. A case in point, shelterbelts and windbreaks are best when integrated across the landscape. This can be much more than countering adverse air movements.

In Figure 17.4, the left plot is further protected by seasonal wind structure that can double as an insect movement corridor or crop-precluding feed source (e.g., as with sunflowers) for birds that eat insects and grains.

The center plot, Figure 17.4, is wind protected through a parkland. A secondary purpose can be as a roost for rodent-seeking or insect-hunting birds.

The right plot, again Figure 17.4, contains a center area of what should be a wind-resistant species, for example, one that does not lodge and is not subject to sandblasting, such as artichoke. A center plot design may also be selected and formulated for an insect-countering SIZ, that is, where the crops near the wind structures are with the insect-protecting SIZ.

The above case represents only a few of the many options. Much of this goes to emphasis, ranking the ability of an agrotechnology to counter multiple threats, picking first those that are strongest against the threat or threats of most concern.

These associations are not only plot focused but also should constitute a landscapewide system of defensive bulwarks. Figure 17.6 (right) shows interconnected auxiliary structures, that is, boundary corridors.[8] This may be a good starting point; others may not depend as much upon auxiliary structures but still be ecologically viable.

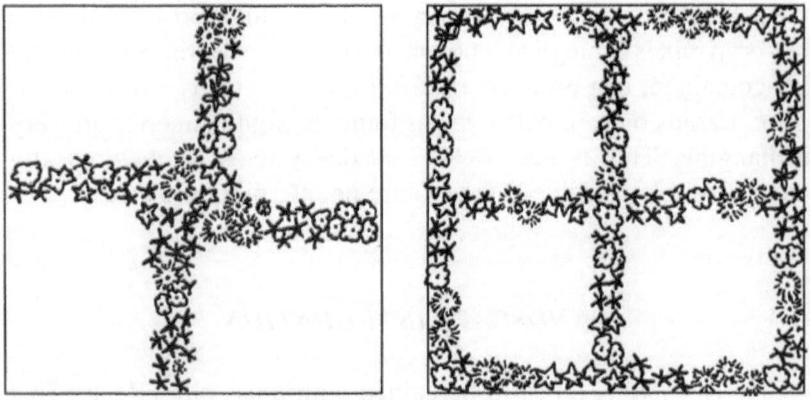

FIGURE 17.6. Two habitat/corridor designs. On the left is a habitat center with extending corridors. At right, the habitat is peripheral with corridors reaching into the center. These layouts allow protective forces, for example, predator insects and/or birds, to flow to the crop plots.

STARTING POINTS

As with multispecies plots, landscape patterns bring to the fore many ecological possibilities.[9] Each agroecosystem offers DAPs and, in the right location, these can be as facilitative as dedicated auxiliary systems. The questions are how this can be accomplished.

Given the myriad of possibilities in overall cross effects, it can be difficult to formulate a landscape without a clear starting point. Current land-use patterns suggest some.

Immediate Associations

Ecological gains can occur by placing one system near another, such as the DAPs that expand or augment. Positioning systems for the best complementarity, without any grand or overriding pattern, may work surprisingly well. The end result would be fragmentation and/or agroforestry coupled with an uneven texture.

Circular

A circular landscape has parallels with concentric plot planting, only the scale is much larger. Although all the ecological advantages of circular plot layouts do always translate to the the larger scale, some do and these are the rationale for use.

Plot Centrific

The plot-centrific approach starts with the one or two key areas, usually land with the best soils, good microclimate, etc. From this, subsequent additions ecologically strengthen the center plots. This can be a mixed, custom-designed landscape, semicircular, or purely circular.

Many landscapes are outwardly ranked in terms of intensity. This model has a very intense set of center plots becoming less intense away from the center. This can be based on an economic rationale, as when the more intense plots are close at hand.

Plot-centrific landscapes can be topographical, and the intense areas are often on agriculturally friendly bottomland. The surrounding, less intense plots are on the riskier hillsides. In regions with rolling hills, the center plot can be a hilltop or small valley. In mountains, a level area might constitute the center.

Ecological Pivot

An ecological pivot pattern places a permanent less or nonproductive ecosystem at the center of a crop ring or plots arranged in a circle. This is shown in Figure 17.6 (left).

The center pivot should be ecologically well endowed, providing facilitative services to circular-patterned agriculture. Forest gardens, forest fragments, mixed-species orchards, or a natural grasslands, those that encourage habitats, serve well. Other center options are for cut-and-carry biomass, either nutrient-laden vegetation and/or repellant plants.

The along-the-rim crop plots might incorporate a rotational sequence either clockwise or counterclockwise around the pivot. Accompanying and expanding the ecological pivot and the habitat function are movement (ecological) corridors. These will radiate outward from the center pivot and can serve multiple purposes as, for example, windbreaks, waterbreaks, infiltration zones, and other auxiliary structures.

Structure-Village Centers

Single farms, multiple farms, or villages may follow the pivot model. Instead of a productively inert auxiliary ecosystem at the center, a house or village may be so positioned.

In tropical regions, it is common to find center households located amid agroforests. The agroforests provide disposal for organic house waste, the garden benefiting from these nutrients. Fruit trees shade the house while offering minor foodstuffs, herbs, medicinal plants, and other useful vegetation. Being ecologically very active, this is a form of ecological pivot. Plate 10.3 shows this center form.

The first crop ring of high value, intensely managed agricultural species may be the most pest susceptible and gain from the bird and bat habitat that the agroforest provides. Subsequent concentric rings are of lesser value and in less immediate danger.

Much the same happens with temperate farms, only the habitat is different. Barns host insect-eating fauna, for example, barn swallows, chimney swifts, and bats. In past eras, free-ranging fowl would have foraged the immediate fields for insects (a two-pronged effect). Rodents are controlled through barn owls and house cats.

A larger-scale version is the strong village layout where instead of one farm, a small village is the ecological pivot. This is unique in that it crosses ownership boundaries. As with center farmstead, the resulting ecological landscape may be informal, rather than a conscious effort.

Each farmer, in owning land at various distances from the village, may choose to plant crops such that the more intensely managed are close at hand and those requiring less effort are at greater distance. Surviving examples are found in Europe. These have inner rings of household gardens, and outside this are mixed-species orchards, grain crops, grazing, and, in the outermost ring, foraged forest.

Peripheral Layouts

Rather than an ecologically endowed center, a land user may choose to locate auxiliary structures at the periphery of fields. This may be an advantage where streamside riparians are needed or where forests surround a cropping area. Shown in Figure 17.6 (right) is where boundary-corridors are joined.

These two strategies, center and peripheral, are far from exclusive. Many landscapes utilize a mix.

Scattering

Another landscape model is scattering. This often finds use where topography is dramatic and, because conditions vary so much across the area, weather can have an uneven impact. With this, plots of like species are scattered across the landscape such that the weather conditions that affect one are less likely to affect all.

As previously mentioned, having like crops grown in various locations and under variety of climates favors multiple varieties. The disadvantage is that each plot may be monitored less and arising threats may not be addressed in a timely manner.

Plot scattering is found in East Africa and Peru where farmers own plots in different locations.[10] In addition to countering counter-negative weather events (where one plot is affected but the one far away is not), this strategy equitably divides the landscape. Individual farmers, each facing more or less the same degree of risk, may have a greater willingness to cooperate in landscape-improving projects.[11]

In New Guinea, topography provides the motive. Farmers, in a search for level areas, place plots in diverse locations.

Scattering is also associated with slash-and-burn agriculture in nonhilly tropical forests. Here having agriculture scattered throughout a forest can reduce some threats, for example, agricultural weeds have trouble crossing unreceptive forest ecosystems. Also, crop-eating birds may not be able to locate these new food sources.

CULTURAL MOTIFS

Throughout this text, numerous cultural influences have been pointed to. Part of cultural agroecology is in how peoples adapt agriculture to a friendly or a reluctant landscape. For the trained eye, differences are apparent at the landscape level. As mentioned in the introductory chapter, endless fields of ripening grain are more a cultural expression than agroecological ideal; others might revel in a limitless expanse of agrobiodiversity.

Agrotechnological Landscapes

A lot of agriculture utilizes set agrotechnologies. An agrotechnological approach, if well mixed, can encourage a wide diversity in landscape forms. This does not preclude cultural expression.

Figure 17.7 shows two landscape interpretations based upon academic or cultural values. Using two plot landscapes, the lower is of farm forestry in which trees and crops are separate entities with no in-

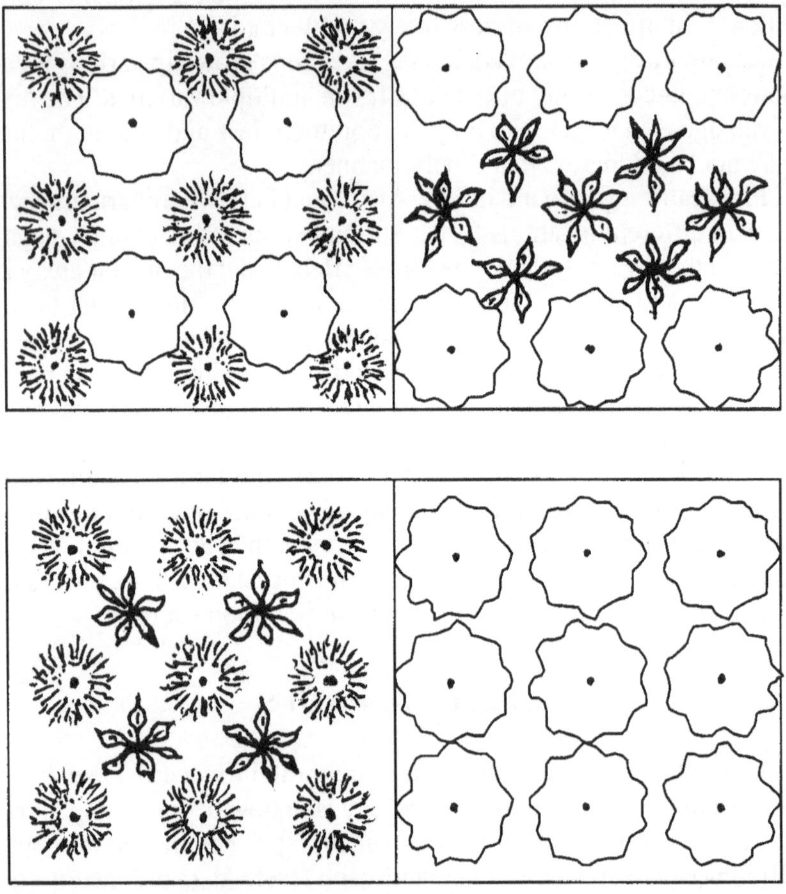

FIGURE 17.7. Two landscape views: the bottom two plots are farm forestry (a single-species forestry plot and bicultural agriculture); the top two mix crops and trees (agroforestry). These layouts also apply to orchards or tree crop plantations.

tended ecological interactions. The upper farm expresses agroforestry where the trees and crops are agriculturally integrated.[12]

Figure 17.8 continues along the cultural diversity path. The bottom farm is a two-plot landscape with some degree of fuzziness, in this case, overlapping with less than clearly defined plots. In contrast, Figure 17.8 (top) shows two plots, each integrated and disarrayed.[13]

FIGURE 17.8. Two further displays of cultural agroecology. The two-plot farm on the bottom exhibits fuzziness in design. The two-plot farm at the top is disarrayed.

Figures 17.7 and 17.8 carry certain cultural assumptions and values. In brief, these are as follows:

1. farm forestry or agroforestry,
2. plot type (geometric or terrain—as in Figure 17.3),
3. plot fuzziness (plot bounds are clearly defined or overlapping and indistinct), or
4. whether plots are internally ordered or disarrayed

These four variables can be combined in any order to achieve the desired landscape motif.[14]

Freeform Landscapes

The agrotechnology, as the basis for agricultural progress, provides contained set units of predetermined origin with clearly defined input and labor needs, known risk parameters, and, in the suggested form, predictable yield potential.

The alternative, roughly labeled nonagrotechnological, is driven by historic and cultural norms of society and a derived view of the nature-people interface. This approach offers yet another set of options.[15]

Guidelines

In discarding set agrotechnologies, one embarks on the equivalent of a journey of exploration without maps, compasses, and land references. Few, without prior knowledge, can survive such a trip. Landscape agrodesign without guidelines, especially within a world of extreme biocomplexity, is equally perilous.

As with the agroforest, operational guidelines are critical in allowing these freeform systems to be established and managed with relative ease.[16] A guided landscape without plots or agrotechnologies is quite doable. In basic form, these guidelines have

1. plant species located where they (one or more) grow best provided each does not substantially interfere with a more valuable species and
2. the populations of each plant species correlate with their relative value as long as the first rule is not violated.

These rules relate to the value-productivity of individual plants. The first locates plants where they grow best. A high-valued species, for example, a staple crop, is planted on the best site. If land is limiting, this may require growing the high-valued species on a less than suitable site. If another crop species will produce better on a site and is the best productivity-value option, then a changeover is in order.[17]

Vegetative structures, such as live fencing, despite their lack of market value, may be the most valuable planting for an immediate site. This is not true a few centimeters away.

What complicates these rules is that they must also apply to polycultural associations between a primary and secondary species. It is easy to see that, lacking a local knowledge on crops and sites, an agrotechnology approach is far less complex.

Tacked on to these is a generalized SIZ rule. This does not repeal complementarity, only refines and focuses it. This has

3. plants or groups of plants placed such that their defense influences (SIZs) overlap, reinforce, or augment across a range of threats.

SIZs, through reinforcement and augmentation, address concerns (soils, rainfall, etc.) or threats (insects, diseases, damaging weather, etc.). The latter involves the number of incidences, their magnitude, and, where applicable, their duration. Basic rules carry into field practice through local rule refinement and mode of application all coupled with strong regional knowledge (climate, soils, etc.).

Applications

As a refinement of these rules, the mode of application can be (1) semiplot or (2) fully freeform. The first of these, the semiplot, has fields of crops, usually a primary species. Within these areas, secondary species are planted. The rules hold that if secondary species is competitive, it will be interplanted in a course pattern, but if complementary, a fine pattern will be utilized.

This can make for spotty placement as one locale can be moist and support an intercrop, but a nearby drier area may be more competitive. Differences in in-field soil structure can also be exploited.

A rocky outcrop may not be the best site for the primary species and another crop may be substituted.

The type of landscape that results is demonstrated in Figure 17.8. Small plots and trees, each seeming randomly placed, are observable in traditional small holder farms of East Africa (see Plate 17.2).

The freeform version (see Figure 17.9) does not overlay a secondary species but relies totally on matching species with site. This can be a more difficult undertaking, made easier if a practitioner has extensive knowledge of crops and their relation to the local conditions.

In a more visually extreme form, fuzziness may not be just an edge feature but may extend across entire plots. For the fully fuzzy farm, examples lie in landscapes dominated entirely by agroforests or

PLATE 17.2. A East African farm landscape exhibiting a semiplot freeform motif.

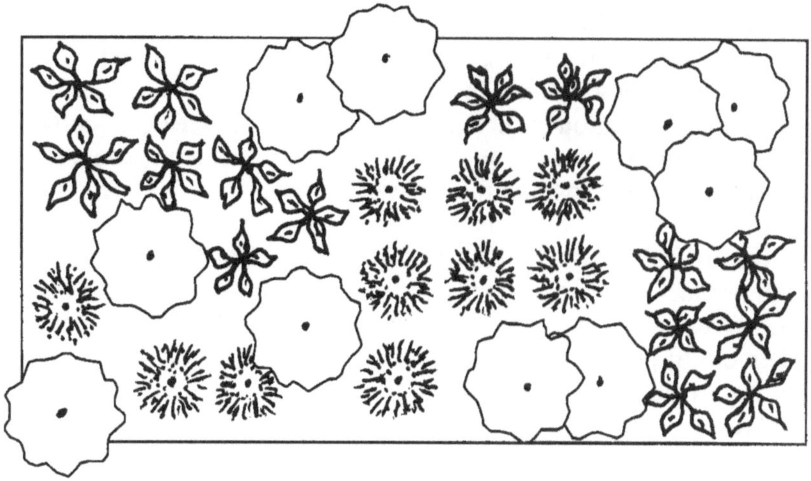

FIGURE 17.9. Semiplot, freeform placement where the local planting rules determine plant placement. In comparing this against Figure 17.8, it can be difficult to visually separate a loose agrotechnological from a rule-guided, freeform landscape.

farms of shaded cocoa and coffee. The lack of definition is noticeable in long-term coffee and cocoa plantations where, because of length of rotation (at times, over 100 years), the individual plots tend to merge.

CHANGE

Where and when to make the jump to less self-contained, more integrated plots (and agrotechnologies) and landscapes is one question. An implementation opening comes as agricultural landscapes change. This can be economic or, as stated in the chapter introduction, driven by other forces.

With a static land-use and marketing situation, there is little opportunity to introduce new or revised agroecosystems. When change is in progress, land users, facing a switch, may be more willing to alter the course of their land-use practice.

Change comes in two obvious forms: expansion or contraction. This may be an oversimplification but useful in describing trends.[18]

The first of these has land is in short supply and population or a constriction in agricultural potential forces more, and often less suitable, land into production. This is all too common in poorer countries where the population is increasing faster than per area productivity. The opportunity comes in being able to increase per area productivity through intercropping and the like while addressing short- and long-term sustainability concerns.

The other situation is the diametric opposite. This is where agricultural productivity has reached a point where land can be taken out of production. Ongoing in parts of Europe and North America, land is taken out of production. This can be for more valuable, nonagricultural activities (e.g., urban expansion) or, lacking another use, the land resorts to forest.[19] An alternative course is to reorder agriculture, maintaining profitability while changing to less intense, more environmentally friendly forms.

Chapter 18

Summary

A Chinese curse suggests "you will live in interesting times." Rather than wishing someone a life of turmoil and upheaval, agroecology can bring interesting and, in the challenges presented and results obtained, more salutary times.

In the first chapter, some lofty goals were posed. Restated, these are

- providing the basics of human existence;
- enhancing the quality of life and improving the human experience;
- ensuring that natural flora and fauna can survive and thrive;
- maintaining the land and the climate; and
- having a clear and ecologically sound vision that persists into the distant future.

These remain, although now somewhat overshadowed by the mechanisms, approaches, and agricultural alternatives.

Economic measures often connect agriculture with the human experience. Nevertheless, agroecology is more than an economic endeavor and the connection can also be through social, community, and cultural issues. This includes quality of life with all accompanying blessings.

NATURAL MANDATES

What nature intends for a site may differ or correspond roughly to that which people desire. Keeping in mind the argument that, in some

Introduction to Agroecology: Principles and Practices

doi:10.1300/5553_18

active natural landscapes, wild animals consume up to 80 percent of the biomass, the farm landscape is a comparable, output-oriented entity.[1]

In the broad continuum of agriculture, some systems closely follow a natural prescription, while others may be at odds with and clash with the climate and site. There are countless examples of either occurrence.

One goal of agroecology is to put a nature-friendly face on agriculture. There is an imperative to adhere to natural prescriptions and avoid, if at all possible, fighting nature for control of crops or the cropping area. In a more accommodating form, adherence has wanted plants grown in a sustainable manner, including being in harmony with natural flora and fauna.

A yet stronger version utilizes, as much as possible, natural dynamics to achieve management objectives. Although seldom realizable in its absolute form, steps can be taken in this direction. Clearly, letting nature provide or replace costly agricultural inputs makes economic sense. Birds and bats that consume crop-eating insects are part of this.

Further along this road may be called the Garden of Eden approach. This satisfies all the nuances of sustainability, is productive, and demands few inputs, but does require cultivation within narrow, nature-defined principles. The traditional agroforests of the South Pacific exemplify this form of agriculture. Less of an ideal and more of a practical expression of natural goals, elements are found throughout agroecologically whole agriculture.

CULTURAL MANDATES

Throughout this text, there are numerous mentions of cultural influences. This is for good reason: culture dominates on a number of fronts. Staple crops are often handed down from generation to generation and, in being set by heritage, are difficult to change. What results sets the stage for utilization of crop and plants. A case in point, French cooking is quite different from Chinese. Although there are foodstuff commonalities, a large proportion of what is eaten or ignored are food-construed cultural comments.

At the larger scale, the types of agriculture practiced may also be ingrained in culture through established field practice. Rooted in past perceptions are a willingness to accept intercropping or the view that trees are part of agriculture. Societies can perceive the nonharvest option as the wise recycling of low-valued outputs or the wasting of hard-gotten gains.

Under cultural agroecology comes culturally influenced landscape motif. Figures 17.7 through 17.9 sample some of what is encountered and culturally reflected in the people-nature interface.

MAINSTREAM AGRICULTURE

Out of default, mainstream agriculture has been intrinsically defined. This is usually monocropping, operating from a fairly narrow genetic base, supported and actively defended from immediate threats (climate, insects, etc.) by outside, often agrochemical, inputs. The mainstream also has little or no rotational planning and is short on ecologically useful land modifications. In many aspects, this is unfortunate as this also defines the agroecological worst-case scenario.

Most variations add agroecological or other elements in response to current views and requirements, that is, the core approaches. In some regions, the sole vehicle for agricultural advancement, outside or in addition to agrochemicals, is the push for genetically modified super plants. Again, this is not an environmental gain unless other agricultural approaches (core or composite) are utilized and agroecosystems put into a nature-conforming posture.

AGROFORESTRY

From the practices described in earlier chapters, much originates from and can be classified as agroforestry. Maintaining agronomic purity is not a mandate of agroecology. Nature, in a quest for or the attainment of ecological perfection, employs trees in wide abandon. In agroecology, trees are ignored only with ecological peril. Farmers should not forgo the ecological and economic possibilities.

In the simplest view, that of ecologically independent farm forestry, trees are only an income addition. Trees do this and a lot more. Through tree-crop and tree-farm integration, these become part of the ecological aggregate.

If there is any negative aspect, it lies with the vast number of agrotechnological options presented. Bodies of thought, some offered in previous chapters, are devoted to making sense out of the conceptional tangle of tree-crop associations. Prominent are DAPs and agrotechnological groupings. Once mastered, good things happen.

AGROECOLOGY

Some confine agroecology to a singular ecological phenomenon (e.g., insect control, organic fertilizers, etc.). Although readers are free to focus on one or more topics, it is the full breath of agroecology that is advocated here.

A key aspect is the will to embrace complexity. For many, the multispecies plot is the vehicle of choice.

Not all of agroecology is represented by in-field biocomplexity. A well-formulated landscape employing monocultures can be an adequate expression of agroecological thought. This is accomplished not through static, sweeping monocultures but through small, contained monocrops, spatially juxtaposed, rotationally active, ecologically amplified by surrounding, often pronature, ecosystems. There are many variations on these themes.

Field expression can also be an exercise in nature-inspired ecological diversity. Just when one thinks an understanding is at hand, the agroecological rationale may not follow expected lines. It is the unusual circumstances, and the need to quickly adapt, that make agroecology an interesting study.

Although the cardinal study, not all agroecology is harvestable crops. Agroecology affords a laboratory to test and make sense of ecological theories and the functionalities of nature.[2] Other statements of agroecological principles fall outside pure ecology or biology into the spheres of environmental science and sociology. Here agroecology is concerned with the well-being of humankind vis-à-vis useful plants and terrestrial resources.

CONCLUSION

This text looks at the productive options from an agroecological standpoint. Explored are how people view and employ agricultural ecology. As agriculture represents humankind's most intimate association with the terrestrial world, this interface is deserving of attention.

Agroecology can and should be more than the raising of crops. Through the various associations, agroecology can add new insights, promote an interchange of ideas, foster an awareness on how others exploit the land, and enhance the agriculture of one region through an awareness of what others do. There is no one superior agroecosystem and farm landscape: each has uses; each has applications; each can be (re)formulated to be an ecological whole. It is from this that one secures an all inclusive, user-friendly, and environmentally sound agroecology.

Notes

Chapter 1

1. Gold (1994) provides two definitions of agroecology: one broad—"[a] more environmentally friendly and socially sensitive approach to agriculture" and one narrow—"the study of purely ecological phenomena within the crop field." At this stage, agroecology, as a discipline, is still under construction and an unequivocal definition may be more a moving or unfolding target. Therefore, both definitions apply.

There may also be an implicit direction and flow associated with agroecology. Conventional agronomy has become a de facto repositioning toward biosimplicity (as shown by the commonness of and an ever-progressing emphasis on the monoculture). In agroecology, the drift should be in the opposite direction, with an unceasing push toward greater biocomplexity. Hence, a far-reaching agroecological definition may best embody the notion of productivity through escalating biodiversity. A broad view of biodiversity may also underwrite fauna, as well as flora, in the crop-production process (e.g., when land users seek the assistance of weed-eating ducks).

2. As stated in the preface, agronomy and agroecology are both considered subsets (i.e., separate directions) of agriculture. Therefore, mainstream or conventional agriculture and mainstream or conventional agronomy are synonymous terms. There are no profound implications in this other than to provide textual clarity.

3. These agroecological imperatives are from a more detailed list restated by Rigby and Cáceres (2001).

4. A broad definition of trees is utilized throughout this text. Included are bamboos, palms, etc. (Kleinn, 2000). The term tree crop applies when trees offer an output other than wood. This can be fruits and nuts as well as sap, buds, flowers, leaves, bark, or some root extract.

5. Some may consider the agroecological goals idealistic, out of reach, or something not to be mused over, but given the influence and similarly eloquent objectives of environmental science, they seem entirely appropriate.

6. A slew of factors come under the totality of land and climate heading; Agroecology, if well applied, can also mitigate the damage from loss of forests (higher temperatures, less rainfall), can slow the extinction of species, sequester carbon, and, if on a wide enough scale, confer other positive global effects. These are discussed by Vitousek et al. (1997) and, in an agroecological context, by Matson et al. (1997). The land and climate gains, although innate in agroecology, are not

Introduction to Agroecology: Principles and Practices

doi:10.1300/5553_19

specifically discussed in the text. Instead the focus is on those factors of most concern to a farmer, a farm family, and the neighbors. The supposition is that solutions to global issues start by improving practice on individual farms.

7. Some of the historic arguments for the productivity of North African agriculture are found in Hughes (1994). Although the ancient Romans made land improvements, they also oversaw productivity losses through poor soil management (Jackson, 2002). Similar methods to overcome a hostile desert landscape are found with the Hopi of North America (Gliessman, 1998, p. 82). From the broader perspective, those who study ancient societies can benefit from understanding the strains on and the surges of these societies from good or bad agricultural practice.

8. Throughout this text, the financial definition of profit is subscribed to, that is, profit is calculated as revenue exceeding costs. The economic definition, profit resulting from the greatest investment opportunity, is not used.

9. Tree-based cotton comes from the kapok tree of Central America. This tree has spread to and is common in other tropical regions. The tree-grown tomato relative is the tamarillo of South America (NRC, 1987).

Chapter 2

1. Exceptions to the norm are underscored in the fact that so few scientific laws apply in agriculture. Two come to mind: Beer's Law (discussed in a following endnote) and Yoda's Law (deals with overcrowding). Other laws inflate the nomenclature for what are merely hypotheses. A case in point, the Law of the Minimum is the unproven von Liebig hypothesis (also mentioned in subsequent endnotes in this chapter).

2. The idea that complexity can be overwhelming has been stated by Levins and Vandermeer (1990, p. 349). A reading of Altieri and Nicholls (2004) shows, that unless well presented, intricacy can be the intractable norm. This applies, in particular, to weed, insect, and nutrient relationships that can easily transcend the principles developed in this and the following chapters. Most cases can be understood and managed within known knowledge and application parameters.

3. The trace elements include zinc, calcium, iron, magnesium, boron, sulfur, copper, manganese, and molybdenum.

4. The leaf area index (LAI) is the ratio of total leaf area of plant (sum of leaf areas) to the circumference of canopy. High LAI plants have dense canopies without much light penetration; low LAI plants allow considerable light penetration. The measure may seem daunting, however, in practice, LAI is visually apparent.

5. The basis for this assertion is Beer's Law. This is where

$$I = I_0 [1 - \exp(-kL)] \qquad \text{(N. 1)}$$

and where the amount of light intercepted *(I)* is a function of that available (I_0) and the leaf area index *(L)*. The *k* value is a species-associated constant. A multispecies version has been proposed by Keating and Carberry (1993). The derivation difficulties for these functions have been recounted by Dai et al. (1993) and Paris (1992).

6. The fertilizer capture rate (40 to 50 percent) is a summary by Matson et al. (1997). The biomass figure (20 percent) was estimated by Palm (1995). The assumption of greater capture efficiency, through increased planting density, is based upon generally accepted ecological notions.

7. The resource-yield functions presented in this chapter are hypothetical. For functions of this type, the units along the *x*-axis (from soil moisture to nutrient contents) will vary great but can be reconciled with two reference points: (1) where no resources are available, i.e., the zero point, and (2) where optimal productivity is attained (highest possible yields when all other resources are optimal). This allows for a common conceptional framework (as in Figures 2.1 and 2.2).

8. The function form when a species demands a resource (i.e., yields are low without ample amounts) is shown in Figure 2.1 (top left). Tolerance to a resource shortfall (i.e., yields are high despite a lack of the resource) is shown in Figure 2.1 (bottom left). In practice, the resource need profiles for key crops can be roughly estimated using fertilizer and irrigation recommendations.

9. The functional form and yield losses that occur when there is too much of one resource, for example, with waterlogged maize, is shown in Figure 2.1, top right. This is also illustrated in Figure 3.2 (the rightmost function).

10. The prevailing and unproven hypotheses (for a direct relationship between resource functions) include the von Liebig, Liebscher, and Mitscherlich hypotheses. Each proposes a different functional relationship between the essential resources and the productive outcome. These are functionally very similar; see Wojtkowski (2002, p. 21) or Paris, (1992), with no definitive work on which is correct or even favored; see Dai et al. (1993), Paris, (1992) or, for more theoretical development, Harmsen (2000).

The focus on a single-limiting resource, as presented here, is the von Liebig hypothesis. If there is any overstepping, as suggested by the Liebscher and Mitscherlich hypotheses, the magnitude will not be great, most likely less than 10 percent. Consequently, the single limiting resource hypothesis can still underwrite most assessments.

11. The effect of the limiting resource(s) on growth and yields can be looked at as being the bottom layer in a series of tiered occurrences. Above the single limiting resource core (alternatively, the core may be another resource hypothesis) are the indirect relationships. Added to the mix is the number of possible essential resources and an ever changing situation (i.e., hourly, daily, etc.) with regard to the plant and essential resources.

This and like problems aside, there are published yield equations that cross-link the essential resources (e.g., Foster and Goh, 1976). Computer crop yield prediction models incorporate such or surrogate equations and do so with success. Although too numerous to cite here, models exist for the most common commercial monocrops: wheat, maize, sorghum, rice, etc.

12. Experienced practitioners, even those without access to scientific measurements, acquire a pretty good idea as to which essential resources most affect their crops and apply fertilizers, irrigation, and other inputs accordingly.

13. Various views of the niche concept have been summarized by Liebold (1995). Discussion here, that which applies to agroecology, is an offshoot of this without being definitive on exactly what constitutes a niche.

14. As applied to agroecology, some useful inroads can be made in implementing the concepts of the fundamental and realized niche. A fundamental niche occurs when a plant lacks competition, as occurs with a single plant in a flowerpot. A monoculture is where each species lacks interspecies competition, but each plant is challenged by niche-identical competitor, might be more aptly termed a constrained fundamental niche. The remainder of agroecosystems, that is, the polycultures, fall under the heading of realized niches.

15. Some references for the maize-bean intercrop as these pertain to relative yields, planting densities, and other variables are found Davis et al. (1987) and Davis and Garcia (1983). The observation on minimum rainfall is from Rao (1986).

16. Credit for the LER goes to Mead and Willey (1980).

17. The LER equation is plot based using per area yields. Similarly, this can be row derived where the

$$\text{LER} = (Y_{ra}\, r_a / Y_a) + (Y_{rb}\, r_b / Y_b) \qquad \text{(N.2)}$$

For this, Y_{ra} and Y_{rb} are the average yields for rows of species *a* and *b*, respectively, and r_a and r_b are the number of rows per area for these same species. As with the text equation, Y_a and Y_b are monoculture yields for species *a* and *b*, respectively. This version might find use in an experimental situation where density varies across a single plot.

18. A forward glance at Figure 3.3 may help solidify this calculation. Shown are the denominators, that is, Y_a and $Y_{b'}$ labeled as 1.0. These are the normal yield levels if all available site resources are directed toward the one species, *a* or *b*. The other values, Y_{ab} and $Y_{ba'}$ labeled, respectively, 0.7 and 0.5, complete the calculation.

19. This upper LER figure of 3.5 is based upon the work of Ong (1994), who reported a high LER value of 3.69 with a facilitative agroecosystem (alley cropping). This may be an exception, as this may be from a particularly well-designed and situated agroecosystem. Ong (1994) has also reported lower values for the same agrosystem type, for example, 0.58.

Dangers lurk with LER use. These are mostly due to insufficient data, but also in capturing the essence of the analysis without incurring statistical anomalies (Vandermeer, 1989, p. 26). If very high values are encountered, one should first ensure that this is not a statistical anomaly.

20. Planting ratios are also referred to as the replacement series approach (see Barker and Francis, 1986).

21. Ratio lines can be graphically displayed as shown in Figure N.1.

22. The optimal monocultural density as the density apex in mixed, high interface plantings is an unchallenged assumption. This density is also assumed best for intraspecies spacings (e.g., between *aa* and *bb* in a cross-sectional spatial design of *bbaabb* or *aabbaa*).

23. Many, having studied basic agricultural economics, will be familiar with the production possibilities curve (PPC). The PPC is a simplification of the surface described in the text. Using ratio lines (as above), a curve can be derived connecting the optimal or high points on Figure 2.4. This is shown, in overview, in Figure N.2.

Owing to a dearth of data, LER surfaces and curves (either PPC or ratio lines) are more postulates than finished products. LER surfaces, based on modeling light

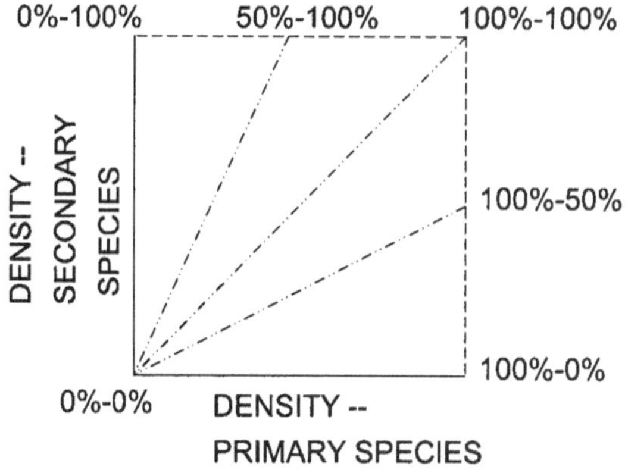

FIGURE N.1. Planting density ratio lines.

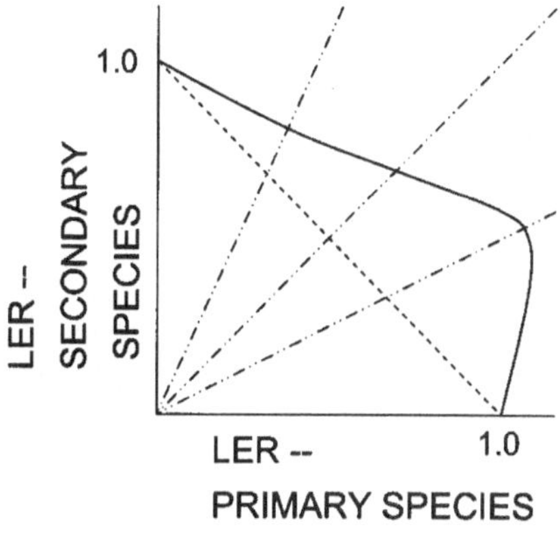

FIGURE N.2. Production possibilities curve based on planting density ratios.

allocation only are in Wojtkowski and Cubbage (1991) and Wojtkowski et al. (1991). An actual, experimentally derived PPC for a sorghum-pigeonpea combination is found in Ranganathan et al. (1991).

24. The 5 to 15 percent value range is from Dai et al. (1993). Studies exist for all common crops, many from earlier periods, which link planting densities with the essential resource content of local soils. When a single figure is suggested, this is for "normal" rainfall and uniform soils (as when the suggested fertilizers are applied).

25. Far from being the only view, Young (1989b) suggests six basic patterns:

1. Borders
2. Rows
3. Individual
4. Blocks
5. Strips
6. Clumps

This is not the only view. Wojtkowski (2002, p. 52) presented five fine and five coarse patterns:

Fine	Coarse
1. Individual	1. Strips
2. Checkerboard	2. Groups
3. Rows	3. Blocks
4. Boundary	4. Boundary strips
5. Subplot	5. Gaps

Mixed- and counterpatterns are also possible. Mixed patterns are combinations of the above (e.g., a second species clustered within a row ordered primary species or a second species as a boundary around a plot containing some other spatial pattern). A counterpattern has the rows of one species perpendicular to the rows of another.

26. The reason for focusing discussion on the bicultural patterns is that as the number of included species increases the pattern possibilities rises geometrically or even exponentially. Therefore, it would be presentationally foolhardy to venture down this road unless needed for immediate discussion.

27. The minimum width of a coarse strip starts at three species. This number is far from definitionally sound. Factors such as relative heights and root spread between unlike species intervene. Therefore, it is the ecology of the situation, not the number, that decides. More precisely, pattern coarseness occurs when similar species interact in more resource-constrained fundamental niches rather than the realized niches associated with interspecies competition and fine patterns. Rows, stated here as being one or two species wide, follow from these rough, but not exacting, guidelines.

28. Inversion cluster gaps are as shown in Figure N.3.

29. In greater detail, the reasons for an *aabaabaa* rather than *ababa* are when (1) species *b* is taller and the *aa* groups create a canopy gap and more lower level light penetration and/or (2) the limiting resource for species *a* is more abundant than for species *b* (in this case, these will not have the same limiting resource). The third reason, a greater value and demand for species *a,* is self-evident.

Where fertilizers are applied, an *aabbaabbaa* pairing may reduce interspecies competition and increase the LER (Schroth, 1999). The theory is that when nutrients are placed only between similar rows, that is, *aa* and *bb,* the roots grow away from

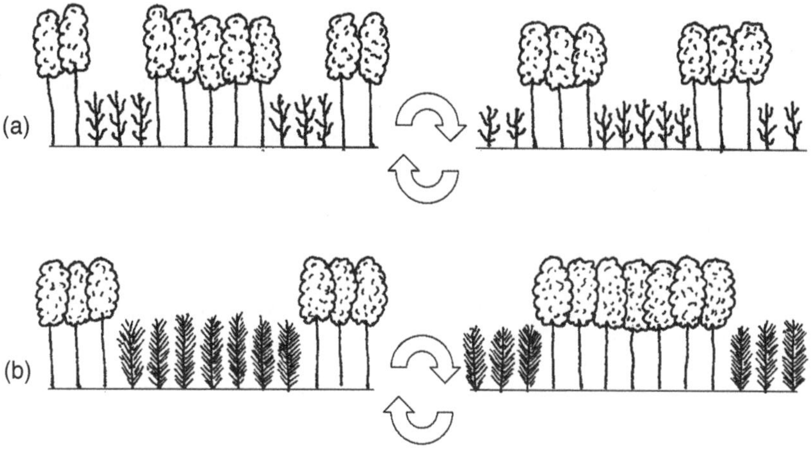

FIGURE N.3. Inversion cluster gaps.

the interspecies interface, reducing the likelihood for productivity sapping competition (e.g., Wijesinghe and Hutchings, 1999).

30. Although not formally studied in this exact form, Malcolm (1994) shows that, if provided with enough light, an understory can compete favorably against and even suppress overstory growth. This relies not upon complementarity but on a competitive relationship where all belowground niches are occupied. In other words, taller species lack, due to competition, the belowground resources to actively compete for light.

31. Schroth (1995) has mentioned buffer plants, which when planted before the primary species and allowed time to develop can force nearby crop roots into a lower soil strata.

32. As seen, disarray does not always equate with randomness. Of all the pattern possibilities, the canopy orderings (as in Figure 2.7) may be the most common with perennial disarray, for example, agroforests. Also encountered are internal clumps (gap or patch). With fine patterns, semi-disarray exists in two versions:

(1) Where the primary species, in the upper or lower canopy, is row ordered while the other species (generally more than two) lack any perceivable arrangement
(2) Where all species are arranged in what might be called sloppy (non-straight) rows where the interface distances are far from uniform

33. The issue of row orientation and light is far from resolved. Beets (1982, p. 54) and Donald (1963) express different views. For this text, relative height difference is the main factor, keeping in mind that, in severe climates, light shading can help understory plants. A documented case is green beans and cauliflower grown in the hot, dry air of Syria (Alhamid et al., 2003).

34. The SIZ concept was introduced by Wojtkowski (2004).

Chapter 3

1. Evelyn (1664, p. 92) commented on good and bad neighbor plants. One does not have to dig deep into the past to find terminology differences. These exist in recent literature and, in a larger content, may express divergent viewpoints. The vocabulary utilized includes relay crops (i.e., rotations) and crops with service plants (i.e., facilitative species). These two word uses come from García-Barrios (2003).

2. As might be expected, the terminology is not universal. One problem is that nature is not accommodating enough to neatly self-classify things. People do this on the basis of preconceptions and on field data which may or may not be complete. As such, differences are to be expected. This is especially true with comparatively new topics that have not stood the test of time, that is, terms and classifications which have not had time to gel.

3. The mechanisms of acquisition are from Vandermeer (1989, p. 78).

4. The timing of maize, millet, and sweet potato is from Brown and Marten (1986).

5. Roots strata diagram (Figure 3.1) is modeled from Weaver (1919, 1920). Much of this work, including vertical stratification, was in grain fields.

6. The Hawaii forest study is from Stratton et al. (2000). Paulownia *(Paulownia elongata)* with wheat has been described by Yin and He (1997), and the root architecture of paulownia by Wang and Shogren (1992). The facilitative gains from deep spreading roots are discussed later in this chapter.

7. The root profile for any one species may range from an inborn trait to a totally freeform condition. Most species lie between, having a predetermined rooting disposition tempered to a large degree by neighboring influences and the location of sought-after in-soil resources. Evidence shows strata separation is found more in water-limiting, rather than nutrient-limiting, situations (Casper et al., 2003). Management has a hand in this. Fertilizer placement can help separate roots, that is, putting nutrients in the center of groups or rows of one species helps keep the roots from interacting with neighboring species (Schroth, 1999).

8. The late planting of barley, rye, or clover with onions, peas, cabbage, or beets comes from first-person field notes (Anon., 1872). There is no accompanying scientific validation, but the logic seems unassailable. Practical application comes when grains and clover are cover crops.

9. The tropical grass species vetiver shows this characteristic, that of vertical, dense roots. It can be planted in long narrow rows that do not spread laterally.

10. Although not discussed here, light wavelength and/or light reflection (as with the capture of light reflected off other leaves) may prove a positive competitive factor. At this stage, it remains only a scientific curiosity.

11. Light gathering and leaf orientation (wheat included) has been discussed by Donald (1963).

12. The mechanisms of partitioning are from Vandermeer (1989, p. 78).

13. Although multiple, cross effect, marginal gains are theoretically sound, this situation may be hard to diagnose and, because such gains may be very site specific, any resulting agroecosystem may be limited in geographic scope. Still, the

possibility for successful application does exist and may underwrite many successful pairings.

14. The millet-groundnut case is from Ong (1991). The reasons why grains respond to shade is not clear. One hypothesis holds that lower, under-shade temperatures slow grain filling, which allows for greater weight accumulation.

15. The data for increased area of leafy crops are from Brenner (1996).

16. The author of this text, from years of informal tasting, can attest to the less sweet tang of shade-grown fruit.

17. The increased nutrient content of shaded pasture grass and sweeter sugar beets is from Brenner (1996), for oats from Nandal and Bisla (1995), while Muschler (2001) undertook the coffee tasting. Larger-sized coffee beans have been mentioned by Muschler (1999, p. 61). Running counter to organic farming, fruit quality is often judged through surface blemishes. Surface dyes and agrochemicals, liberally applied, do produce a blemish-free output.

18. The RVT is from Schultz et al. (1982).

19. Two plants, grown together, exert many influences. Therefore, without manipulation, it can be experimentally difficult to statistically detach competitive partitioning from facilitation. Methods employed include artificial shade (e.g., Early et al., 1966) or manually separating the biomass from shade and roots (e.g., Ong, 1995). With some basic understanding, progress can be made even without statistically separating the underlying causes.

20. One reported gain is the transfer of plant sugars from the roots of one plant to fungi and then to another plant species (Pennisi, 2004).

21. The basis of comparison (as briefly mentioned in Chapter 2) has 40 to 60 percent of the fertilizer applied being utilized by component plants (Matson et al., 1997). The 20 percent biomass figure may be low due to timing (slow decay), location (deep in the soil or otherwise out of reach of roots of the target species), or nutrient balance (little uptake as this is not the limiting resource).

22. Indirect but not conclusive, some evidence for the nutrient pump comes from Velasco et al. (1999), who found more phosphorus in pastures with trees than without. Because phosphorus is often a sought after or an active essential resource, this mechanism is also called the phosphorus pump (Francis, 1986).

23. Bromeliads are not agricultural plants and, to date, no clear facilitative uses have been demonstrated. However, these show that airborne nutrients are a potent source.

24. The timing and release dilemma may be hard to overcome (Zingore et al., 2003), although the application-decay-uptake window may be shortened by promoting, through nutrient balance and favorable microclimate, rapid decay. Kelty (1992) describes where a facilitative species accomplished this in a forest plantation setting.

25. Hydraulic lift has been verified by Emerman and Dawson (1996). For the Hawaii example early in this chapter of mixed shallow and deep root architecture, see Stratton et al. (2000), facilitative water sharing may or may not have been a contributing factor. In all likelihood it exists and, in the process, supports a stronger, more viable ecosystem. The same will hold true in a similarly rooted agroecosystem.

26. Trees with lower-strata roots are some types of oaks and paulownia, mentioned earlier. The paulownia case is documented by Wang and Shogren (1992) who found that 80 percent of roots are between 40 and 100 cm below the soil surface. This tree is utilized in China as a facilitative addition. Oak roots are less studied but do show some form of crop compatibility, for example, see Leuschner et al. (2001). Bainbridge (1988) has argued for the inclusion of oaks in agriculture.

27. Plants that accumulate various nutrients should be more common than the scientific literature suggests. The abilities of *Tithonia diversifolia* to gather nutrients comes from Buresh (1999).

28. Examples where species have buffered, and improved, problematic soils are found in Dagar et al. (1995).

29. Throughout this text, the term biocomplexity is used as a shorthand for an ecosystem with threshold (operational or functioning) levels of density, diversity, disarray, and duration.

30. There seems to be an unfolding view on what is the minimum species number for ecosystem governance. Some of the references are Kareiva (1994), Naeem et al. (1994), Tilman et al. (1997), and Baskin (1994).

31. The fungi-plant diversity relationship is mentioned by Pennisi (2004). A more complete list of microbes as agents of facilitation, along with references, can be found in Chapter 15.

32. The benefits from worms have been studied by Hulugalle and Ezumah (1991), Hauser et al. (1997), Scheu and Parkinson (1994), and Shuster and Edwards (2003). Ants in agriculture have received attention, for example, in work by Risch and Carrol (1982), Stanton and Young (1999), and Mando and Van Rheenen (1998).

33. Evidence on the nontransfer of plants comes from Matlack (1994) and Peterken (1993). The latter study found that 40 percent of the vascular plants seen in the ancient woodlands of England did not spread to newer, detached, and receptive woodlands. This is despite the "newer" woodlands being in place for hundreds of years. Therefore, the corollary of ecosystem governance are transfer points. These may be

1. a spatial neighbor, a nearby, biologically well-endowed, natural ecosystem that has been in place into the distant past;
2. a temporal neighbor where a similar agroecosystem immediately replaces a long-duration natural one, for example, a natural grassland is replaced by a pasture or a natural forest with an agroforest;
3. a system of corridors that naturally convey organisms to fragmented blocks; or
4. failing all, organisms manually and individually imported into receptive ecosystems. Dangers of the latter include spreading unwanted and unintended organisms or upsetting the natural balance. The type of transfer is important if a complex agroecosystem is to be a refuge for endangered plant and animal species.

Chapter 4

1. For a lot of agroecosystems, the immediate threat lies with exposed and erosion-prone soils. The difference in the magnitude of and susceptibility to the threat can be quantified (e.g., for example, Wiersum [1984] and Howeler et al. [2003] or benchmarked through the Universal Soil Loss Equation (e.g., Young [1989a]). Less observed, and more easily repaired, are when particular minerals resources are depleted.

2. Do-no-harm admonitions apply to agroecosystems or the types of component species selected. Schenck (1904) and Schlich (1910) forewarned about selecting trees that impair sites. General discussion on the scope and applications of sustainability is found in Hansen (1996).

3. The *Salix matsudana* study is from Ohte et al. (2003). As an aside, the hydraulic need profile of a plant should match the water availability profile of the region and site. Native species tend to do this, while exotics may not. The exotic and invasive tamarisk (also called salt cedar) in the southwestern United States also hogs scare water (Sink, 2004).

Also mentioned for poor hydrology and other negatives (increased fire danger, interacts poorly with native wildlife, etc.) are species of eucalyptus in California (Williams, 2002). It should be noted that not all subscribe to the negative view of exotic species, as explained in Zobel et al. (1987).

4. As for other reasons, irrigation systems can collapse because of an ebb in cross-landscape societal cooperation. This can be the lack of channel maintenance or the loss of forest watershed and runoff. The overtapping of deep water sources also leads to system failure (Postel, 1999).

5. The idea that organic agriculture is less risky than conventional techniques is generally accepted. Specific references include Lotter et al. (2003) and Dapaab et al. (2003).

With organic and other risk-countering practices, the acceptance of a measured, but direct, crop loss is not a great economic departure from a strategy involving input-secured loss avoidance. Whereas increased outside inputs (e.g., chemical sprays) (and the costs thereof), reduce profits, the crop reductions that can accrue with less severe, multiple ecological defenses (e.g., nonproductive spread barriers and minor populations of plant-eating insects) also reduce profits. The net aftermath of the two strategies, that is, (1) the costs of added inputs that reduce profit or (2) yield reductions that reduce revenue and accordingly profit, should be comparable.

6. The use of an intercropped, secondary species, one that resists drought and yields when the primary output fails, is further discussed in Chapter 8.

7. The acorn as an alternate food source is discussed by Bainbridge (1985). The shea and baobob trees are among 55 Nigerian species listed for famine relief (Harris and Mohammed, 2003).

8. Putting risk into a framework that predicts when farmers will take countermeasures has been highly problematic, for example, Just and Pope (2003) and Rabin and Thaler (2001). Not wanting to enter into endless discussion, the subject is not broached here.

9. The 50 percent land cover figure is minimal, especially when dead biomass (grass, leaves, etc.) is laid directly on the ground. Another factor is the amount of water subtracted, through in-soil infiltration, from the surface flow. A live ground cover, a plant that has deep, soil-penetrating roots, will help in this regard. These same plants keep the soil-protecting biomass in place, even if rainfalls are overpowering. Whenever soils are protected with a vegetative cover, the more the better is always a good general rule.

In a useful aside, large leaves (pumpkin, watermelon, squash, taro, banana, etc.) also aid in suppressing weeds, better yet if slow decaying; these attributes are discussed later in the chapter. Exceptions to the large-leaf rule lie with small-leaf cover crops, those that form thick, dense, close-to-the-ground, vegetative mats (as in Plate 4.1, top).

10. The listing of wind damage is from Cieugh et al. (1998).

11. Frost-countering options are found in Bolivia (Zimmerer, 1999), Pakistan (MacDonald, 1998), New Guinea (Waddell, 1975), China (Wang, 1994), and South America (Caramori et al., 1996). These are discussed in greater detail (as agrotechnologies) in Chapter 14.

12. The idea of integrated weed management is advanced in Liebman et al. (2001), Liebman and Gallandt (1997), and Anderson (2005). For all intent and purpose, this approach is in its infancy. Further conceptional and empirical development is needed for widespread application. The unpretentious steps presented here come with many unanswered questions, including the ability to closely target only the most competitive weed species.

13. Purslane was found to coexist well (i.e., no yield penalty) with broccoli (Ellis et al., 2000).

14. Altieri and Nicholls (2004, p. 69), among others, discuss desirable weed levels, placement, and timing. Outside of pure control, Plate 12.1 shows weed exploitation through purposely placed corridors.

15. As not all can be covered in an introductory text, further enlightenment is found in Liebman et al. (2001), Staver (1999), and Liebman and Gallandt (1997). For a more practical view of the problem and proposed solutions, see McCord (1992). In an integrated end run around the problem, Worlidge (1689, p. 115) suggests lessening soil fertility for grapes. The reason is to improve the harvest index and to reduce the number of weeds.

16. Worsham (1989) discussed how decoy plants can trigger seed germination but not support the growth of the difficult-to-remove weed striga (*Striga* spp., also called witchweed).

17. Delayed sowing as a weed-control measure is mentioned by Jordan et al. (1996).

18. Liebman and Gallandt (1997) point out where weed species are reduced or eradicated by plowing. Past practitioners may have used multiple plowings, presumably optimally timed, for weed control (Worlidge, 1669, p. 32). Anderson (2003, 2005) sheds more light on the till options.

19. Lists of weed-attacking organisms have been compiled, for example, by Julien and Griffiths (1998) and Liebman et al. (2001). A listing of edible weeds is provided by Duke (1992). Taste, rather than user knowledge, is the main obstacle to

these being anything more than a casual or emergency food. Having said this, weeds do find uses as foodstuff, pot and medicinal herbs (Vieyra-Odilon and Vibrans, 2001; von Weizsäcker, 1995).

20. The examples of weed-suppressing mulch come from, respectively, research by Kamara et al. (2000), Chou (1992), and Jobidon (1992).

21. Lots of examples exist where one plant or a mix of plant species control one or more weed taxa, for example, see Rice (1995, p. 111). The mechanisms vary greatly. Not much work has been done synthesizing this information into a set of theories and field-ready practices. The weed exclusion properties of *Peltophorum pterocarpum* mentioned in the text were documented by van Noordwijk et al. (1992) and those on wild mustard by Gliessman (1998, p. 221).

22. Weed control with the cover crop mucuna is discussed by Tarawali et al. (1999).

23. Weed-controlling cover crops of ryegrass and white clover with pepper and okra are examples taken from Biazzo and Masivnas (2000). Mentioned in Chapter 3, the late planting of barley, rye, or clover with onions, peas, cabbage, or beets should confer a similar result (Anon., 1872). Note that all these are aboveground compatible, for example beets are shade tolerant, cabbage can shove aside rye or barley, and so on.

24. The idea that large, slow-decaying leaves are good for weed control comes from Budelman (1988) and Eason (1991). The insertion of squash with maize, as a weed-control measure, was mentioned by Gliessman (1983) and as an insect control by Altieri (1994). Other species with weed-control potential are soybeans and sweet potato (Akobundu, 1987, p. 445).

25. A summary of practical animal counters is found in McCord (1992).

26. The case is presented by Pain (2003) where well-constructed and well-positioned fencing, along with managed herds, offer protection from lions and other dangerous beasts. Agressive and noisy dogs also help. In contrast, Giesser and Reyer (2004) found hunting a better control against European wild boar than fencing or feeding. This study did not look at the deterrent effect of dogs.

27. There are some comparative rankings of plant edibility (from palatable to nonpalatable) based on spines, thorns, taste, and so on, mentioned in Woodward and Coppock (1995).

28. The list of plants that repel small mammals is from Reilly (1993, p. 78). These animals can easily scurry around a pungent plant. This raises questions such as how these plants are best employed as a close companion or plot barrier.

29. The observation on grain awns deterring birds comes from Harrison (1775).

30. Honeybees as an elephant repellant is discussed in Anon. (2002) and O'Brien (2002).

31. The barn owl is a prime hunter where, for gopher control, well-designed bird (owl) houses can encourage large predator populations (Brower, 2004).

32. Chili (hot) peppers against elephants is from Beard (1997). Generally less studied, repellants may be equally effective against other large crop-eating animals.

33. The general practice of semihusbandry is discussed by Linares (1976). A key promotional factor seems to be field-forest interface, especially in uneven edges, that is, the more interface the better (Peterson, 1981). In a possible multipurpose role, that is, as guide species to attract insect-eating birds and for game, Green

(1908, p. 52) has mentioned the planting of fruit trees in forest-field edges. In promoting the wild turkey, Robinson (2005) discusses the need for oak trees, mixed grasses, and habitat-favoring shrubs.

34. The capibara is mentioned by Bruce (1991, p. 26).

35. This breakdown of West African hunting comes from Leach (1994) and the Peruvian breakdown from Bruce (1991, p. 26).

Chapter 5

1. There is some discussion on the original goals of IPM that can be either to reduce or eliminate chemical use (Walter, 2003, p. 59). This may seem minor, but the ramifications, as to direction taken, are significant. The IPM steps presented here are modified from those restated by Walter (2003, p. 62). This is not the last word on this. From the points raised in this chapter, more conceptional development is needed if IPM, especially an agroecological version, is to fully materialize.

2. Question can be raised on what is the minimum damage threshold. African farmers, out of necessity, accept 1 to 40 percent (Abate et al., 2000). The upper value may be unacceptable; a value at the low end of the range may be okay (e.g., ~10 percent). It is entirely possible that given the prey required for a healthy predator-prey relationship nature, rather than farmers, may have the greater say.

3. Hypotheses underwrite some of these countermeasures. Specifically, these are the

1. disruptive crop,
2. repellant plant,
3. trap crop,
4. enemies, and
5. microclimate (e.g., Vandermeer, 1989, p. 94).

Altieri and Nicholls (2004, p. 39) have compiled active cases and dissenting opinions.

4. A short list of environmental and other difficult-to-classify control factors can be found in Colbach et al. (1997).

5. Some references that support the hypothesis that insects can be controlled by microclimate are Koech and Whitbread (2000), Schroth et al. (2000), and Bach (1988). Morales et al. (2001) noted a negative relationship between manure and aphids: not one was found when chemical fertilizers were applied. Waring and Pitman (1985) studied beetle density and resistance in pines.

6. The study where less water equating with less insect damage was by Daane and Williams (2003). García-Barrios (2003) has discussed this mechanism: less essential resources in leaves as an insect deterrent.

7. The reduction of barley disease threats, specifically yellow dwarf virus, powdery mildew, rusts, and sharp eyespot, is mentioned by Jordan et al. (1996).

8. The case of alfalfa weevils comes from Stringer and Alverson (1994).

9. Cases of beneficial insects being harmed are found in Landis et al. (2000).

10. Ants improving tree growth has been documented by Fuente and Marquis (1999). In this study, ant-visited trees were taller than their nonvisited counterparts; no reasons are given. Fuente and Marquis (1999) allude to ants molesting other insects. The caveats on ants are indirectly provided by Davidson et al. (2003).

11. Spiders in agriculture (with over 6,000 potential species to choose from) are championed by Marc et al. (1999). Starcher (1995) mentions the 6,000 species of wasps and gives specific cases for application.

12. A complete list of those insects that can be purchased and introduced is easily obtained through an Internet search. Dangers exist when selecting ecosystem-exotic predator insects unless proper testing protocols and other safeguards are in place, for example, Loomans et al. (2002).

13. Many lure options remain unexplored. Among the possibilities (Suckling and Karg, 1999) are lure and kill (strips containing the pheromone bait and insecticide), lure and infect (strips with the chemical attractant and a targeted pathogen), and lure and contain (the lure entices an insect to enter an unescapable contrivance). The list of susceptible species is from Mancarrow and Taylor (2000, p. 70).

14. Brown and Marten (1986) described the coconut ant trap.

15. Another option is to induce greater repellant abilities into standing crops or repellant plants (Day, 2001; Ruc, 1990). Being labor-intensive, this may be more of an outbreak measure.

16. The passive repellant plant list is condensed from Reilly (1993, p. 78). More examples are given later in this chapter.

17. Biowarfare against a specific insect is the subject of discussion by Smits (1997).

18. Pheromones as a tool in agriculture have been discussed by Kirsch (1988) and Suckling and Karg (1999).

19. A general list of benign insect control chemicals can be found in Weinzierl (1998). Milkweed meal against nematodes and armyworms is summarized by MacKenze (1999), milk against mold has been studied by Bettiol (1999), and diatomaceous earth seems time tested, with published promotions going back to the 1800s.

20. There is a wealth of anecdotal experience where herbivore insects return in greater number after harsh chemicals subside. In Africa, a published study was undertaken by van der Valk et al. (1999). The dangers of poor dosing (too high or low) is based upon statements by Kaiser (2003).

21. This generalist-specialist approach may not be as clear cut as presented. Generalists can eat the specialist (e.g., Snyder and Ives [2001]) or interfere with procrop pollinating insects.

22. Early literature seems to support this. Geese and poultry were allowed a free run of fields, so much so that authors noted only when they should not be included (Gent, 1681). Given the open and accommodating nature of early farm buildings, it is assumed that chimney swifts, barn swallows, and bats were commonplace.

23. The African push-pull arrangement is from Khan et al. (2000) who further noted that the repellent plant silverleaf helps control the weed striga.

24. Citrus ants in China are a well-referenced example. The underlying work is by Huang and Yang (1987).

25. Altieri and Nicholls (2004, p. 47) exhaustively address, through almost countless examples, weed miniecosystems. This includes the statement on the decreased likelihood of pest outbreaks (p. 55).

26. Another category, that of traps, may rise to this level, especially if synthesized pheromones are available that lure whatever specialized pests are targeted at a given moment.

27. The lists of repellant plants, generalist and specific, are from Mancarrow and Taylor (2000, pp. 58 and 62). These are in addition to those listed earlier under passive repellant plants.

28. This can be a mixed bag. In a study from Nigeria, Apantaku (1999) found most plant-derived home remedies to be of dubious value; a few, however, are quite effective.

29. This case is from the Shakers, a religious sect that gained prominence in the eastern United States in the mid-1800s. The remedies given come from Beale and Boswell (1991) and Anon. (1843) with no guarantee of effectiveness. Some current home remedies are found in Philbrick and Philbrick (1974), Reilly (1993, p. 83), and Nancarrow and Taylor (2000, p. 52).

30. These remedies are for North America and Europe. Similar lists exist for other regions. For parts of Africa, Morna et al. (1990) have made suggestions on which remedies work.

31. This statement on ants in Mexico is from Risch and Carrol (1982).

32. The mixing of ants and spiders may or may not be feasible depending on the species involved. In one study, Vandermeer et al. (2002) found exclusivity but did not discount this as an option.

33. Altieri and Nicholls (2004) present a large sample of the literature regarding entomo-control.

34. The lists of specialist insects is from Mancarrow and Taylor (2000, p. 42).

35. Clues on the how birds can be utilized, that is, their habitat and diet, can be found in Pearson (1917), Forbush and May (1955), and Adams (1994). Other specifics, as given here, are from personal accounts: growers explaining how they have accomplished set avian tasks.

36. The use of mulberry trees is anecdotal; the reason may be in satisfying the 5 percent of the housewren diet that is vegetative.

37. Information on woodpecker allurement is from firsthand accounts. Published observations on woodpecker effectiveness can be found in MacLellan (1970).

38. The example of integrated countermeasures in grape cultivation comes from Ebersole (2004). Note that the natural manure, provided by the sheep, may also be an indirect gain, through the environmental setting (see Chapter 15, "Agricultural Alternatives").

Chapter 6

1. The origins of maize, from Central America, and the earliest domestication efforts, be they accidental or purposeful, are discussed by Fedoroff (2003) and Zohary (2004); theories behind this are in Rindos (1984) and Hawkes (1969).

2. An account of the mid-twentieth-century Green Revolution can be found in Stakman et al. (1967). A counterargument on effectiveness is offered by Shiva (1991).

3. Bananas and breadfruits, in contrast to recent seedless fruits, have a long history. Producing seedless varieties is based on varying underlying motives. For bananas, fruits packed with hard seeds are difficult to consume. For breadfruits, why the seedless type dominates is not apparent, as a species with edible seeds (breadnuts) is still grown (Zerega, 2003). For oranges and watermelons, this is a consumer choice, as seeded varieties are widely available.

4. There is an illustrative case of recovered shade resistance in cassava (de Foresta et al., 1994). A shade-tolerant stem and canopy from a related species is grafted onto the productive rootstock of a high-yielding variety. Although grafting is not applicable to all crops, this shows that high yields do not require high levels of sunlight (i.e., an efficiency gain). The same may be true with water capture (Fahl et al., 2001). Where mineral nutrients are part of the output (rather than the process as with potassium in banana fruit), there may be less opportunity for engineering shortfall resistance. This is not true where minerals are part of structure (as with a shorter stalk), rather than the output.

5. The need to specifically develop varieties for low-input agriculture and intercropping has been discussed by Atlin and Frey (1989) and Smith and Francis (1986), respectively.

6. The earliest agriculture is attributed to New Guinea, the Near East, China, Central America, South America, and eastern North America (Newmann, 2003). Each of these centers had various crops. Hence, there began a process that resulted in a plenitude of varieties.

7. The tally of wheat varieties is from Percival (1922). This spans numerous species, for example, *Triticum dicoccum, T. orientale, T. durum, T. polonicum, T. turgidum, T. pyramidale, T. vulgare, T. compactum, T. sphaerococcum,* and *T. spelta.* The number of rice varieties may be even higher, although no definite figure is advanced (as in Grist, 1965, p. 81). For rice, most of the varieties are associated with *Oryza sativa,* and far fewer with *O. glaberrima.* There are similar counts for other cops. At the regional level, Pearce (2005) mentions the 60 varieties of almond found in Afghanistan and the 1,300 varieties of the common bean *(Phaseolus vulgaris)* seen in Rwanda.

8. Advocacy of heirloom varieties and explications on why these are important from a quality-of-life perspective are provided by Weaver (1997).

9. To be more elaborate, there is cabbage (*Brassica oleracea* var. *capitata*), Brussels sprouts (var. *gemmifera*), broccoli (var. *italica*), cauliflower (var. *botrytis*), kohlrabi (var. *gongylodes*), and collards (var. *viridis*). The amaranth example is

from Mapes et al. (1997) wherein the two divisions represent many varieties of this plant (Espitia-Rangel, 1994).

10. The visual changes in vegetable varieties that have occurred since Roman times is from Weaver (1997).

11. The dynamic nature of domestication for traditional farmers has been discussed by Perales et al. (2003). Mention has been made of the expanded opportunities to informally exploit the genetic resources of wild relatives when these are found near their domestic counterparts (Johns, 1990, p. 138). The latter may be a more common force than supposed as others; for example, Alhamid et al. (2003), have noted the presence of wild relatives as nearby weeds. Others have questioned how much, if anything, domestic crops can gain from close-at-hand wild relatives (Jackson and Koch, 1997).

12. The lists of wild plant species with cropping potential, that is, currently not receiving attention, are quite large. Lists have been compiled by Styger et al. (1999), Aman (1998), NRC (1996), Van den Eynden et al. (2003), Mertz (1998), NRC (1987), and a host of others. In total, there are thousands of edible plants still in the wild that could be part of human diets.

13. The concept of desirable plant characteristics (DPCs) derives from that of the multipurpose tree (e.g., Huxley, 2001).

14. A more formal and comprehensive ranking procedure has been proposed by Wojtkowski (1998, p. 99). Matrix based, this does not delve into the problems of assigning each of the DPCs an importance ordering. Bellon (1996) and Franzel et al. (1995) have looked into field-level selection, Rhodes and Nazarea (1999) into the societal aspects.

15. The Uganda banana study was conducted by Gold et al. (2002).

16. It is implied that the one-plot–one-ecosystem model underlies this discussion of agroecosystems. In simplifying the presentation, an inexperienced reader is not plunging into a morass of ideas without a solid footing. The one-plot–one-ecosystem model provides a solid footing to later venture from.

17. Gliessman (1998, p. 304) has presented a somewhat different view. Here change (1) seeks to increase the input efficiency of a system by, for example, changing types of inputs applied; (2) substitutes one type of input for another, for example, biomass for fertilizer; and/or (3) can be a redesign in which the current agroecosystem is not in line with the changing economic or land-use situation, and the introduction of an entirely new system outweighs any disadvantages of a complete change.

18. The statement that each agrotechnology is a solid, well-defined, landscape addition may be overblown; some are far less defined than others. Conceptional promotion is the reason for any exaggeration. It should also be noted that pure examples are rare; most have been modified. The photos are, in large part, of agrosystems that have drifted from their technically unsullied state.

19. The preface to this book contains a brief rationale for agrotechnology as the practical foundation for this text.

20. The list of agrotechnologies described in subsequent chapters follows those categories listed here. Conspicuously missing are those associated solely with silvi-

culture. The absent categories includes multispecies and/or multiage forest tree plantations.

21. Techniques such as plowing, weed reduction through plastic sheets, etc., are management inputs. These do not establish an agroecosystem but do affect the internal environment.

22. The notion of understandability should not be understated. The supremacy of the monoculture may be based in very large part, if not solely, on understandability. Other practices, such as biocomplex agroforests (Chapter 10) and nonagrotechnologial landscapes (Chapter 17), would not be utilized if they did not come with rudimentary operational guidelines. Explicitly stated, these allow inexperienced practitioners to make positive inroads into biocomplexity in the face of changing species mixes and elaborate user objectives.

23. The African case study on bringing alley cropping design and management in line with local conditions is from Kanmegne and Degrande (2002).

24. Agrobiodiversity headlines a large and growing body of theory. The question is how this is best done. For presentational simplicity, most agroecosystems are discussed as bicultures. As stated in the text, three-plus polycultural versions are entirely possible.

Given two distinct bicultural systems, one of *ababa* and a second of *edede* where species *a* and *e* are the primary species while *b* and *d* are secondary, the routes or approaches to bioamplification are as follows:

- *Expanded primary:* Two systems come together as *abebabeba* with the assumption that species *a* and *e* are complementary with species *b.* There is no requirement that *a* and *e* should be complementary.
- *Expanded secondary:* Two systems merge into *abadabada* where species *a* is complementary with species *b* and *d* while *b* and *d* can be competitive.
- *Merger:* Two systems do not share a common spatial pattern, as with expanded and primary and secondary systems, but overlay one above the other. This can be a parkland over hedgerow alley cropping or light shade over a vine support system. Other combinations are incompatible, for example, heavy shade over vine support may not gain traction.
- *Reenforcing:* The two systems can exist separately, but one can be a subset of another. This is course pattern where, for example, one is a cluster or wide strip within another. For example, this may be a light shade system as clumps within a parkland system. In cross section this can be as follows:

  abababa(edede)abababababa(edede)ababababa
- *Transitional:* One agroecosystem can be a subset of another, but without clear, distinct boundaries. Instead, these transition into each other in fuzzy form, i.e., some species in each intruding or transitioning into another system. In cross section, this may be as follows:

  abababeba(edade)adababababada(edade)abebababababa

There is no need for spatial likeness with the reinforcing or transitional approaches.

Chapter 7

1. There is no question that the few principles and practices behind the monoculture, those few that do exist, are much easier to grasp. As for the problems of fostering or advancing biocomplex agriculture, one should look beyond this text. For a full understanding of the ramifications, one should read books and articles on the subject. As for the economic superiority of high-input monocultures, this premise is challenged throughout this text.

2. This statement on the preferences of banks and other financial or lending institutions for biosimplicity is from the findings of Godoy and Bennett (1991) and McNeely (1993).

3. Advertising, promises, and peer pressure help in the push to purchase and become dependent on outside farm inputs. Monocultures are also encouraged by an abundance of government and/or corporate mobilized research. Governments benefit through increased tax revenue (income, sales, and/or land) with less noncash subsistence and more commercial farming. Although seldom overtly mentioned, land taxes (as a fixed cost that sets a lower limit to cost-oriented agriculture) may motivate farmers toward revenue-oriented cash crops. Although commonly perceived, this may be false, as the quest for more revenue need not translate into greater profits (see Economics, Chapter 14).

4. The previous discussion of heritage varieties (Chapter 6) should have debunked the myth of special purpose "improved" varieties. The quotation marks stress that improvement here is narrowly defined, in a monocultural rather than a polycultural setting.

5. The concept and use of the EOR are far more important than the actual calculation. If needed, measurement is achieved through the economic orientation ratio (EOR):

$$\mathrm{EOR} = (P_a Y_{ab} / P_a Y_a) - (C_a / C_{ab}) \quad \text{(N.3)}$$

For this, Y_a and C_a are the yields and costs associated with a monoculture of species a, while Y_{ab} and C_{ab} are those associated with an intercrop of species a and b.

The EOR does not indicate productive or economic superiority; it measures only how the economic gains or losses are achieved. A monoculture of the primary crop can be compared against a biodiverse alternative. In the other direction, the comparison uses an intercrop as the base, checking the economic orientation of monocultural inputs.

When revenue increases are greater than the cost increases, the EOR value is greater than zero. When cost decreases are greater than the loss in revenue, the EOR values are less than zero. Revenue orientation is indicated when the EOR exceeds zero.

6. Cost-oriented monocultures come to the fore with rotations and/or through small, high-interface plots that achieve agroecological character through an association with and spillover from ecologically well-endowed neighboring systems. These options are discussed in later chapters.

7. The reference to mixed varietal rice in China is from Yoon (2000), wheat in Pakistan from MacDonald (1998), wheat in Syria from Plucknett and Smith (1986), and maize in Mexico from Altieri and Trujillo (1987). These may be the tip of the iceberg, as multiple varieties may go unobserved in many regions. Markets in many regions show that locals accept multiple varieties, for example, sweet potatoes *(Impomea batatas)* in New Guinea, potatoes *(Solanum tuberosum)* in Peru and Bolivia, and rice in many areas.

8. The reason goes back to understandability (see Chapter 6). By viewing and managing the system as a monoculture, users do not have to enter directly into the realm of biocomplexity and adopt a more intricate set of operational guidelines. Of course, much can be lost in agroecological terms in doing so.

9. In a perceived monoculture, weeds can contribute forage, potherbs, and medicinal and ornamental plants. This view, that of the useful weed, is an operational guideline in some cultures, for example, Vieyra-Odilon and Vibrams (2001) and von Weizsäcker (1995).

Chapter 8

1. The list of tricultures and some of the common garden bicultures comes from Wallace (1980). Other garden mixes are from Dennis and Miller (1997). The staple and commercial intercrops are from Vandermeer (1989), Altieri (1994), Rao (1986), and Okigbo and Greenland (1976), with a few, personally observed, added by the author. The four- and five-crop mixes are from Beets (1982, p. 44) and Okigbo and Greenland (1976).

2. This listing of common California covercrops is from Altieri (1995).

3. The lower limit for rainfall in maize-bean intercrop has been set at 400 mm (Rao, 1986). At this level, a monoculture of either species produces well, whereas an intercrop will not.

4. The difference in internal and edge-row yields, as water moved from being limited to abundant, was documented by Ghaffazadeh et al. (1994).

5. The intercrop LER values and the LER/weed reductions were reported by Altieri (1995, pp. 207 and 228). The weed-reduction figure was the percentage of the ground area covered. There is no conversion figure for percent area covered and cost of removal.

6. A published example, where cassava yielded but maize failed under drought conditions, is from Nigeria (Mutsaers, 1991). From anecdotal information obtained by this author, it seems that sorghum intercropping in India may be similarly formulated.

7. A number of as of yet unexplored spatial options or changes in harvest machine design would simplify harvesting. The lack of viable machine harvesting may be the greatest economic impediment to the adoption of commercial intercropping. The discussion is ongoing with seemingly little progress. Among those entering the discussion are Erbach and Lovely (1976) and Vandermeer (1989, p. 199).

8. This 3 m recommendation seems the norm with many unknowns and little formal guidance. The strip contents may well prove more important than the width. This applies to both generalist and specialist predators, for example, Altieri and Nicholls (2004, p. 148).

9. The 90 percent erosion-reduction figure is provided by ICRAF (1996). This work also found that with alternating seasonal strips erosion was 40 to 60 percent less than a fully plowed area. Alternating active crop strips with permanent strips achieved the 90 percent mark.

10. The buckwheat barrier surrounding cucumber is from Platt et al. (1999).

11. Tall annual species and their effectiveness as windbreaks is discussed by Beets (1982, p. 26). As an added note, annual windbreaks may be integrated into a cross-landscape series of wind defenses, that is, shelterbelts → perennial windbreaks → annual windbreaks, each being a part of or an extension to the SIZ of the taller, larger structure.

12. Sunflowers protecting tomato and grass strips guarding cotton are from Ferber (1969). The aforementioned use of sorghum as a border was documented by Capistrano and Marten (1986); no agroecological reason is given. As a clarifying point, Jerusalem artichoke is considered seasonal because it sprouts anew each spring.

13. The shyness of birds and the protection afforded by field designs (layouts) has been described by O'Conner and Shrubb (1986). The use of concentric cropping in Peru was reported by Beckerman (1983, 1984) and Stocks (1983). The latter reference lists the agroecological gains from this field pattern arrangement. In addition to the risk factors stated in the text, these may include the better apportionment of sunlight and improved nutrient flows.

14. Mechanized agriculture with cover crops is a very unsettled topic. A sampling of studies, for example, Ashford and Reeves (2003), Teasdale and Rosecrance (2003), Mallory et al. (1998), and Creamer and Dabney (2002), shows unresolved issues and economic problems. Among the obstacles, the DPC list for cover crops do not currently offer enough detail to address the options.

15. The mucuna with maize case is from Tarawali et al. (1999). Another example, barley with a forage legume, offers, in addition to nutrient gains and soil and weed protection, the option of grazing after the barley harvest (Zewdu, 2004).

16. Suppression through closer than normal spacing was found effective by Fischer and Burrill (1993). This was with maize over clover.

17. Some descriptions of complex gardens are, from Nigeria, by Okigbo and Greenland (1976) and, from the South Pacific, by Siwatibau (1984).

18. Complex homegardens, as a feature of European farming, may have been first described by Stevens and Liebault (1616, p. 287). These continue to this day in France, Spain, and many other countries. A more contemporary study of Spanish homegardens is by Agelet et al. (2000), where 50 percent of medicinal species in the overall Spanish landscape come from homegardens.

19. The microenvironment as an in-field positioning guide is from Capistrano and Marten (1986).

Chapter 9

1. For those that rank nitrogen fixation a more desirable characteristic than fast growth, or wish both, regional lists of common N-fixing species are available. For Africa, such a list is found in Brewbaker (1989). Further listing is found in Nair et al. (1999).

2. The whorteberry or pine grass example was mentioned by Koch (1996), the pine-poplar sequence was presented by Peterson and Squiers (1995), and the oak-shrub combination is from Callaway (1992).

3. The comparison figures for total per area biomass and in-leaf nutrients come from Young (1989a, pp. 122 and 136). Matching leaf nutrient content with crops is problematic, requiring more work and some rethinking if nutrients are to attune with and to flow smoothly, without a temporal lag, between facilitative trees and crops (Nair et al., 1999).

4. The increased sorghum yield attributable to tree root decay is from Korwar and Radder (1994). Associated issues in nutrient recovery are discussed by Schroth (1995).

5. The high LER value for alley cropping comes from Ong (1994). This was for alley cropping in East Africa where the trees enriched poor soils.

6. Multiparticipant systems of this and other types are further discussed (and references provided) in Chapter 14.

7. The term *horqueta tree* applies where trees are used to stash forage, in branches or high platforms, out of reach of animals. The term comes from South America. Although the practice is widespread, no corresponding English word was identified.

8. A single large tree can be home and cohabitat to any number of bird species, for example, Herzog and Oetmann (2000).

9. The old German practice of placing T-shaped roosting poles in farm fields was encountered by the author and verified by a number of observers as once commonplace but now an increasingly unused practice.

10. How far the natural savanna–parkland agrotechnology analogy goes, at least toward explaining direct nutrient gains, is the subject of some dispute. Studies in natural savannas show the trees contributing toward soil fertility (Belsky, 1992). Some show that these gains transmit to crops, for example, Kho et al. (2001); others are less sure, for example, Wilson et al. (1998) and Nair et al. (1999). Moisture-demanding crops and water as the limiting resource may explain the nonpositive yield outcomes (Ong and Leakey, 1999). This may not be the case with pasture grasses, for example, Bolívar et al. (1999).

11. The expanded pruning options include branch cutting (where main branches remain untouched, the secondary branches alone are removed) to complete pollarding (almost all branches are pruned away). As a practical warning, to reduce tree mortality, not all branches and leaves should be cut; a few must remain. In some cases, severe pruning might be economically feasible only if the pruned branches have economic value or if crop growth is increased dramatically.

12. Oak parklands are observable in the lower Midwest and have been actively advocated, for example, Bainbridge, 1988. This list applies to cropping parklands. With less intense grazing, other less crop-compatible species are possible, for example, de Oliveira and de Oliveira (1994) and Harvey et al. (1998).

13. The water and soil runoff figures are from Pratap et al. (1998). This is from India in a region with 1,000 mm rainfall and on a site with a 4 percent slope. Soil loss reductions with alley cropping were slightly better than contour planting without a hedge (48 versus 45 percent), but the alley system was superior in reducing water runoff over nonhedge contour planting (40 versus 27 percent). The LER was slightly higher with the alley system.

14. The design package elements for maize in a hedgerow alley system were compiled from Sanchez (1995) and Reynolds (1991).

15. The nutrient gains from alley cropping, especially with nitrogen-demanding maize, have been discussed by Kang (1997) and Shepherd et al. (1997). Despite the nutrient gains, slope and erosion control may turn out to be the principal selling point. This is supported by author-acquired anecdotal evidence and from Ehui et al. (1992).

16. This list was modified from the works of Young (1989b), Rachie (1983), and Beer (1987).

17. Rao et al. (1991) and Nair et al. (1999) reviewed the literature, offering insight into the crops and pairings utilized in hedgerow alley cropping.

18. On the subject of hedge height, Versteeg et al. (1998) found that the tree species *Acacia auriculiformis* required a pruned height of at least 0.8 m to insure survival. This species would not find favor where a lower height is preferred, for example, the situation described by Kanmegne and Degrande (2002).

19. Cooper et al. (1996) have discussed the minihedge.

20. The mixing of hedge species follows from the fact that the leaves are important. Of the commonly utilized hedge species, leucaena is plagued by the leucaena psyllid (Liping, 1991), and sesbania leaves are eaten by the sesbania beetle (Wale et al., 1996). If one is attacked, another can thrive.

21. A case study with an inclusive strip counterpattern is by Versteeg et al. (1998).

22. Descriptions and the yield gains from plantain-tree intercrops in West Africa are presented by Banful et al. (2000) and Ruhigwa et al. (1994). In Guam, the raising of jackfruit with various trees has been described by Anon. (2003).

23. Roman agroforestry, natural trellises included, has been reexamined by Lelle and Gold (1994). From what is assumed to be first-person observation, Chaucer (1382, line 177) alluded to vines raised on elm trees. A contemporary example from Bolivia, grapes in trees, has been described (FAO, 1994).

24. Of the climbing plants listed, guaraná may be the lesser known. This plant, found in Brazil, provides a high-caffeine-content fruit. This is a shrub that, if given the opportunity, will climb.

25. Salam et al. (1991) advocate pepper on tree trunks. Also under this heading are rattan as a supplementary vine addition (Mohd Ali and Raja Bariza, 2001). Rattan can be raised above any tall, perennial ecosystem, for example, orchards and treecrop or forest plantations. These are destructively harvested, that is, physi-

cally pulled from the canopy, but, as with all supplementary additions, do not require any design accommodation within the recipient ecosystem.

26. Hops can fall into the seasonal category if winter killed, requiring spring replanting.

27. This refers to in-field shade trees employed in agriculture (see heavy shade, this chapter), not those in an urban setting or around houses. This list was modified from Beer (1987).

28. The methods for establishing shade include (from Muschler, 1999, p. 107)

1. thinning a natural forest;
2. permanently planting the desired species;
3. planting temporary followed by permanent shade; and
4. planting one permanent species, pruning it after some growth, and then planting a second shade species.

29. The role of shade in weed control is discussed by Staver (1999) and Muschler (1999, p. 26). Soto-Pinto et al. (2000) found yields in shaded coffee down about 50 percent with about 40 to 50 percent shade. In addition to reductions in operating costs, there can be quality gains. Muschler (2001) observed that shade caused more uniform ripening and a tastier coffee.

30. The DPCs of shade trees, those presented here, are from Akyeampong et al. (1995). An expanded list is found in Wojtkowski (1998, p. 99).

31. There is a fair bit of documentation on the ecological benefits, facilitative and otherwise, of heavy shade systems, for example, Perfecto and Armbrecht (2003), Perfecto et al. (1996), Rice and Greenburg (2000), and, in a more formal assessment, Mas and Dietsch (2004). The 94 to 97 percent bird figure is from Tolme (2004).

Chapter 10

1. Most of the tree-with-crop list was compiled by Vandermeer (1989), a few pairings are from Plucknett and Smith (1986), and some supplied by this author.

2. The intercropping of citrus with cucumber was seen by this author. The brazilnut over rubber example was also personally encountered. The Kenyan fruit mix comes from Aiyelaagbe (1992). Another case, pecans with pasture, has been reported as treerow alley cropping (Stuhr, 2001), but, because the trees do not touch, this system may be better categorized as agroforestry intercropping.

3. Mango trees in Mexican and other regional farm fields were observed by this author on numerous occasions.

4. Palms do grow to considerable height. In doing so, these change the ecological uniqueness of treerow alley cropping and make these into light shade systems. Dwarf varieties of oil and coconuts palm avoid this while making fruit harvest much easier.

5. Whereas row cropping is only row based, the same species can be employed in nonalley forms.

6. Row orientation in alley cropping is briefly discussed in Chapter 2. The argument for a north-south orientation with treerows is obvious. Further advancing this point, a north-south orientation is the norm for the successful commercial systems observed by this author.

7. The mix of coffee and banana along with shade, forest, and fruits trees is from Escalante (1995).

8. A sample study of a light shade, coconut-cacoa system can be found in Mialet-Serra et al. (2001).

9. The heavy shade example with breadfruits is from Ashley (1986). The triple canopy, brazilnuts-rubber-cacao, was encountered in Brazil by the author. This is a more mixed system, the brazilnut and rubber being an agroforestry intercrop association, and the mix of brazilnut over the rubber puts the cocoa in heavy shade.

10. Temperate examples are far from common, but have been observed by this author in south-central Chile. These are located in smaller towns and villages near, but not in, the center. Not yet studied, the purpose may be to maximize output from a comparatively small area. These are absent in larger farms.

11. The overall economic or absolute productivity of agroforests has yet to be measured. Those studies that do exist, based upon contributions to family diet, find these to be comparatively high yielding, for example, Cooper et al. (1996). Others, for example, Soemarwoto (1987) and Torquebiau (1992), note the stable yields and low level of labor inputs. Normally, no agrochemicals are utilized.

12. The large percentage of tall woody perennials separate agroforests from ecosystem-governed pastures and complex gardens (Chapter 8).

13. In tropical agroforests, the potential biodiversity can be exceedingly large, although the actual biodiversity per agroforest is less. Table N.1 expresses some of the variability.

14. The management (rule-of-thumb) guidelines for agroforests have been modified slightly from those presented by Wojtkowski (1993). Operational guidelines, that is, understandability (see Chapter 6), either formally stated or ingrained in local

TABLE N.1. Biodiversity variability in tropical agroforests.

Location	Total number	High-low range for individual gardens	Source
Mexico	325	21-37	Cooper et al. (1996)
Costa Rica	55	6-29	Zaldivar et al. (2002)
South India	60	NA	Babu et al. (1992)
Honduras	240+	30-130	House and Ochoa (1998)
Cuba	101	18-24	Wezel and Bender (2003)
Peru	25	5-25	Coomes and Ban (2004)

Note: NA = not available.

lore, are, without exception, an agricultural prerequisite. This certainly applies to agroforests.

15. In addition to the yet unknown productivity levels of agroforests, the reductions in risk, as a theory applied through spatial disarray (biodisarray, see Figure 2.9), may be better understood by practitioners. Optimal implementation requires a level of knowledge well beyond current agricultural science. However, users often have a good understanding of agroforest plants and how they ecologically fit within a complex ecosystem (Teklehaimanot et al., 2001). Therefore, it is conceivable that very experienced practitioners, those with may years in the field and with an intimate knowledge of various species, through informal guidelines, might make a good stab at an optimal layout.

16. The exploitation of the edge effects through greater biodiversity has been studied by Jensen (1993), who found 50 percent more species at the edge. Internal subdivisions (gaps included) have been noted across regions by Gillespie et al. (1993), Aumeeruddy and Sansonnens (1994), Michon and Mary (1994), Gamero et al. (1996), and Méndez et al. (2001). These zonal differences increase biodiversity.

17. Variations on tropical homegardens within the same region (Central America) can be assessed by comparing the work of Nasser et al. (1994), Víquez et al. (1994), and Gillespie et al. (1993).

18. With oaks and beeches, this might qualify as a hog feed system. The difference is when the trees are primary crop; a feed system has fauna in this role. In this case, the hogs were of enough importance to be an economic determinant, not an economic bystander. This suggests a feed system, a point made by Woolsey and Greeley (1920, p. 70).

19. The list of botanicals that can be raised in North American forest garden setting is from Dix et al. (1997). Mushrooms that fit are listed in Hill (1999).

20. Some of the roles where the agroforest substitutes for a specialized, nonproductive auxiliary ecosystem are in adding frost resistance, for example, Wang (1994).

21. A look at the present uses and future potential for animals in agroforestry (semihusbandry included) is provided by Redford et al. (1992).

22. Dehesa systems also provide a less-mentioned output: the cork oak *(Quercus suber)* provides cork bark. The other common species are *Q. faginea* and *Q. ilex* that produce only acorns.

23. Traditional German streuobst, without animals recycling the nutrients, might also be delineated as varietal and perceived monocultures. Multiple apple varieties provide in-product genetic variation. A few interspersed pear, cherry, and other fruit trees add the noninterference component. A description of these systems is provided by Vieyra-Odilon and Vibrams (2001) and Herzog and Oetmann (2000).

24. In the South Pacific, Raynor (1992) noted that traditional forest animals, mainly pigs and chickens, are common in local agroforests.

25. In an aquaforestry example from China (Gongfu, 1982), it is observed that carp fill most below-water spatial niches. There are upper zone, grass and bighead carp; middle zone, silver carp; and lower zone, mud and black carp. Among all these niche feeders, most of the leaf fall is consumed. Very little is wasted, and the system is very feed efficient.

26. Various species of fruit- and nut-eating fish are found in the Amazon Basin. These have not yet attracted much notice as a regional aquaforestry resource.

27. Salmon with willow, employed on Chilean farms as was observed by this author.

28. Insects that are found in diets and that could be raised in mixed tree systems are in Australia, the witchetty and green grubs; in Mexico, the sago grubs, maguey worms (in some brands of tequila these are put in the bottle), and mealworms; and in the United States, crickets (Menzel and D'Alvisio, 1998). Worldwide, there are over 3,000 eatable insect species, making this an underutilized food resource.

Chapter 11

1. Water is generally not included in the carryover. There can be exceptions, as with a healthy water table or through moist, water-holding, in-soil biomass.

2. The –6 percent to +7 percent range for water capture is from Morris and Garrity (1993). The literature does mention any indirect effect (e.g., reduction in transpiration). From a case of sequentially planted groundnut and maize in West Africa (Versteeg et al., 1998), this seemly sound intercrop, on a nutrient basis, had a problem after a few seasons. The water-holding capacity of the soil fell below that required for the maize. It is thought that this occurred because neither the maize nor the groundnut imbued the soil with enough moisture-holding organic material. Facilitative intercropping, formulated to address the situation, rectified the problem.

3. Barley-clover is a potential intercrop that does well in a rotational sequence. This seems time tested, for example, Harrison (1775).

4. The list of common rotational sequences is from Alteiri (1995).

5. The same logic holds if species *a* and *b* had a relatively low LER potential when mixed, only elevated through a wide interspecies interface. Strengthening a low LER pairing in a sequential situation may require inserting a seasonal fallow or adding the appropriate nutrients.

6. The marigold-nematode relationship is described by Michalak and Gilkeson (1994). Wallace (1980, p. 171) states that the temporal SIZ (residual effect) can last several years.

7. The references for medieval agricultural rotations are Vasey (1992) and Gras (1940).

8. Using soil and site variables, crop-specific temporal sequences are possible, for example, Jordan et al. (1996) and Jordan et al. (1997a, b). Although agrochemicals were employed, a DPC-based study can help usher in fully organic, cross-seasonal rotations.

9. The weed and barley virus example is from IACPA (1998).

10. Suggestions have been made on how to string together spatial agrotechnologies as the ecological (essential resource) situation mutates, for example, Wojtkowski et al. (1991).

11. Questions abound on nutrient loads in organic systems and whether to plow. Among those asking the questions are Ankens et al. (2003); Calanca et al. (2001); and Zihlmann et al. (2000).

12. The dangers of a bareground phase are well documented. However, if needed, see Biglow et al. (2004).

13. More can be found on these two slash options, that is, slash and mulch (Thurston, 1997) and slash and burn (Peters and Neuenschwander, 1988).

14. Weed suppression comes to the fore and even drives the type of fallow used. When farmers face hard to eradicate weed species, suppression has been found to tip the weed balance in favor of easy to control species. Speargrass, a dreaded weed found in Southeast Asia, can be controlled through such a fallow-cover-suppression strategy (van Noordwijk, 1996).

15. Postfallow burns, often undertaken after extensive drying of the vegetation, are not the only option. Cold burns, those that result from flaring wet biomass, allows some of the biomass to decay naturally. With some systems (e.g., hedgerow alley cropping), the woody component can be cut and left until the leaves shed, and the woody branches are subsequently piled and burned. In this case, the ability to readily shed leaves is a DPC.

16. The bamboo fallow-regeneration sequence is from Indonesia (Christanty et al., 1997). The kudzu seed regeneration fallow is from Central America (Kass and Somarriba, 1999). These do not exhaust the list of possibilities. Kass and Somarriba (1999) discuss some other extremely novel options. Gliessman (1998, p. 76) mentions a grass fallow with a unique cut-plant-burn sequence (not the usual cut-burn-plant).

17. The question of NPV is less the methodology, more the discount figure utilized. High discount values bias things toward an initial period. Zero values are time-period neutral.

18. The debate on the best measure is far from resolved and may never yield an answer. LER trendlines are good but of little credence when business decisions are paramount. Net present value evaluations, with their wide-ranging business and financial applications, have a strong presence but, in being front loaded, they tend to bias decisions against long-term investments (Price, 1995). In addition to universal computational familiarity, more of the lofty agroecological goals (see Chapter 1) can be quantified and evaluated (e.g., Hansen and Jones, 1996). Indices (not mentioned in the text) lack intuitiveness (a strength of the LER), lack widespread acceptance (as with the NPV), and are highly subjective. However, these can be formulated to appraise intangibles, for example, clean water, wildlife accountability, and so on (Stockle et al., 1994).

19. In Chapter 3, barley, rye, or clover with onions, peas, cabbage, or beets illustrate root strata partitioning. The explanation continues here, offering a reason for the late planting: to pre-start, through a temporal overlap, a second rotation.

20. The El Salvador overlapping crop sequence is from Beets (1982, p. 129). Okigbo and Greenland (1976) afford examples of other overlapping sequences.

21. The word taungya originated in Burma in the 1840s. For this simple form, teak trees were mixed with upland (nonpaddy) rice and other crops (described by Blanford, 1925). Before this, the history of the taungya can be traced through Evelyn (1664), where turnips first inaugurate the planting of hollies, laurels, yews,

and junipers; Browne (1832), where carrots, wurzel (a type of beet), and potatoes initiated various tree planting; Brown (1878), the mixing pines and broomstraw; and Schenck (1904), where white oak was planted with barley, oats, or summer rye. Although less documented, there is parallel development in Asia (Menzies, 1988) and certainly in other regions as well.

22. In a useful aside, there are six tree planting methods;

1. *Seeds:* These are seldom utilized in forestry because of the longer time frame, danger to young vulnerable trees, and no assurance that the resulting stock will be of good quality.
2. *Seedlings:* The common planting method where trees 10 cm to 1 m high are transplanted from a nursery.
3. *Striplings:* This is where trees 2 to 3 m tall are transplanted. Normally, all branches, except those at the very top, are removed (this helps with survival). Striplings are often utilized when the trees are to be located in grazed areas.
4. *Stumps:* This is when small cut sections containing 10 cm of root and 5 cm of stem are transplanted. Stump planting finds application when many trees must be hand carried and planted in a remote area. Stump planting does not work with coniferous species, for example, pines.
5. *Sprouts:* This is when an existing plant is bent to the ground and held in place to root. This finds use to fill gaps in existing hedges.
6. *Stems:* These are long sections of tree branches or stems that are stripped of leaves and planted. Branch planting is employed as a low-cost method for establishing hedge rows. Long stems (5 to 10 m) are used in tree plantations or with agroforestry when a long clear trunk is needed, a fast start is desired, and/or to shorten the tree rotational period.

23. The maize-papaya-orange taungya was reported by Villachica et al. (1990). The walnut–Christmas tree case is in Merwin (2001).

24. These mixes follow guidelines for forestry bicultures, in this case, requiring the second tree to be shade resistant (Blanford, 1925).

25. The LER calculation for forest trees is based on annual growth, either yearly average diameter growth or as the overall per plot wood volume added during one year or one growing season.

Chapter 12

1. Numerous studies support the basic premise that wild or semi-wild areas offer habitat for farm-beneficial creatures. A sample includes Alomar et al. (2002), Vickery et al. (2002), Bäckman and Tianen (2002), Maudeley et al. (2002), Groppali (1993), and Ma et al. (2002).

2. The concept of economic orientation does not apply unless these offer some output. If this occurs, such auxiliary systems should be cost-oriented.

3. The argument that auxiliary systems predate agriculture comes from Pringle (1998).

4. Often overlooked are habitats as refuge for pollination insects, for example, Bohart (1970). More subtle, but equally potent, are habitats as a sanctuary for beneficial microflora (e.g., mycorrhizae) and slightly larger organisms (e.g., earthworms) that travel poorly. If present, these do lend an ecologically helping hand to neighboring ecosystems.

5. The importance of corridors in allowing native plants, animals, and microorganisms to travel and colonize reception ecosystems is discussed under ecosystem governance (Chapter 3). The Chapter 3 endnotes provide references. Corridors also aid larger creatures, for example, Driscoll (2004).

6. This grass strip design is from a comprehensive study on ecologically sound landscapes (IACPA, 1998). The resulting economic impact is mentioned under economic gains, see Chapter 14.

7. Shelterbelts and windbreaks are a well-studied topic, encompassing much more than can be presented here (e.g., the internal design, cross landscape spacing, etc.). A good starting point for further study is Caborn (1965).

8. Pollination gains in the absence of winds have been mentioned by Stuhr (2004).

9. The ability of a windbreak to contain spray drift is well-known. Also possible is to utilize these structures to contain bad odors, such as those from poultry and hog farms, see for example, Malone (2002) and Colletti and Tyndall (2002).

10. Except where noted, this list of windbreak gains is based mostly on one compiled by Prinsley (1992).

11. The magnitude of protection on maize, wheat, and bean yields is from Brenner (1996), the benefits in water retention in soybean from Miller et al. (1973). Another reference, which set the pattern for Figure 4.1, is Sun and Dickinson (1994).

12. The uses of shelterbelts and windbreaks go beyond that suggested here. Besides being habitat corridors, waterbreaks, and so on, these can be snow fences. Well designed and well placed, these pile winter snow in or near fields. This can be an advantage as more moisture, from slowly melting snow, is available to plants in the spring (Strange, 1998).

13. In greater detail, the wind reduction possibilities, expressed in percent reduction in wind speed and distance from the structure, is, on average, for a 10 m high windbreak as shown in Table N.2. The source of the data in Table N.2 is Caborn (1965, p. 226). Others have more conservative view where effect of a wind structure is from 4 to 10 "heights" (Wight and Studr, 2002).

14. An abbreviated list of wind-resistant and wind-susceptible plants is found in Peri and Bloomberg (2002).

15. The use of shelterbelts for secondary outputs is briefly discussed by Wight (2003) who, in relaying firsthand accounts, suggested the species listed.

16. Another role of a riparian buffer may be streambank stabilization. A well-established buffer should do this without problem. If banks are in need of stabilization, any number of techniques can be used. These are outlined in Wells (2002).

TABLE N.2. Wind reduction possibilities (in percent).

	Distance away			
Structure	**50 m**	**200 m**	**150 m**	**300 m**
Open canopy structure	–54	–46	–37	–20
Medium structure	–60	–56	–48	–28
Dense structure	–66	–55	–44	–25

17. There is a lot of published testimonial on the effectiveness of riparian buffers (one such study is by Fail et al., 1987), with less agreement on the design parameters. Management and vegetative content are important. Fast-growing, nutrient-demanding plants represent a good starting point (e.g., Perry et al., 2001). Ecological theory suggests that a mix of species is best, as these trap a mix of nutrients. If the runoff contains a set, predetermined nutrients mix, a specialist plant, one that demands the specific mix, might be employed (Wojtkowski, 2002, p. 233).

18. Peruvian streamside farming has been described by Pinedo-Vasdquez et al. (2003) and, in the Sudan, by Simpson and Simpson (1991).

19. Special purpose structures may be best when floods are a yearly event. Otherwise, an integrated waterbreak system may serve best. More on integrated landscape and cross purpose structures can be found in Nebel (2001). An in-stream variation of the waterbreak, the gabion (rock barriers placed a cross seasonal streams), promotes infiltration and prolongs flows; a first-person account of effectiveness is found in Cohn (2005).

20. The living fence DPCs come from Ayak (1997).

21. There is a long history of living fences in Europe. Thorny species are found in the distant past to the present day, for example, Anderson (1800, Vol. I, p. 19) and Gregg (1988, p. 97).

22. The suggested DPCs for barrier species are expanded from a list supplied by ICRAF (1993). The cacti examples are from Wojtkowski (1998) and Soorae (2002). The African tree, *Acacia drepanolobium,* has a symbiotic relationship with many ant species. Although not yet common as a barrier, some of these insects viciously defend the tree from uninvited and unwanted visitors (Stanton and Young, 1999).

23. Pleaching is a time-tested practice, once common in ancient Europe, where, through intent, the branches of different plants of the same species actually grow or fuse together. This was once common enough that the verb, to pleach, was needed.

24. The statement that soil improvement favors high-value crops is from Jama et al. (2000).

25. Some historic, long-evolved catchments are in the Middle East (Nevo, 1991) and India (Gunnell and Kishnamurthy, 2003). These are often found, but not always documented, even when agriculture depends upon their existence. A contemporary example is found in Lövenstein et al. (1991).

26. Empirical studies are far from clear on the advantages of one type of vegetation over another. Paved parking lots produce a lot of runoff but are not considered serious runoff alternative, as pavement and other impermeable surfaces contribute little, if anything, to water cycling. As a result, many consider highly vegetated areas best for gathering water (e.g., Sandström, 1998). The indirect evidence (e.g., Ataroff and Rada, 2000) has widespread forest removal, weakening the overall rainfall-evaporation-rainfall cycle, lessening regional precipitation.

Chapter 13

1. A description of nonvegetative infiltration structures to encourage water retention, prolong the growing season, and make agriculture possible is found in Nevo (1991).

2. The historic references to floating gardens are not very precise. One case, the Aztec practice, often referred to as floating gardens, was the farming of channel banks and built-up shallows. The local name is chinampa farming. A true example, crops grown on floating, soil-covered rafts, is in Bangladesh (Rezaul Haq et al., 2004)

3. Figures on the effectiveness of mounds as a frost countermeasure are from Waddell (1975). Photographs and layout descriptions of New Guinea mounds are in Gardner and Heider (1974, p. 48).

4. It can be difficult to assign a use rationale to many nonecological descriptions. The account (Carrier, 1923) of pre-Colombian maize cultivation in New England fails to indicate whether the mounds counter frosts, waterlogged soils, or both. Period explanations (e.g., Lathrop, 1825) suggest that maize was more cold susceptible, and mounds more necessary, than with current varieties.

5. The other gains from mounds (i.e., root health and tuber productivity) come from Sillitoe (1996, p. 382). These, along with ease of harvest, may be important. In New Guinea (sweet potato) and Bolivia (regular potato), tuber crops are the staple food in the high mountains.

6. Theory suggests an optimal range with regard to the low temperate and to rock size, spacing, and numbers, that is, smaller rocks, because of their greater surface area gathers more heat but releases it quicker; fewer larger rocks collect less but release heat at a slower rate.

Old tires serve a similar purpose, that of holding heat. Tires also repel squash and cucumber beetle (Mancarrow and Taylor, 2000, p. 62). While a means to recycle this blight on the landscape, the disadvantages include labor costs and, in the water collected, old tires are a good breeding ground for mosquitoes.

7. Fish (2000) lists the advantages of rock clusters. Gonzáles (2001, p. 142) mentions rocks and maize in Mexico. This duplicates what occurs in nature. Steenbergh and Lowe (1969) showed that, in natural situations, survival rates of cacti are better in rock clusters, as opposed to seemly more inviting and more fertile alluvial sites.

8. An example of seasonal cropping utilizing water-collecting holes is Nyakanda et al. (1998).

9. Adsorption zones filled with organic plant matter have been described by McIntyre (1999), and their use in the Sahel has been mentioned by Glausiusz (2003).

10. The design and layout parameters for the adsorption zone microcatchment combination can be found in Weber (1986, p. 106).

11. Guidelines or starting values for the spacing of infiltration structures are found in Valdéz and Abastidas (1993). Some recommended interditch distances are as follows:

Land slope (percent)	Interditch distance (m)
12	10.5
20	9.5
30	8.0
40	6.5
50	6.0

12. The placing of rocks along the uphill side of an infiltration bund comes from CEPIA (1986) without a technical explanation. The ability to provide a more porous soil surface may be one reason.

13. The study on ridge erosion effectiveness is reported in ICRAF (1996).

14. Cajete use in Mexico has been described by Mountjoy and Gliessman (1988). The description may be more of a subdivided or segmented infiltration ditch than pure cajetes with their unique separate or spaced-ditch design.

15. Terraces are usually assumed to be purely contour. This is a strong assumption. If one accepts the notion that terraces serve to reduce the slope (not in providing a flat working area), cross-slope layouts are possible and the design options expand. Under the slope reduction assumption, the five terrace options are shown in Figure N.4. The

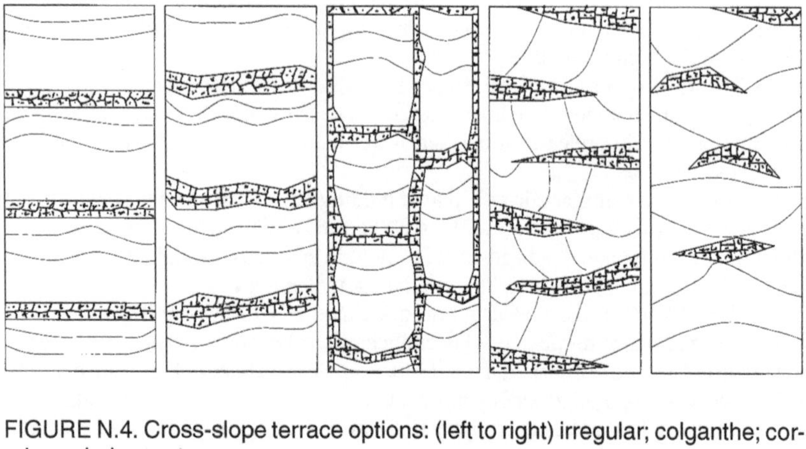

FIGURE N.4. Cross-slope terrace options: (left to right) irregular; colganthe; corral; graderia; contour.

top is the highest elevation. The contour lines indicate the resulting slope. These are from Schulte (1996).

16. Tree-stone terraces and similar Andean structures have been documented by Sorgedrager et al. (1991) and FAO (1994). The FAO study has maize, potatoes, and wheat benefiting from terraces and 20 or so purposely selected tree species. MacDonald (1998) studied the farm landscape and stone terraces in northern Pakistan.

17. Grass-covered terraces are found in the Philippines (Stark et al., 2000) and in Honduras (Walle and Sims, 1998). Napier grass is mentioned in the Philippine example with the positive insect DPCs re-cited by Altieri and Nicholls (2004, p. 95). The other candidate, vetiver, is discussed by Hengchaovanich (2003). Grasses have an advantage over hedges in that less labor is needed.

18. The progress terraces described are vegetative. The grass version adds a new upper contour as the soil collects against the grass strip. For the tree-based type, seasonally cut branches are piled on the uphill side of each contour row. This damming effect appends layers. Although no descriptive studies were located, a stone version is quite possible, a new layer of piled rocks being added as the contained soil rises.

19. The outcome tallies for the progressive terrace are from Banda et al. (1994).

20. Wojtkowski (2004, p. 125) presents three paddy-channel layouts:

1. linear, where each paddy borders and is fed off a single extended channel;
2. branching or nodal, where a group of paddies are fed through a water dividing node at the end of the channel; or
3. progressive fill, where, as a downhill series, the overflow from an uphill paddy fills the next lower paddy.

21. The fish-nutrient-crop nutrient cycle has been discussed by D'Silva and Maughan (1994). The broader role of ponds in a larger integrated landscape has been portrayed by Attenborough et al. (2000).

22. Downhill shoveling of the sludge from channel bottoms keeps it from washing back into the channel. This is also applied in infiltration ditches.

23. A brief description of Aztec chinampa farming is found in Ayres (2004).

Chapter 14

1. Small (2003) presents the case of multiple fathers.

2. For the Viking colonization of Greenland, the pastoral agriculture introduced had far less climatic resilience than the gathering strategy (fish, seals, etc.) of the native Inuit. The colony died off, unable to survive a series of bad years (McGovern, 2000).

3. The superiority of local agriculture is demonstrated (1) in the Amazon, where high populations were once sustained, higher than is currently possible (Mann, 2002b), and (2) from Fiji, where sustainability is better with traditional biodiversity than with simpler, imported practices (Siwatibau, 1984).

4. Without question, governmental policy is a factor in agricultural direction and outcome, for example, Altieri and Rojas (1999) and Ascher (2000). Due to uncertainties

and the amount of discussion needed to bring this topic to a meaningful conclusion, it is not broached here.

Some place gender as an influencing topic. The arguments for this are well founded: women may even farm better than men (Anon., 2004).

Irrespectively, in some cultures, men and women drive tractors, while in others, only men do this. In some cultures, men harvest crops, and in others, this is done by women. Similar statements can be made of most farm activities. It is quite possible that the gender-established duties as assigned by one culture may be countered by what is done in another. Also problematic is that this is a moving target, for example, Gurung and Lama (2004).

It is therefore difficult to universally state, without being contradicted through example, the sway that men and women, individually, not collectively, have on agroecology. Among the issues, summed linkages with agroecological principles and practices are troublesome, especially how gender influences agroecosystem design. Without a definitive resolution, or part thereof, a separate heading may be inaccurately counterproductive.

5. In this and other texts, the degree of land control is reflected in the terminology utilized. The word *farmer* denotes an individual with some control over land-use decisions and the motivation for a long-term planning horizon. The term *land user* tends to apply to those more at the destiny of others.

6. The multiparticipant taungya is a forestry fixture in land-scarce regions. Discussion of the multiparticipant taungya is found in Agyeman et al. (2003). Outside the taungya, there may be other multiparticipant opportunities. This can be in dividing along species lines. This can be part of a rotational scheme that crosses farm boundaries. For this, each farmer specializes in one crop (of a rotational sequence). Gains come (1) because each knows a lot about the one crop and (2) specialized farm machinery is used to full utility.

7. With multiparticipant systems, the labor and cost savings are far from uniform and the tradeoff not linear (where one loses as much as the another gains). Instead, there may exist a spacing and planting density where both participants stand to gain. Rudimentary analysis has shown that each may receive 90 percent of anticipated yields for their crop while the inputs for each are reduced 50 percent or more (Wojtkowski et al., 1988).

8. From this author's experience, islanders in the South Seas tend to take a utilitarian view. In contrast, the French are famous for their culinary take on food.

9. These two adages come from Tenney (1811, pp. 22 and 50).

10. As endnoted in Chapter 5, Apantaku (1999) found that home remedies were not all that effective, and a few were of value. As for practicality, this is addressed by Kaval (2004), who observed that on-farm alternative methods tend to be more profitable; when studied on research experimental stations, these exhibited less profitability.

11. The relationship between clean water and human health is beyond question. However, if a reference is needed, try Nyong and Kanaroglou (1999).

12. Many agroecosystem content studies list species that are purely ornamental (nice foliage and/or flowers, no useful output), for example, Raynor (1992) and Babu et al. (1992).

13. Household gardens, as an aesthetic feature, come in different forms: with herbs (Beale and Boswell, 1991; Wilson, 1994), with vegetables (MacDonald, 1994), or other plant combinations.

14. The aesthetic but functional (edible) landscape is well established. There are French (Racine et al., 1987), ancient Roman (Jashemski, 1979 and 1987), early English (McLean, 1980), Persian (Moynihan, 1979), and early Islamic (Husain, 2000) versions with varying edible components.

15. The practice of planting apples and pears along public ways is mentioned by Emerson (1846, p. 437). Although there are safety issues (slippery fruit on the road) and the inadvisability of attracting and putting wildlife in harm's way (as road kill), this practice can still enhance public spaces. There may be another dimension as public plantings can preserve and promote heritage fruit varieties.

16. The rudiments of aesthetics have been discussed by Litton et al. (1974), Porteous (1996), and Bell (1999). Despite discussions, beauty is still in the eye of the beholder and how best accomplished is an undefined process.

17. The economic confusion and mixed results have been summarized by Kaval (2004) and Lynam et al. (1986).

18. As one affirming conclusion in a long string of taungya studies, Esekhade et al. (2003) show a good economic aftermath.

19. For those who have studied the full farm, cost-oriented changeover, Jordan et al. (1996), Jordan et al. (1997a, b), and IACPA (1998) reported productively 10 to 15 percent less; gross margins (profits) declined over conventional farming by about 4 percent. Mäder et al. (2002) found yield 20 percent lower and gross margins about equal to a conventional approach. Charles and Clover (1993, p. 120) recount the most dramatic change, yield reductions of approximately 40 percent, increases in gross margins of 40 percent with organic versus conventional farming. Some of the yield losses are from utilizing less potent natural nutrient gains (e.g., rotations) instead of chemical fertilizers. Others are from subdividing fields with auxiliary travel corridors. The gains come from reduced agrochemical inputs.

Chapter 15

1. The de facto base may be looked at as the fundamental agroecosystem, equivalent to the fundamental niche in ecology, but on an ecosystem level. This is also a de facto base because it underwrites the purest version of the LER.

2. There is the view that complex ecology cannot be analytically rendered into a single equation but is the product of numerous mechanisms. How these interrelate determines the outcome (Grimm et al., 2005). Through vectors, this applies to agroecology.

Questions of terminology are appropriate here. The milder phrasing, development path or approach, is used, although these may be more fittingly called research vectors, development groupings, or development vectors. As the basis for research (as vectors), each path has direction and magnitude (in countering threats, etc.). As groupings, each of the paths or headings is subject to branching.

3. The Congo cassava example is from personal communications with those familiar with the problem. As with all things undertaken without understanding the potential effects, unanticipated side effects perverted a well-intended intervention. Indirect effects aside, a lot of good can result in more thought-out applications (DeVries and Toenniessen, 2001).

4. The argument made is that a reliance upon a single crop can cause poor health (Pollan, 2003). This presents a link between farm policy, overproduction, cheap maize, poor diet, obesity, and poor health in the United States. There is also belief that static diets, those with a long history of use, tend to be balanced whereas change upsets the nutritional balance (Johns, 1999). Adding to this is the argument that the consumption of multiple varieties is better for human health (Raloff, 2005).

5. The connection between maize and environmental shortfall is from Williams (2004). This need not be the case; industrial raw materials and animal feeds do not rely on consumer partiality. These can just as easily encompass a greater amount of varietal diversity without value loss.

6. Wilson (2003) found that disease- and pest-resistant cassava varieties are not those first chosen if natural processes are available to ensure minimal yields. This criteria may not apply where the output is sold and the decisions purely financial.

7. The risks of overextended genetics are hotly debated; some of this argument can be found in Pollack (2004) and Cox (2002).

8. The genetic and varietal categories could be considered a single class in that both seek suitable genetic material. However, genetics, especially through plants manufactured by inserting genes, some foreign to a species, unveils a far different course than the finding the DPC, labeling, and the putting to use the existing, nature-sanctioned, genetic resources.

9. The wheat varieties, as mentioned in an endnote in Chapter 6, are described by Percival (1922). The figure, over 600 varieties, is not set in stone and most likely changes (+ or –) yearly.

10. Harrison (1775, p. 212) provides a fairly robust description of how varieties of barley yield better and can counter diseases better if ecologically in tune with the site. This follows along with the idea that healthier plants are better at resisting diseases and insects. The large number of varieties found in some countries, and the potential for healthier plants through better site matching, is indicative of possibilities, for example, Pearce (2005).

11. The matching of maize varieties with site soils is from Castelán-Ortega et al. (2003).

12. The use of existing, within-species biodiversity to realize specified goals may be best accomplished utilizing the full varietal spectrum, for example, Goff and Salmeron (2004). Nature can be very imaginative, for example, in countering hungry insects and fast-spreading diseases. The internal mechanisms can be multitudinous (layered), targeted (effective against only categories of insects or diseases), and/or dynamic (springing into action only when imperilment strikes). It is a shame to not use, and lose through the extinction of promising varieties, what nature worked hard and long to achieve.

There are other provarietal arguments to be made. The alternative, genetic modification through the insertion of a species-foreign gene, although potent, is, as mentioned in the text, the equivalent of putting all eggs into one basket. A one-size-fits-all defense is generally not in nature's plan, whereas well-chosen genetic diversity, displaying many defensive fronts, many be an overall less risky approach. Risk reduction comes through an ability to counter evolving threats (e.g., insecticide-resistant insect species) and in being kinder to the overall environment (e.g., fewer slow-to-breakdown toxins remaining in the soil).

13. At times, the term *Kyusei nature farming* is found. It is not clear what this encompasses.

14. Some of the facilitative gains of in-soil microbes are documented by Setälä (2000) and Statälä and Huhta (1991). A wider discussion of mycorrhizae is found in Picone (2003). There is evidence that soil inoculation is better if the organisms have more material to work with, for example, microorganism-gentle fertilizers, mulch, and/or manure (Hussain et al., 1999).

15. Although some lament the lack of studies on microbial interactions, for example, Klironomos (2002), enough is known so that this becomes a viable alternative. There is the known ability of microbes to increase available water (e.g., Statälä and Huhta, 1991), to avert harmful diseases (e.g., Milius, 2003) and insects (e.g., Ruc, 1990), to make nutrients available (e.g., Setälä, 2000; Reynolds et al., 2003), and to counter climatic stresses. Pennisi (2003) offers a large window into this promising alternative.

16. The endophytic possibilities in countering climate have been summarized by Pennisi (2003). This stems, in part, from work by Redman et al. (2002) and others. Endophytes have also been mentioned in connection with disease resistance (Milius, 2003). The reduction in light needs is an indirect mechanism where sugars are borrowed from root-clinging fungi (Pennisi, 2004).

17. Somewhat speculative at this point, the scattered research (some referenced in the above notes) comes close to supporting a fully microbial-based approach where agroecosystems benefit on many counts from the unseen labor from millions of microorganisms.

18. Some facilitative gains come from earthworms that improve general soil conditions (e.g., Hulugalle and Ezumah, 1991; Hauser et al., 1997) and from ants and termites that do much the same (e.g., Risch and Carrol, 1982; Stanton and Young, 1999; Molles, 1999, p. 338; Mando and Van Rheenen, 1998).

19. Some definitions of agrobiodiversity are quite broad. These can include all organisms, both flora and fauna, that influence crop growth (e.g., McNeely and Scherr, 2003) or all useful plants (e.g., Brookfield, 2001). As used here, agrobiodiversity is only those organisms, mainly plants, that provide a useful harvest. This follows from Almekinders and de Boef (1999).

20. Wojtkowski (2002, p. 307) has suggested four categories or directions to general biodiversity:

1. *Coverage:* Many species with no particular object other than to capture those environmental gains that come with more plants per area
2. *Intensity:* Adding biodiversity with the express purpose of furthering the productive potential (the added species are facilitative or provide a useful output)

3. *Eloquence:* Adding locational complementarity (plant-plant and/or topographical) such that more is better and both environmental or productive gains are solidified
4. *Holism:* When biodiversity benefits production as well as broader landscape objectives (such as promoting one or more native wildlife species)

The topic of direction in farm- or landscape-wide biodiversity is not carried beyond this brief look.

21. Numerous examples of microlocational placements are given in Chapter 13. These exploit the areas around and within land-modification features such as farmed channel banks where different crops are located set distances from the standing water.

Restating the premise using essential resources. If a site is too wet for normal wheat, the varietal approach would seek a moisture-tolerant variety. Under a location approach, a moisture-loving crop, such as rice, is substituted. This concept applies to shortfalls in essential mineral resources.

22. Scattering in East Africa is from den Biggelaar (1996, p. 46), in Peru from Zimmerer (1999). As a landscape topic, scattering and the ramifications of a plot or site-first approach are discussed in greater depth in Chapter 17 under freeform landscapes.

23. Ex-farm inputs are not directly from organic (vegetative) sources. Some commercial insecticides derive directly or indirectly from plants. Pyrethrum, from the dried flower of *Chrysanthemum cinerariaefolium,* is one case. Synthetic fertilizers, but not most manures, are included as ex-farm inputs.

24. Herbicide resistance in weeds and the negative consequences are from Bradford (2004).

25. The manures and aphids case is from Morales et al. (2001), upheld by Brown and Tworkoski (2004). In comparing fertilizer application against fallow usage, Njoku et al. (2004) attributed the productive gains to better physical soil characteristics and improved micronutrients. These results may well be an offshoot of the ecosystem or environmental setting subsequently engendered.

26. The notion that composing is a good thing has been around, with excellent results, for a long time. The idea of underlying, mini-ecosystem is more of a logical offshoot.

27. The fertilizer shock is not an empty assertion (Pennisi, 2004).

28. Given a three-level approach, that of a primary, secondary, and trenary ordering, there are 1,331 (11 × 11 × 11) possibilities with only three core elements; more if permutations are added; even more if additional levels are added. And this only starts the process.

29. The variations on alternative agriculture presented derive from a scan of the literature, but mostly from the list supplied by Gold (1994). Not included are exotic, often nonfarm, methods, for example, hydroponics, greenhouses, etc., where practice is less allied with the land.

30. Case studies on organic farms show that much variation exists (NRC, 1981), cementing the idea that rankings are a burdensome undertaking. The conclusion from all of this, if any, is that there are many ways to reach organic objectives. Un-

derlying much of this is understandability (discussed in Chapter 6) and having practical operation guidelines that help one thread through the complexity of options (in this case, the core approaches) to reach stated goals.

31. Permaculture is exhaustively explained by Mollison (1992). The complexity of the argument, along with peripheral thoughts on a sustainable lifestyle, shows the difficulty in precisely, or even roughly, determining the underlying, core approach mix.

32. Natural ecosystem mimicry has been advocated by Lefroy et al. (1999), Jackson (2002), and others. Although in rudimentary form, the concept can be progressed in any number of directions. An agroecosystem could be measured by comparing the types of organisms hosted by a natural ecosystem against that of the mimic. Also possible is to gauge the depth of ecological interactions supported by the agroecosystem imitator and those of an equivalent natural system.

33. Mimicry in a prairie landscape can bring farm ecology closer to natural state through simple replacement, as has been described by Jackson (2002). Robinson (2005) discusses how a forest tree plantation can be modified, through mimicry-promoting biodiversity, to favor wild turkeys as a secondary, semihusbandry, output.

34. In Europe, the now extinct mega-fauna are elephants, rhinoceros, and aurocks (an early form of cattle). Together with large populations of deer, horses, boars, bison, and beavers, these would have had a profound ecosystem-altering effect (see the next endnote).

35. In the savannas of Africa, grazing animals can eat up to 80 percent of the aboveground biomass (Vera, 2000, pp. 53 and 91). In Europe, portions of this landscape type still survive (Rackham, 1998; Joffee et al., 1999) as do now-rare plant species that once depended on now-extinct, animal-driven dynamism (e.g., Charles, 1997).

In North America, there are similar indications of hungry grazers and extensive savannas (e.g., Mann, 2002a). As with the European situation, plants still subsist that once depended upon large animals for seed spread (Barlow, 2001). Some altering forces remain. Fire, now missing in eastern North America, is still an active influence in western regions. As with Europe (Vera, 2000, p. 72), human-induced change in landscape types occurred in North America, for example, Foster and Zebryk (1993).

The record for Asia paralleled that of Europe (Elvin, 2004). In other parts of the world, the past is less known. Landscapes could have bounced between gap and patch types depending upon the level of human activity, for example, Mann (2002b); Bush and Colinvaux (1994); Stokstad, E. (2003); Willis et al. (2004).

Chapter 16

1. The market summarily lost was for bananas, the site was in Costa Rica, and the study author Soluri (2001).

2. Although points can be raised, much of the design discussion continues on without a satisfying and definitive outcome. This reinforces the view that agroecology is more an art than a science.

3. Work has been done on formal design methodologies, especially in defining the steps for sorting through or enhancing the numerous agroforestry options leading to farm-suitable systems. The steps are prediagnostic (defining goals) → diagnostic (matching the goals with an agrotechnology) → modification (changes in the standard agrotechnology) → planification (on-farm testing) → implementation. Each has feedback loops if the proposed system fails to live up to expectations. References for this are in Raintree (1983, 1990). The processes presented in this chapter are less formal, more generic, and bear little resemblance to the efforts of Raintree (1983, 1990).

4. The groupings here (adding productivity, trimming costs, and reducing risk threats) are a distillation of work by Gliessman (1998, p. 304). Gliessman separates ecosystem change into three orders of magnitude: (1) efficiency gains, (2) input substitution, and (3) a total system redesign.

5. Hydroponics are where crops are grown sans natural soils, sunlight, and rainfall. All inputs are artificially provided. Although an extremely expensive way to grow things, this finds application with very high-value crops (e.g., illegal drugs) that must be hidden. Decorative plants inside buildings are a lesser example, as they are grown in real soil.

6. Agroforests are discussed in Chapter 10. In the most hands-off versions, nature provides all inputs. Although nature does much, harvesting and some minor management must be applied to obtain the fruits of the system. The penalty for the free nature-provided services is having to accept nature-dictated outputs in nature-dictated quantities.

7. This argument goes back to the concept of marginal gains (see Chapters 2 and 3) and the comparatively large amounts of inputs that must go into obtaining very high per area rates of productivity.

8. Some of the argument for retaining weeds, and the benefits provided, are found in Altieri and Nicholls (2004, p. 47). In Chapter 5, this is provided as an example of a miniecosystem friendly to insect-eating insects.

9. The cataloguing of agrotechnologies is comparatively recent and the list is still expanding. Traditional and lost practices, some under study, for example, Mann (2002b), may prove unique enough to be added to this list. There might also be changes as this concept matures, as with a subsequent dividing of categories.

10. The process of selecting and customizing an agrotechnology has been formalized by Wojtkowski (2004, p. 51). This is a four-phase process:

1. *Draft,* where a need (site and socioeconomic) is identified and appropriate options proposed
2. *Agrotech,* where a candidate agrotechnology emerges as the best
3. *Refinement,* where the agrotechnology is implemented but needs fine-tuning
4. *Expansion,* where the chosen system broadens in scope (DAPs) and receives wider cognizance

11. The list of maize-compatible agrotechnologies assumes that maize is light demanding, an assertion by Earley et al. (1967) and Early et al. (1966).

Chapter 17

1. Much of that presented here derives from Wojtkowski (2004). If one needs practical justification for a landscape approach, one need not look beyond De Loach (1970). This is a key, well-referenced study with a finding of reduced insect pests through landscape redesign.

2. Cases where early ownership patterns trace to an early period are not uncommon. Paoletti (2001) found fields that have remained in continued use without substantial change from the time of the ancient Romans.

3. The difficulties in formally determining land-use intensity have been discussed by Shrair (2000). Among the problems are in determining the degree of intensity (labor) per area and in assigning classifications to features such as auxiliary systems and forest fragments. This is, in part, attributed to lack of knowledge on local (agro)ecosystems, for example, Fairhead and Leach (1998).

4. Marginal gains in connection to plot inputs and outputs figure prominently in previous discussions (Chapters 2 and 3, specifically Figures 2.1, 2.2, and 3.3). The theory here has scarce inputs allocated first to those plots that will benefit most, that is, the added resources overcome the limiting resource at the least cost. If inputs are still available after addressing the easy-to-rectify plots, inputs go to the less needy (less resource constrained) plots.

5. In a quick review of agricultural economics, the operative equation is, for a three-plot example,

$$(\text{outputs/inputs})_{(\text{plot } 1)} = (\text{outputs/inputs})_{(\text{plot } 2)} = (\text{outputs/inputs})_{(\text{plot } 3)} \qquad (\text{N.4})$$

The outputs and inputs are normally expressed in monetary terms.

The supposition is that the first inputs should give a large boost in productivity. Later inputs are not as yielding, that is, the input gains recede as lower per unit output is obtained from each added input. Eventually, a point is reached where the value of inputs exceeds the value of outputs (when this happens, the output-input ratio, in monetary terms, falls below one; this is also referred to as negative marginal gains).

Commercial farmers should not seek negative marginal gains in equating plots. Subsistence farmers (those facing starvation if the land does not yield) or hobbyists (those deriving pleasure from backyard gardens) are generally exempt from these economic rules (that of equating interplot marginal gains and/or, in seeking very high yields, that the output/input monetary ratio does not fall below one).

6. Ranking, as to DPCs, is discussed in Chapter 6. Ranking can also be used with agrosystems concerns and threats. Although the basic methodologies are in

place, a large conceptional gap remains on how these apply to DAPs and the locational possibilities. This limits discussion to a few illustrative examples.

7. A large body of literature fronts the agricultural-natural ecosystem dilemma when native plant and animal populations become isolated from each other. There is far less on how this change influences nearby agriculture.

In an intensity reverting landscape, the need to restock flora, both micro and macro, within the fragments is problematic (Matlack, 1994 and Peterken, 1993). Well-endowed, ecologically fully functioning forest fragments are an agroecological plus as reservoirs/habitats (especially when incorporation of bottom-up ecosystem governance is part of the ecological scheme—see Chapter 3). Strategically placed corridors greatly help with this.

8. The ecological reasons for a system of interlocking corridors and transfer points are attested to in previous chapters, that is, Chapter 2, under implications for ecosystem governance, and Chapter 12, under corridors.

9. There is a line of development that advocates the use of ecological ecotones (the zones of transition from one ecosystem to another), the natural biodiversity contained, and the opportunities for agroecological exploitation. This topic, as applied more to hunter-gatherer situations, is introduced by Turner et al. (2003). Except the occasional and brief mention (e.g., under agroforests), the possibilities are not delved into in this text.

10. Plot scattering in Peru dates to the Inca period or before. This has been described by Zimmerer (1999). The East Africa case, from Rwanda, is by den Biggelaar (1996, p. 46). The purpose may be in not having all eggs in the same basket, that is, all crops exposed to the same risk threats and threat levels. As previously mentioned, there are also productive advantages in being able to place crops on a close to ideal site.

11. Equitable scattering as an incentive to cooperate on joint projects is mentioned by Trawick (2002). In this account, irrigation maintenance chores are embraced by an Andean community because each has a stake in (and scattered plots throughout) the whole system. Other accounts support this. In India, change to a more uniform, less scattered, ownership pattern reduced the level of community involvement in a common irrigation system (Gunnell and Kishnamurthy, 2003).

12. The crop-tree dichotomy (as in farm forestry) has been referred to as college detachment (Wojtkowski, 2004, p. 284) in that programs of higher education are structured in this way (with separate departments or schools for forestry and for agriculture). This extends to research and extension groups and on to farm layouts. There is no other rationale for this.

13. Plot bounds can be clear or, lacking definition, fuzzy. Most are clear, but fuzzy plot edges have been documented. In studies of Central American agroforests, Méndez et al. (2001) and Gamero et al. (1996) noted fuzzy plot constructs. With crop plots, this can occur through boundary encroachment, for example, Padoch and de Jong (1987). Overlapping (i.e., fuzzy) boundaries as an ecological strategy are discussed by Wojtkowski (2004, p. 189).

14. Using the four criteria listed, sixteen landscape formulations are possible ($2 \times 2 \times 2 \times 2$). Avoiding a full listing, these derive as (farm forestry-agroforestry) × (geometric-terrain) × (ordered-disarrayed) × (clear-fuzzy plot fringes). Wojtkowski

(2004, p. 280) has carried this further with many more variations on these and similar themes. Some arise from subdividing the four categories.

15. Agrotechnologies need not be as set in place as previous chapters indicate. Wojtkowski (2002, p. 152) breaks this into three agrotechnological groupings:

1. *divisional* (set agrotechnologies with predetermined DAPs),
2. *formulation* (visually similar, but with a wider DAP range), or
3. *placement* (as described under the freeform landscape heading).

A freeform, nonagrotechnological landscape is a step beyond the agrotechnological variations, but, despite the difference in approach, the outcomes can coincide. The placing of crops among widely scattered trees can be indistinguishable from a disarrayed parkland system. An examination of the full landscape often reveals the underlying approach.

16. As a conceptional thread that crosses different practices, operational guidelines, i.e., understandability (first mentioned in Chapter 6), are a prerequisite for simple and for biodiverse systems. Another thread is the core development path (as discussed in Chapter 15) where the rules for freeform landscape, as presented here, enshrine a plot-first, locational approach.

17. In a study of Rwanda farming, den Biggelaar (1996, p. 69) found crop location is based on

1. whether soils or nearby crops are damaged (the underlying and not to be ignored do-no-harm admonition, see Chapter 4);
2. where they fit within farming priorities;
3. where they grow best; and
4. their potential uses (e.g., food crop, medicinal, etc.).

These have been distilled to the two factors presented in the text.

18. As stated, Gliessman (1998, p. 304) has presented change in terms of (1) efficiency gains, (2) substitution, and (3) redesign. The last, redesign, drives major landscape change where farmers shift cropping systems (a cost-oriented design to one that is more revenue oriented or a change in the level of agrobiodiversity).

19. Land resorting, when cultivated lands become natural areas, is often looked at as a recent occurrence. Not new, for thousands of years the ebbs and flows of civilizations and populations have ecologically changed regions (Willis et al., 2004). Present changes have policy implications. One question: Are (1) smaller areas of high-intensity farming with accompanying untouched nature reserves better than (2) large areas of low-intensity agriculture? Green et al. (2005) (and subsequent commentary by Vandermeer and Perfecto [2005]) suggests that the best way to encourage nature is the high-intensity option.

Digging deeper into this, the tenor can lie with (1) static ecosystems that favor certain natural plant and animal species versus (2) dynamic versions that encourage more transitional forms of flora and fauna. The latter includes event-driven ecosystem transfigurations, those that mimic nature-induced calamities, for example, wildfires, floods, high winds, and, in past epochs, herds of hungry grazers (Vera,

2000, p. 55). Because of strong equivalency, that is, high-intensity ecosystems tend to be static; this suggests a mix of types may be best. In forestry, the analogous debate is by Attiwill (1994) and Wohlgemuth et al. (2002).

Chapter 18

1. The argument for dynamism in natural ecosystem is made in Chapter 15 under mimicry. The 80 percent figure on biomass consumption is from Vera (2000, pp. 53 and 91).

2. The assertion that agroecology is a vehicle to test ecological theory comes from Vandermeer (1989, p. x).

References

Abate, T., van Huis, A., and Ampofo, J.K.O. (2000). Pest management strategies in traditional agriculture: An African perspective. *Annual Review of Entomology 45*(1):631-660.

Adams, G. (1994). *Birdscaping Your Garden.* Rodale Press, Emmaus PA.

Agelet, A., Bonet, M.A., and Vallès, J. (2000). Homegardens and the role as a main source of medicinal plants in mountain regions of Catalonia (Iberian Peninsula). *Economic Botany 54*(3):295-309.

Agyeman, V.K., Marfo, K.A., Kasanga, K.R., Danso, E., Asare, A.B., Yeboah, O.M., and Agyeman, F. (2003). Revising the *taungya* plantation system: New revenue-sharing proposals from Ghana. *Unasylva 212*(54):40-43.

Aiyelaagbe, I.O.O. (1992). Fruit crops: Are they relevant for agroforestry development? *Agroforestry Today 4*(3):12-13.

Akobundu, O. (1987). *Weed Science in the Tropics: Principles and Practices.* John Wiley and Sons, New York.

Akyeampong, E., Hitimana, L., Franzel, S., and Munyemana, P.C. (1995). The agronomic and economic performance of banana, bean and tree intercropping in the highlands of Burundi: An intern assessment. *Agroforestry Systems 31:*199-210.

Alhamid, S.K., Håkasson, B., and Gustafsson, M. (2003). Economic viability of the traditional farming system in the Ghouta Oasis of Damascus Syria. *American Journal of Alternative Agriculture 18*(4):196-205.

Alomar, O., Goula, M., and Albajes, R. (2002). Colonization of tomato fields by predatory mirid bugs (Hemiptera: *Heteroptera*) in northern Spain. *Agriculture Ecosystems and Environment 89:*105-115.

Altieri, M.A. (1994). *Biodiversity and Pest Management in Agroecosystems.* Food Products Press, Binghamton, NY.

Altieri, M.A. (1995). *Agroecology: The Science of Sustainable Agriculture.* Westview Press, Boulder, CO.

Altieri, M.A. and Nicholls, C.I. (2004). *Biodiversity and Pest Management in Agroecosystems,* Second Edition. Food Products Press, Binghamton NY.

Altieri, M.A. and Rojas, A. (1999). Ecological impacts of Chile's neoliberal policies with special emphasis on agroecosystems. *Environment, Development and Sustainability 1:*55-72.

Altieri, M.A. and Trujillo, J. (1987). The agroecology of corn production in Tlaxcala, Mexico. *Human Ecology 15*(2):189-220.

Aman, R. (1998). Rare and wild fruits of Peninsular Malaysia and their potential uses. In Nair, M.N.B., Sahri, M.H., and Asharri, Z. (Eds.), *Sustainable Manage-*

Introduction to Agroecology: Principles and Practices

doi:10.1300/5553_20

ment of Non-Wood Forest Products (pp. 204-211). Proceedings of an International Workshop Universiti Putra Malaysia, October 14-17, 1997, Universiti Putra Malaysia Press, Serdang.

Anderson, J. (1800). *Essays Relating to Agriculture and Rural Affairs,* Vol. 1, C.G. and J. Robinson, London.

Anderson, R.L. (2003). An ecological approach to strengthen weed management in the semiarid Great Plains. *Advances in Agronomy 80:*33-62.

Anderson, R.L. (2005). A multi-tactic approach to manage weed population dynamics in crop rotations. *Agronomy Journal 97*(6): 1579-1583.

Ankens, T., Walther, U., Rishner, W., Weisskopf, P., Nievergelt, J., Stamp, P., Schmid, O., and Mäder, P. (2003). Nitrogen dynamics, nitrate leaching and plant development using mouldboard ploughing and no-tillage cultivation techniques. In *Proceedings of the Conference of the International Soil and Tillage Research Society* (pp. 2-6). July 13-19, Brisbane, Australia.

Anon. (1843). *The Gardener's Manual.* United Society [Shakers], New Lebanon, NY.

Anon. (1872). Gardeners' Journal [Manuscript]. United Society [Shakers], New Lebanon, NY.

Anon. (2002). Bee off with you. *The Economist 365*(8297):78.

Anon. (2003). Alley cropping in Guam; Hedgerows benefit crops and soil nitrogen. *Inside Agroforestry* (Summer/Fall):7-8.

Anon. (2004). The new land girls. *The Economist 373*:32.

Apantaku, S.O. (1999). Indigenous technical knowledge and use of forest plant products for sustainable control of crop pests in Ogun State, Nigeria. *Journal of Sustainable Agriculture 14*(2/3):5-13.

Ascher, W. (2000). Understanding why governments in developing countries waste natural resources. *Environment* (March):8-18.

Ashford, D.L. and Reeves, D.W. (2003). Use of a mechanized roller-crimper as an alternative kill method for cover crops. *American Journal of Alternative Agriculture 18*(1):37-45.

Ashley, M.D. (1986). *Agroforestry in Haiti.* University of Maine Press, Orono, ME.

Ataroff, M. and Rada, F. (2000). Deforestation impact on water dynamics in a Venezuelan Andean cloud forest. *Ambio 29*(7):440-444.

Atlin, G.N. and Frey, K.J. (1989). Breeding crop varieties for low input agriculture. *American Journal of Alternative Agriculture 4*(2):53-58.

Attenborough, D., Hogarth, J., and Saito, S. (2000). *Japan's Secret Garden.* NOVA Broadcast No. 2716, PBS, air date December 19.

Attiwill, P.M. (1994). The disturbance of forest ecosystems: The ecological basis for conservative management. *Forest Ecology and Management 63:*247-300.

Aumeeruddy, Y. and Sansonnens, B. (1994). Shifting from simple to complex agroforestry systems: An example for buffer zone management from Kerinci (Sumatra, Indonesia). *Agroforestry Systems 28:*113-141.

Ayak, E.T. (1997). Adoption of agroforestry technologies: The case of live hedges in the central plateau of Burkina Faso. *Agricultural Systems 54*(2):189-206.

Ayres, A. (2004). Xochimilco's sunken treasure. *New Scientist 182*:50-51.

Babbar, L.I. and Zak, D.R. (1994). Nitrogen cycling in coffee agroecosystems: Net N mineralization and nitrification in the presence and absence of shaded trees. *Agriculture, Ecosystems and Environment 48:*107-113.

Babu, K.S., Jose, D., and Gokulapalan, C. (1992). Species diversity in a Kerala home garden. *Agroforestry Today 4*(3):15.

Bach, C.E. (1988). Effects of host plant patch size on herbivore density patterns. *Ecology 69*(4):1090-1102.

Bäckman, J.-P. C. and Tianen, J. (2002). Habitat quality of field margins in a Finnish farmland area for bumblebees (Hymenoptera: Bombus and Psithyrus). *Agriculture Ecosystems and Environment 89:*53-68.

Bainbridge, D.A. (1985). The rise of agriculture: A new perspective. *Ambio 14*(3): 148-151.

Bainbridge, D.A. (1988). The oaks: A neglected multiuse treecrop. In Allen, P. and van Dussen, D. (Eds.), *Global Perspectives on Agroecology and Sustainable Agricultural Systems: Proceedings of the Sixth International Scientific Conference of the International Federation of Organic Agriculture Movements* (pp. 601-657). The Ecology Program, Santa Cruz, CA.

Banda, A.Z., Meghembe, J.A., Ngugi, D.N., and Chome, V.A. (1994). Effect of intercropping maize and closely spaced *Leucaena* hedgerows on soil conservation and maize yield on a steep slope at Ntcheu, Malawi. *Agroforestry Systems 27*:12-22.

Banful, B., Dzietror, A., Ofori, I., and Hemeng, O.B. (2000). Yield of plantain alley cropping with *Leucaena leucocephala* and *Flemingia macrophylla* in Kumasi, Ghana. *Agroforestry Systems 49:*189-199.

Barker, T.L. and Francis, C.A. (1986). Agronomy of multiple cropping. In Francis, C.A. (Ed.), *Multiple Cropping Systems* (pp. 161-182). Macmillian Publishing Co., New York.

Barlow, C. (2001). Ghost stories from the ice age. *Natural History 110*(7):62-67.

Baskin, Y. (1994). Ecologists dare to ask: How much does diversity matter? *Science 264*:202-203.

Beale, G. and Boswell, M.R. (1991). *The Earth Shall Blossom.* The Countryman Press, Woodstock, VT.

Beard, J. (1997). Marauding elephants feel the heat. *The New Scientist 154*:25.

Beckerman, S. (1983). Barí swidden garden: Crop segregation patterns. *Human Ecology 11*(1):85-101.

Beckerman, S. (1984). A note on ringed fields. *Human Ecology 12*(2):203-206.

Beer, J. (1987). Advantages, disadvantages and desirable characteristics of shade trees for coffee, cacao and tea. *Agroforestry Systems 5*:3-13.

Beets, W.C. (1982). *Multiple Cropping and Tropical Farming Systems.* Westview Press, Boulder, CO.

Bell, S. (1999). *Landscape Patterns, Perceptions and Process.* Routledge, New York.

Bellon, M.R. (1996). The dynamics of crop intraspecific diversity: A conceptional framework at the farmer level. *Economic Botany 50*(1):26-39.

Belsky, A.J. (1992). Effects of trees on nutrient quality of understory gramineous forage in tropical savannas. *Tropical Grassland 26:*12-20.

Bettiol, W. (1999). Effectiveness of cow's milk against zucchini squash powdery mildew *(Sphaerotheca fuliginea)* in greenhouse conditions. *Crop Protection 18*(8):489-492.

Biazzo, J. and Masivnas, J.B. (2000). The use of living mulches for weed management in hot pepper and okra. *Journal of Sustainable Agriculture 16*(1):59-79.

Biglow, S.W., Ewel, J.J., and Hagger, J.P. (2004). Enhancing nutrient retention in tropical tree plantation: No short cuts. *Ecological Applications 14*(1):28-40.

Blanford, H.R. (1925). Regeneration with the assistance of taungya in Burma. *The Indian Forest Records 11*(3):1-39.

Bohard G.E. (1970). Management of habitat for wild bees. In *Proceedings of the Tall Timbers Conference on Ecological Animal Control by Habitat Management,* Volume 2, February 25-27 (pp. 185-195). Tall Timbers Research Station, Tallahassee, FL.

Bolívar, D., Ibrahim, M., Kass, D., Jiménez, F., and Camargo, J.C. (1999). Productividad y calidad forrajera de Brachiaria humidicola en monocultivo y en asocio con Acacia mangium en un suelo ácido en el trópico húmedo. *Agroforestería en las Americas 6*(23):48-50.

Bradford, S. (2004). Argentina's bitter harvest. *New Scientist 182*: 41-43.

Brenner, A.J. (1996). Microclimatic modifications in agroforestry. In Ong, C.K. and Huxley, P. (Eds.), *Tree-Crop Interactions: A Physiological Approach* (pp. 159-188). CAB International, Wallingford, UK.

Brewbaker, J.L. (1989). Nitrogen-fixing tree species for fodder and browse in Africa. In Kang, B.T. and Reynolds, L. (Eds.), *Alley Farming in the Humid and Subhumid Tropics* (pp. 55-62). Proceedings of an International Workshop, Ibadan, Nigeria, March 10-14, 1986.

Brookfield, H. (2001). *Exploring Agrobiodiversity.* Columbia University Press, New York.

Brower, K. (2004). The proof is in the pellet. *Audubon 106*:78-83.

Brown, B.J. and Marten, G.G. (1986). The ecology of traditional pest management in Southeast Asia. In Marten, G.G. (Ed.), *Traditional Agriculture in Southeast Asia: A Human Ecology Perspective* (pp. 241-272). Westview Press, Boulder, CO.

Brown, J.C. (1878). *Pine Plantations on the Sand Waste of France.* Oliver and Boyd, Edinburgh, UK.

Brown, M.W. and Tworkoski, T. (2004). Pest management benefits of compost much in apple orchards. *Agriculture, Ecosystems and Environment 103*(3):465-472.

Browne, P.J. (1832). *The Silva Americana.* William Hyde and Co., Boston.

Bruce, J.W. (1991). *Securidad Alimentaria Familiar y Silvicultura: Análisis de los Problemas Socioeconómicos.* FAO, Rome.

Budelman, A. (1988). The performance of leaf mulches of *Leucaena leucoephala, Flemingia macrophylla* and *Gliricidia septium* in weed control. *Agroforestry Systems 6:*137-146.

Buresh, R.J. (1999). Phosphorus management in tropical agroforestry: Current knowledge and research challenges. *Agroforestry Forum 9*(4):61-65.

Bush, M.B. and Colinvaux, P.A. (1994). Tropical forest disturbance: Paleocological records from Darien, Panama. *Ecology 75*(6):1761-1768.

Caborn, J.M. (1965). *Shelterbelts and Windbreaks.* Faber and Faber, London.

Calanca, P., Neftel, A., and Fuhrer, J. (2001). N management of European grasslands: Can the exchange of N species be influenced at the operational level? *The Scientific World Journal 1*(2):652-657.

Callaway, R.M. (1992). Effects of shrubs on recruitment of *Quercus douglasii* and *Quercus lobata* in California. *Ecology 73*(6):2118-2128.

Capistrano, A.D. and Marten, G.G. (1986). Agriculture in Southeast Asia. In Marten, G.G. (Ed.), *Traditional Agriculture in Southeast Asia: A Human Ecology Perspective* (pp. 6-19). Westview Press, Boulder, CO.

Caramori, P.H., Androcioli Filho, A., and Leal, A.C. (1996). Coffee shade with *Minosa scabrella* Benth. for frost protection in southern Brazil. *Agroforestry Systems 33:*205-214.

Carrier, L. (1923). *The Beginning of Agriculture in America.* McGraw-Hill Book Co., New York.

Casper, B.B., Schenk, H.J., and Jackson, R.B. (2003). Defining a plant's belowground zone of influence. *Ecology 84*(9):2313-2321.

Castelán-Ortega, O., González E.C., Arriaga J.C., and Chávez M.C. (2003). Mexico. In Brookfield, H., Parsons, H., and Brookfield, M. (Eds.), *Agrodiversity: Learning from Farmers Across The World* (pp. 249-269). United Nations University Press, Tokyo.

CEPIA (1986). *Technologias Campesinas de los Andes del Sur.* Centro el Canelo de Nos, Perú.

Charles, D. (1997). Can bison replace engines of war? *New Scientist 154:*10.

Charles (Prince of Wales) and Clover, C. (1993). *Highgrove: An Experiment in Organic Gardening and Farming.* Simon and Schuster, New York.

Chaucer, G. (1382). Parliament of Foules.

Chou, C.H. (1992). Allelopathy in relation to agricultural productivity in Taiwan: Problems and prospects. In Rizvi, J.H. and Rizvi, V. (Eds.), *Allelopathy: Basic and Applied Aspects* (pp. 179-203). Chapman and Hall, London.

Christanty, L., Kimmins, J.P., and Mailly, D. (1997). "Without bamboo, the land dies": A conceptual model of the biogeochemical role of bamboo in an Indonesian agroforestry system. *Forest Ecology and Management 91:*83-91.

Cieugh, H.A., Miller, J.M., and Böhm, M. (1998). Direct mechanical effects of wind on crops. *Agroforestry Systems 41:*85-112.

Clason, T. and Robinson, J.L. (2000). From a pine forest to a silviculture system. *Agroforestry Notes 18:*2.

Cleveland C.C., Townsend, A.R., Schmidt, S.K., and Constance, B.C. (2003). Soil microbial dynamics and biogeochemistry in tropical forests and pastures of southwestern Costa Rica. *Ecological Applications 13*(2):314-326.

Cohn, J.P. (2005). A new breed of ranchers. *Americas 57*: 6-13.

Colbach, N., Licas, P., Cavelier, N., and Cavelier, A. (1997). Influence of cropping system on short eyespot in winter wheat. *Crop Protection 16*(5):415-422.

Colletti, J. and Tyndall, J. (2002). Shelterbelts: An answer to growing odor concerns. *Inside Agroforestry* (Spring):3.

Coomes, O.T. and Ban, N. (2004). Cultivated plant species diversity in home gardens of an Amazonian peasant village in northeastern Peru. *Economic Botany 58*(3):420-434.

Cooper, P.J.M., Leakey, R.R.B., Rao, M.R., and Reynolds, L. (1996). Agroforestry and the mitigation of land degradation in the humid and sub-humid tropics of Africa. *Experimental Agriculture 32:*235-290.

Cox, T.S. (2002). The mirage of genetic engineering. *American Journal of Alternative Agriculture 17*(1):41-43.

Creamer, N.G. and Dabney, S.M. (2002). Killing cover crops mechanically: Review of recent literature and assessment of new research results. *American Journal of Alternative Agriculture 17*(1):32-40.

Daane, K.M. and Williams, L.E. (2003). Manipulating vineyard irrigation amounts to reduce insect pest damage. *Ecological Applications 13*(6):1560-1660.

Dagar, J.C., Singh, N.T., and Singh, G. (1995). Agroforestry options for degraded and problematic soils of India. In Singh, P., Pathak, P.S., and Roy, M.M. (Eds.), *Agroforestry Systems for Sustainable Land Use* (pp. 96-120). Science Publishers, Enfield, NH.

Dai, Q., Fletcher, J.J., and Lee, J.G. (1993). Incorporating stochastic variables in crop response models: Implications for fertilizer decisions. *American Journal of Agricultural Economics 75:*377-386.

Dapaab, H.K., Asafu-Agyei, J.N., Ennin, S.A., and Yamoah, C. (2003). Yield stability of cassava, maize, soya bean and cowpea intercrop. *Journal of Agricultural Science 140:*73-82.

Davidson, D.W., Cook, S.C., Snelling, R.R., and Chua, T.H. (2003). Explaining the abundance of ants in lowland tropical rainforest canopies. *Science 300:*969-972.

Davis, J.H.C. and Garcia, S. (1983). Competitive ability and growth habit of indeterminate beans and maize for intercropping. *Field Crops Research 6:*59-75.

Davis, J.H.C., Roman, A., and Garcia, S. (1987). The effects of plant arrangement and density in intercropped beans *(Phaseolus vulgaris)* and maize: II. Comparison of relay intercropping and simultaneous planting. *Field Crops Research 16:*117-126.

Day, S. (2001). Fight the blight. *New Scientist 171:*36-39.

de Foresta, F., Basri, A., and Wiyono (1994). A very intimate agroforestry association cassava and improved homegardens: The Mukibat technique. *Agroforestry Today* 6(1):12-14.

De Loach, C.J. (1970). The effects of habitat diversity on predation. In *Proceedings of the Tall Timbers Conference on Ecological Animal Control by Habitat Management,* Volume 2, February 25-27, 1970 (pp. 223-241). Tall Timbers Research Station, Tallahassee, FL.

de Oliveira, M.A. and de Oliveira, L. (1994). *The Cork.* ICEP, Portugal.

den Biggelaar, C. (1996). *Farmer Experimentation and Innovation: A Case study of Knowledge Generation Process in Agroforestry Systems in Rwanda.* Community Forestry Case Study Series 12. FAO, Rome.

Dennis, C.T. and Miller, L. (1997). *If You Like My Apples.* Avery Publishing Group, Garden City Park, NY.

DeVries, J. and Toenniessen, G. (2001). *Securing the Harvest: Biotechnology, Breeding and Seed System for African Crops.* CABI Publishing, Wallingford, UK.

Dix, M.E., Hill, D.B., Buck, L.E., and Rietveld, W.J. (1997). Forest farming: An agroforestry practice. *Agroforestry Notes* 7(November):1-4.

Donald, C.M. (1963). Competition among crop and pasture plants. In Norman, A.G. (Ed.), *Advances in Agronomy*, Volume 15 (pp. 1-118). Academic Press, New York.

Driscoll, D.A. (2004). Extension and outbreak accompany fragmentation of a reptile community. *Ecological Applications 14*(1):220-240.

D'Silva, A.M. and Maughan, O.E. (1994). Multiple use of water: Integration of fish culture and tree growing. *Agroforestry Systems 26:*1-7.

Duke, J.A. (1992). *Handbook of Edible Weeds.* CRC Press, Boca Raton FL.

Earley, E.B., McIlrath, W.O., Seif, R.D., and Hageman, R.H. (1967). Effects of shade applied at different stages of plant development on corn *(Zea mays)* production. *Crop Science 7*(2):115-156.

Early, E.B., Miller, R.J., Reichert, G.L., Hageman, R.H., and Seif, R.D. (1966). Effects of shade on maize production under field conditions. *Crop Science 6*(1):1-7.

Eason, W.R. (1991). The effect of leaf litter on sward botanical composition and growth. *Forest Ecology and Management 45:*165-172.

Ebersole, R. (2004). Drink responsibly, choose eco-friendly wines. *National Wildlife 42:*18-19.

Ehui, S.K., Kang, B.T., and Spencer, D.S.C. (1992). Economic analysis of soil erosion: Effects in alley cropping, no-till, and bush fallow system of southwestern Nigeria. In Moock, J.L. and Rhoades, R.E. (Eds.), *Diversity, Farmer Knowledge and Sustainability* (pp. 225-243). Cornell University Press, Ithaca, NY.

Ellis, D.R., Guillard, K., and Adams, R.G. (2000). Purslane as a living mulch in broccoli production. *American Journal of Alternative Agriculture 15*(2):50-59.

Elvin, M. (2004). *The Retreat of the Elephants: An Environmental History of China.* Yale University Press, New Haven, CT.

Emerman, S.H. and Dawson, T.E. (1996). Hydraulic lift and its influence on the water content of the rhizophere: An example from sugar maple *Acer saccharum. Oecolgia 108:*273-278.

Emerson, G. (1846). *Report on the Trees and Shrubs Growing Naturally in the Forests of Massachusetts.* Dutton and Wentworth, Boston.

Erbach, D.C. and Lovely, W.G. (1976). Machinery adaptations for multiple cropping. In Papendick, R.I., Sanchez, P.A., and Triplett, G.B. (Eds.), *Multiple Cropping* (pp. 337-346). ASA Publication Number 27. American Society of Agronomy, Madison, WI.

Escalante, E.E. (1995). Coffee and agroforestry in Venezuela. *Agroforestry Today* 7(3/4):5-7.

Esekhade, T.U., Orimoloye, J.R., Ugwa, I.K., and Idoko, S.O. (2003). Potentials of multiple cropping systems in young rubber plantations. *Journal of Sustainable Agriculture 22*(4):79-94.

Espitia-Rangel, E. (1994). Breeding the grain amaranth. In Paredes-López, O. (Ed.), *Amaranth: Biology, Chemistry and Technology* (pp. 23-38), CRC Press. Boca Raton, FL.

Evelyn, J. (1664). *Silva.* A. Ward, York, UK.

Fahl, J.I., Carelli, M.L.C., Menezes, H.C., Gallo, P.B., and Trivelin, P.C.O. (2001). Gas exchange, growth, yield and beverage quality of *Coffea arabica* cultivars grafted on the *C. canephora and C. congensis. Experimental Agriculture 37:* 241-252.

Fail, J.L., Jr., Haines, B.L. and Todd, R.L. (1987). Riparian forest communities and their role in nutrient conservation an agricultural watershed. *American Journal of Alternative Agriculture* 2(3):114-121.

Fairhead, J. and Leach, M. (1998). *Reframing Deforestation Global Analysis and Local Realities: Studies in West Africa.* Routledge, London.

FAO (1994). Prácticas Agroforestales en el Departmento de Potosí Bolivia: Documemto de Trabajo No 1. FAO/HOLANDA/CDF, Proyecto "Desarrollo Forestal Comunal el Altiplano Boliviano," Potosí, Bolivia.

Fedoroff, N.V. (2003). Prehistoric GM corn. *Science 302:*1158-1159.

Ferber, A.E. (1969?). *Windbreaks for Conservation.* USDA Agriculture Information Bulletin 339, 25 pp.

Fischer, A. and Burrill, L. (1993). Managing interference in a sweet corn-white clover living mulch system. *American Journal of Alternative Agriculture 8*(2):51-56.

Fish, S.K. (2000). Hohokam impacts on Sonoran Desert environment. In Lentz, D.L. (Ed.), *Imperfect Balance Landscape Transformations in the Precolumbian Americas* (pp. 250-280). Columbia University Press, New York.

Forbush, E.H. and May, J.B. (1955). *A Natural History of American Birds of Eastern and Central North America.* Bramhouse, New York.

Foster, D.R. and Zebryk, T.M. (1993). Long-term vegetation dynamics and disturbance history of a tsuga-dominated forest in New England. *Ecology 74*(4):982-998.

Foster, H.L. and Goh, H.S. (1976). Fertilizer requirements of oil palm in Western Malaysia. Reprint from the Malaysian International Agricultural Oil Palm Conference.

Francis, C.A. (1986). Future perspectives of multiple cropping. In Francis, C.A. (Ed.), *Multiple Cropping Systems* (pp. 251-370). Macmillan Publishing Co., New York.

Franzel, S., Hitimana, L., and Akyeampong, E. (1995). Farmer participation in on-station tree species selection for agroforestry: A case study from Burundi. *Experimental Agriculture 31:*27-38.

Fuente de la, M.A.S. and Marquis, R.J. (1999). The role of ant-tended extrafloral nectaries in the protection and benefit of a neotropical rainforest tree. *Oecologia 118*(2):192-202.

Gamero, E.M., Lok, R., and Somarriba, E. (1996). Análisis agroecológico de huertos caseros traditionales en Nicaragua. *Agroforestería en las Americas 3*(11/12):36-40.

Garbaye, J., Delwaulle, J.C., and Duangna, D. (1988). Growth response in eucalyptus in the Congo to ectomycorrhizal inoculation. *Forest Ecology and Management 24:*151-157.

García-Barrios, L. (2003). Plant-plant interactions in tropical agriculture. In Vandermeer, J.H. (Ed.), *Tropical Agroecosystems* (pp. 12-58). CRC Press, Boca Raton, FL.

Gardner, R. and Heider, K.G. (1974). *Gardens of War: Life and Death in the New Guinea Stone Age.* Penguin Books, Harmondsworth, UK.

Gent, F.W. (1681). *Systema Agriculturæ: The Mystery of Husbandry Discovered.* Harrow, London.

Ghaffazadeh, M., Préchac, F.G., and Cruse, R.M. (1994). Grain yield response of a corn, soybean, and oat grown in a strip intercrop system. *American Journal of Alternative Agriculture 9*(4):171-177.

Giesser, H. and Reyer, H.U. (2004). Efficacy of hunting, feeding and fencing to reduce crop damage by wild boar. *The Journal of Wildlife Management 68*(4): 939-946.

Gillespie, A.R., Knudson, D.M., and Geilfur, F. (1993). The structure of four homegardens in the Petén, Gauatemala. *Agroforestry Systems 24:*157-170.

Glausiusz, J. (2003). A green renaissance for the Sahel. *Discover 24*(1):13.

Gliessman, S.R. (1983). Allelopathic interactions in crop-weed mixtures: Applications for weed management. *Journal of Chemical Ecology 9*(8):991-999.

Gliessman, S.R. (1998). *Agroecology: Ecological Processes in Sustainable Agriculture.* Ann Arbor Press, Ann Arbor, MI.

Godoy, R. and Bennett, C.P.A. (1991). The economics of monocropping and intercropping by small holders: The case of coconuts in Indonesia. *Human Ecology 19*(1):83-98.

Goff, S.A. and Salmeron, J.M. (2004). Back to the future of cereals. *Scientific American 291:*42-49.

Gold, C.S., Kiggundu, A., Abera, A.M.K., and Karamura, D. (2002). Selection criteria of Musa cultivars through a farm participatory appraisal survey in Uganda. *Experimental Agriculture 38:*29-38.

Gold, M.V. (1994). *Sustainable Agriculture: Definitions and Terms.* Special Reference SRB 94-05. National Agricultural Library, Betsville, MD.

Gongfu, Z. (1982). The mulberry dike—fish pond complex: A Chinese ecosystem of land-water interactions on the Pearl River Delta. *Human Ecology 10*(2):191-202.

Gonzáles, R.J. (2001). *Zapotec Science: Farming and Food in the Northern Sierra of Oaxaca.* University of Texas Press, Austin.

Gras, N.S.B. (1940). *A History of Agriculture in Europe and Americas.* F.S. Crofts and Co., New York.

Green, R.E., Cornell, S.J., Scharlemann, J.P.W., and Balmford, A. (2005). Farming and the fate of wild nature. *Science 307:*550-554.

Green, S.B. (1908). *Principles of American Forestry.* John Wiley and Sons, New York.

Gregg, S.A. (1988). *Foragers and Farmers: Population Interaction and Agricultural Expansion in Prehistoric Europe.* University of Chicago Press, Chicago.

Grimm, V., Revilla, E., Berger, U., Jeltsch, F., Mooij, W.M., Railsback, S.F., Thulke, H.-H., Weiner, J., Wiegand, T., and DeAgelis, D.L. (2005). Pattern-oriented modeling of agent-based complex systems: Lessons for ecology. *Science 310:*987-991.

Grist, D.H. (1965). *Rice.* Longmans, Bristol, UK, 548 pp.

Grodzinsky, A.M. (1992). Allelopathic effects of cruciferous plants in crop rotation. In Rizvi, J.H. and Rizvi, V. (Eds.), *Allelopathy: Basic and Applied Aspects* (pp. 77-85). Chapman and Hall, London.

Groppali, R. (1993). Breeding birds in traditional tree rows and hedges in the central Po valley (Province of Cremona, Northern Italy). In: Bunce, R.G.H., Ryszkowski, L., and Paoletti, M.G. (Eds.), *Landscape Ecology and Agroecosystems* (pp. 153-158). Lewis Publishers, London.

Gunnell, Y. and Kishnamurthy, A. (2003). Past and present status of runoff harvesting system in dryland peninsula India: A critical review. *Ambio 32*(4):320-324.

Gupta, G.N., Mutha, S., and Limba, N.K. (2000). Growth of *Albizia lebbeck* on micro-catchments in the Indian arid zone. *International Tree Crops Journal 10:*193-202.

Gurung, J.D. and Lama, K. (2004). Incorporating gender into a male domain: The approach of a forest project in Nepal. *Unasylva 55*(216):50-51.

Hansen, S.W. (1996). Is agricultural sustainability a useful concept? *Agricultural Systems 50:*117-143.

Hansen, S.W. and Jones, J.W. (1996). A system framework for characterizing farm sustainability. *Agricultural Systems 51:*185-201.

Harmsen, K. (2000). A modified Mitscherlich equation for rainfed crop production in semi-arid areas: 1. Theory. *Netherlands Journal of Agricultural Science 48:*237-250.

Harris, F.M.A. and Mohammed, S. (2003). Relying on nature: Wild foods in Northern Nigeria. *Ambio 32*(1):24-29.

Harrison, G. (1775). *Agriculture Delineated.* J. Wilkie, London.

Harvey, C.A., Haber, W.A., Mejias, F., and Jolano, R. (1998). Remnant trees in Costa Rican pastures. *Agroforestry Today 10*(3):7-9.

Hauser, S., Vanlauwe, B., Asawalam, D.O., and Norgrove, L. (1997). Role of earthworms in traditional and improved low-input agricultural systems in West Africa. In Brussaard, L. and Ferrara-Cerrato, R. (Eds.), *Soil Ecology in Sustainable Agricultural Systems* (pp. 113-136). Lewis Publishers, New York.

Hawkes, J.G. (1969). The ecological background of plant domestication. In Ucko, P.J. and Dimbleby, G.W. (Eds.), *The Domestication and Exploitation of Plants and Animals* (pp. 17-29). Aldine Publishing Co., Chicago.

Hengchaovanich, D. (2003). Vetiver system for slope stabilization. In *Vetiver and Water.* Proceedings of the Third International Conference on Vetiver (part 4[1]), October 6-9, Guangzhou, China.

Herzog, F. and Oetmann, A. (2000). Communities of interest and agroecosystem restoration: Streuobst in Europe. In Flora, C. (Ed.), *Interactions Between Agroecosystems and Rural Communities* (pp. 85-102). CRC Press, Boca Raton, FL.

Hill, D.B. (1999). Farming exotic mushrooms in the forest. *Agroforestry Notes 13*(July):1-4.

House, P. and Ochoa, L. (1998). La diversidad de especies útiles en diez huertos en la aldea de Camalote, Honduras. In Lok, R. (Ed.), *Huertos Caseros Traditionales de América Central* (pp. 66-84). Centro Agronómico Tropical de Investigacíion y Enseñanza, Turrialba, Costa Rica.

Howeler, R., Watananonta, W., Vongkasem, W., Klakhaeng, K., Jantawat, S., Randaway, S., and Vankaew, B. (2003). Working with farmers: The key to adoption of vetiver grass hedgerows to control erosion in cassava fields in Thailand. In *Vetiver and Water* (part 1[3]). Proceedings of the Third International Conference on Vetiver, October 6-9, Guangzhou, China.

Huang, H.T. and Yang, P. (1987). The ancient cultured citrus ant. *BioScience 37:*665-667.

Hughes, J.D. (1994). *Pan's Travail: Environmental Problems of the Ancient Greeks and Romans.* John Hopkins Press, Baltimore, MD.

Hulugalle, N.R. and Ezumah, H.C. (1991). Effects of cassava-based cropping systems on physio-chemical properties of soil and earthworm casts in a tropical alfisol. *Agriculture, Ecosystems and Environment 25:*55-63.

Husain, A.A. (2000). *Scent in the Islamic Garden: A Study of Deceani Urdu Literary Sources.* Oxford University Press, Oxford.

Hussain, T., Javaid, T., Parr, J.F., Jilani, G., and Haq, M.A. (1999). Rice and wheat production in Pakistan with effective microorganism. *American Journal of Alternative Agriculture 14*(1):30-38.

Huxley, P.A. (2001). Multipurpose trees: Biological and ecological aspects relevant to their selection and use. In Last, F.T. (Ed.), *Tree Crop Ecosystems* (pp. 19-74). Elsevier, Amsterdam.

IACPA (1998). *Integrated Farming: Agricultural Research into Practice.* A Report from the Integrated Arable Crop Production Alliance for Farmers, Agronomists and Advisors. Ministry of Agriculture, Fisheries and Food. MAFF Publications, London.

ICRAF (1993). *Annual Report.* International Center for Research in Agroforestry, Nairobi, Kenya.

ICRAF (1996). *Annual Report.* International Center for Research in Agroforestry, Nairobi, Kenya.

Jackson, L. and Koch, G.W. (1997). The ecophysiology of crops and their wild relatives. In Jackson, L.E. (Ed.), *Ecology in Agriculture* (pp. 3-37). Academic Press, New York.

Jackson, L.L. (2002). Restoring prairie processes to farmlands. In Jackson, D.L. and Jackson, L.L. (Eds.), *The Farm As Natural Habitat* (pp. 137-154). Island Press, Washington, 297 pp.

Jackson, W. (2002a). Farming in natures' image. pp. 41-45. In: Kimbrell, A. (Ed.), *Fatal Harvest: The Tragedy of Industrial Agriculture.* Island Press, Washington, DC.

Jackson, W. (2002b). Natural ecosystems agriculture: A truly radical alternative. *Agriculture, Ecosystems and Environment 88:*111-117.

Jama, B.A., Palm, C.A., Buresh, R.J., Niang, A., Gachengo, C., Nziguheba, G., and Amadalo, B. (2000). *Tithonia diversifolia* as a green manure for soil fertility improvement in western Kenya: A review. *Agroforestry Systems 49:*201-221.

Jashemski, W.F. (1979). *The Gardens of Pompeii.* Caratzas Brothers, New Rochelle, NY.

Jashemski, W.F. (1987). Recently excavated gardens and cultivated land of the villas at Boscoreale and Oplontis. In MacDougall, E.B. (Ed.), *Ancient Roman Villa Gardens* (pp. 33-75). Dumbarton Oaks Research Library, Washington, DC.

Jensen, M. (1993). Soil conditions, vegetation structure and biomass of a Javanese homegarden. *Agroforestry Systems 24:*171-186.

Jobidon, R. (1992). Allelopathy in Quebec forestry—Case studies in natural and managed ecosystems. In Rizvi, J.H. and Rizvi, V. (Eds.), *Allelopathy; Basic and Applied Aspects* (pp. 341-356). Chapman and Hall, London.

Joffee, R., Rambal, S., and Ratte, J.P. (1999). The dehesa system of southern Spain and Portugal as a natural ecosystem mimic. *Agroforestry System 45:*57-79.

Johns, T. (1990). *The Origins of Human Diet and Medicine.* University of Arizona Press, Tucson, AZ.

Johns, T. (1999). Plant constituents and the nutrition of indigenous peoples. In Nazarea, V.D. (Ed.), *Ethnoecology: Situated Knowledge and Located Lives* (pp. 157-174). The University of Arizona Press, Tucson, AZ.

Jordan, V.W.L., Hutcheon, J.A., and Donaldson, G.V. (1997). The role of integrated arable production systems in reducing synthetic inputs. *Aspects of Applied Biology 50:*419-429.

Jordan, V.W.L., Hutcheon, J.A., Donaldson, G.V., and Farmer, D.P. (1997). Research into and development of integrated farming systems for less-intensive arable production: Experimental progress (1998-1994) and commercial implementation. *Agriculture, Ecosystems and Environment 64*:141-148.

Jordan, V.W.L., Hutcheon, J.A., Glen, D.M., and Farmer, D.P. (1996). *Technology Transfer of Integrated Farming Systems: The LIFE Project,* Third Edition. Integrated Arable Crops Research, Bristol, UK.

Julien, M.H. and Griffiths, M.W. (Eds.) (1998). *Biological Control of Weeds: A World Catalogue of Agents and Their Target Weeds,* Fourth Edition. CABI Publishing, Wallingford, UK.

Just, R.E. and Pope, R.D. (2003). Agricultural risk analysis: Adequacy of models, data, and issues. *American Journal of Agricultural Economics 8*(5):1249-1256.

Kaiser, J. (2003). Sipping from the poisoned chalice. *Science 302:*376-379.

Kamara, A.Y., Akobundu, I.O., Chikoye, D., and Jutzi, S.C. (2000). Selective control of weeds in an arable crop by mulches from some multi-purpose trees in Southwestern Nigeria. *Agroforestry Systems 50:*17-26.

Kang, B.T. (1997). Alley cropping—Soil productivity and nutrient recycling. *Forest Ecology and Management 91:*75-82.

Kanmegne, J. and Degrande, A. (2002). From alley cropping to rotational fallow: Farmers' involvement in the development of fallow management techniques in the humid tropics. *Agroforestry Systems 54:*115-120.

Kareiva, P. (1994). Diversity begets productivity. *Nature 368:*686-687.

Kass, D.C.L. and Somarriba, E. (1999). Traditional fallows in Latin America. *Agroforestry Systems 47:*13-36.

Kaval, P. (2004). The profitability of alternative cropping systems: A review of the literature. *Journal of Sustainable Agriculture 23*(3): 47-65.

Keating, B.A. and Carberry, P.S. (1993). Resource capture and use in intercropping: Solar radiation. *Field Crops Research 34:*273-301.

Kelty, M.J. (1992). Comparative productivity of monocultures and mixed-species stands. In Kelty, M.J., Larson, B.C., and Oliver, C.D. (Eds.), *The Ecology and Silviculture of Mixed Species Forests* (pp. 125-141). Kluwer Academic Publishers, Dordrecht, the Netherlands.

Khan, Z.R., Pickett, J.A., van den Berg, J., Wadham, L.J., and Woodcock, C.M. (2000). Exploiting chemical ecology and species diversity: Stem borer and striga control for maize and sorghum in Africa. *Pest Management Science 56:*957-962.

Kho, R.M., Yacouba, B., Yayé, M, Katkoré, B. Moussa, A., Iktam, A., and Mayaki, A. (2001). Separating the effects of trees on corn: The case of Faidherbia and millet in Niger. *Agroforestry Systems 52:*219-238.

Kirsch, P. (1988). Pheromones: Their potential role in control of agricultural insect pests. *American Journal of Alternative Agriculture 3*(2/3):83-97.
Kleinn, C. (2000). On large-area inventory and assessment of trees outside forests. *Unasylva 51:*3-10.
Klironomos, J.N. (2002). Another form of bias in conservation research. *Science 298:*749.
Koch, P. (1996). *Lodgepole Pine in North America,* Volume 1. Forest Products Society, Madison, WI.
Koech, E.K. and Whitbread, R. (2000). Disease incidence and severity on beans in alleys between leucaena hedgerows in Kenya. *Agroforestry Systems 49:*85-101.
Korwar, G.R. and Radder, G.D. (1994). Influence of root pruning and cutting interval of *Leucaena* hedgerows on performance of alley cropped rabi sorghum. *Agroforestry Systems 25:*95-109.
Landis, D.A., Wratten, S.D., and Gurr, G.M. (2000). Habitat management to conserve natural enemies of anthropoid pests in agriculture. *Annual Review of Entomology 54*(1):175-203.
Lathrop, L.E. (1825). *The Farmer's Library.* William Fay, Rutland, VT.
Leach, M. (1994). *Rainforest Relations, Gender, and Resource Use Among the Mende of Gola, Sierra Leone.* Smithsonian Institute Press, Washington, DC.
Lefroy, E.C., Hobbs R.J., Connor, M.H., and Pate, J.S. (1999). What can agroforestry learn from natural ecosystems? *Agroforestry Systems 45:*423-436.
Lelle, M.A. and Gold, M.A. (1994). Agroforestry systems for temperate climates: Lessons from Roman Italy. *Forest and Conservation History 38:*118-126.
Leuschner, C., Hertel, D., Coners, H., and Büttner, V. (2001). Root competition between beech and oak: A hypothesis. *Oecologia 126:*176-184.
Levins, R. and Vandermeer, J.H. (1990). The agroecosystem embedded in a complex ecological community. In Carroll, C.R., Vandermeer, J.H., and Resset, P.R. (Eds.), *Agroecology* (pp. 341-362). McGraw- Hill, New York.
Liebman, M. and Gallandt, E.R. (1997). Many little hammers: Ecological management of crop-weed interactions. In Jackson, L.E. (Ed.), *Ecology in Agriculture* (pp. 291-343). Academic Press, New York.
Liebman, M., Mohler, C.L., and Staver, C.P. (2001). *Ecological Management of Agricultural Weeds.* Cambridge University Press, Cambridge, UK.
Liebold, M.A. (1995). Mechanistic models and community context. *Ecology 76*(5): 1371-1382.
Linares, O.F. (1976). Garden hunting in the American tropics. *Human Ecology 4*(4):331-349.
Liping, Z. (1991). Biological control of insect pests and plant diseases in agroforestry systems. In Avery, M.E., Cannell, M.G.R., and Ong, C.K. (Eds.), *Biological Research for Asian Agroforestry* (pp. 73-87). Winrock International and Oxford and IBH Publishing Co., New Delhi.
Litton, R.B., Tetlow, R.J., Sorensen, J., and Beatty, R.A. (1974). *Water and Landscape: An Aesthetic Overview of the Role of Water in the Landscape.* Water Information Center, New York.
Loomans, A.J.M., van Lenteren, J.C., Bigler, F., Burgio, G., Hokkanen, H.M.T., and Thomas, M.B. (2002). Evaluating environmental risks of biological control in-

troductions: How to select safe natural enemies. *IOBC/WPRS Bulletin 23*(1): 147-150.

Lotter, D.W., Seidel, R., and Liebhardt, W. (2003). The performance of organic and conventional cropping system in a extreme climate year. *American Journal of Alternative Agriculture 18*(3):146-154.

Lövenstein, H.M., Breliner, P.R., and van Kuelen, H. (1991). Runoff agroforestry in arid lands. *Forest Ecology and Management 45:*59-70.

Lynam, J.K., Sanders, H.H., and Mason, S.C. (1986). Economics and risk in multiple cropping. In Francis, C.A. (Ed.), *Multiple Cropping Systems* (pp. 250-266). Macmillian Publishing Co., New York.

Ma, M., Tarmi, S., and Helenius, J. (2002). Revisiting the species-area relationship in a semi-natural habitat: Floral richness in agricultural buffer zones in Finland. *Agriculture, Ecosystems and Environment 89:*137-148.

MacDonald, J. (1994). *The Ornamental Kitchen Garden.* David and Charles, Devon, UK.

MacDonald, K.I. (1998). Rationality, representation, and the risk mediating characteristics of a Karakoram mountain farming system. *Human Ecology 26*(2): 287-317.

MacKenze, D. (1999). Good ol' milkweed. *New Scientist 164:*24.

MacLellan, C.R. (1970). Woodpecker ecology in the apple orchard environment. In *Proceedings of the Tall Timbers Conference on Ecological Animal Control by Habitat Management,* Volume 2, February 25-27 (pp. 273-284). Tall Timbers Research Station, Tallahassee, FL.

Mäder, P., Fliessbach, A., Dubois, D., Gunst, L., Fried, P., and Higgli, U. (2002). Soil fertility and biodiversity in organic farming. *Science 296:*1694-1697.

Malcolm, J.R. (1994). Edge effects in central Amazonian forest fragments. *Ecology 75*(8):2438-2445.

Mallory, E.B., Posner, J.L., and Baldock, J.O. (1998). Performance, economics and adoption of cover crops in Wisconsin cash grain rotations: On-farm trials. *American Journal of Alternative Agriculture 13*(1):2-11.

Malone, G. (2002). Odor: Can trees make a difference? *Inside Agroforestry* (Spring):1.

Mancarrow, L. and Taylor, J.H. (2000). *Dead Daisies Make Me Crazy: Garden Solutions Without Chemicals.* Ten Speed Press, Berkeley, CA.

Mando, A. and Van Rheenen, T. (1998). Termites and agricultural production in the Sahel: From enemy to friend. *Netherlands Journal of Agricultural Science 64:*77-85.

Mann, C.C. (2002a). 1491. *The Atlantic Monthly 289*(3):41-52.

Mann, C.C. (2002b). The real dirt on rainforest fertility. *Science 297:*920-923.

Mapes, C., Basurto, F., and Bye, R. (1997). Ethobotany of quintinil: Knowledge and management of edible greens *Amaranthus* spp. (Amaranthaceae) in the Sierra Norte de Puebla, México. *Economic Botany 51*(3):293-306.

Marc, P., Canard, A., and Ysnel, F. (1999). Spiders (Araneae) useful for pest limitation and bioindication. *Agriculture, Ecosystems and Environment 74:*229-273.

Mas, A.H. and Dietsch, T.V. (2004). Linking shade coffee certification to biodiversity conservation: Butterflies and birds in Chiapas, Mexico. *Ecological Applications 14*(3):642-654.

Matlack, G.R. (1994). Plant species migration in mixed-history forest landscapes in eastern North America. *Ecology 75*(5):1491-1502.

Matson, P.A., Parton, W.J., Power, A.G., and Swift, M.J. (1997). Agricultural intensification and ecosystem properties. *Science 277:*504-509.

Maudeley, M., Seeley, B., and Lewis, O. (2002). Spatial distribution patterns of predatory arthropods within an English hedgerow in early winter in relation to habitat variables. *Agriculture, Ecosystems and Environment 89:*77-89.

McCord, N. (1992). *Please Don't Eat My Garden.* Sterling Publishing Co., New York.

McGovern T.H. (2000). The Vikings' silent saga. *Natural History 109:*50-56.

McIntyre, R. (1999). Four-level forest farming operation making the most of agroforestry. *Inside Agroforestry* (Fall):10.

McLean, T. (1980). *Medieval English Garden.* The Viking Press, New York.

McNeely, J.A. (1993). Economic incentives for conserving biodiversity: Lesson from Africa. *Ambio 22*(2-3):144-150.

McNeely, J.A. and Scherr, S.J. (2003). *Ecoagriculture: Strategies to Feed the World and Save Biodiversity.* Island Press, Washington, DC.

Mead, R. and Willey, J. (1980). The concept of "land equivalent ratio" and the yield advantages from intercropping. *Experimental Agriculture 16:*217-228.

Méndez, V.E., Lok, R., and Somarriba, E. (2001). Interdisciplinary analysis of homegardens in Nicaragua: Micro-zonation, plant use and socio-economic importance. *Agroforestry Systems 51:*85-96.

Menzel, P. and D'Alvisio, F. (1998). *Man Eating Bugs: The Art and Science of Eating Insects.* Ten Speed Press, Berkeley, CA.

Menzies, N. (1988). Three hundred years of taungya: A sustainable system of forestry in South China. *Human Ecology 16*(4):361-376.

Mertz, O. (1998). Wild vegetables as potential new crops in farming systems of Serwak, Malaysia. In Nair, M.N.B., Sahri, M.H., and Asharri, Z. (Eds.), *Sustainable Management of Non-Wood Forest Products* (pp. 70-83). Proceedings of an International Workshop Universiti Putra Malaysia, October 14-17, 1997. Universiti Putra Malaysia Press, Serdang.

Merwin, M. (2001). Oregon family farm mixes Christmas trees and black walnut. *Inside Agroforestry* (Fall/Winter):4.

Mialet-Serra, I., Bonneau, X., Mouchet, S., and Kitu, W.T. (2001). Growth and yield of coconut-cacao intercrops. *Experimental Agriculture 37:*195-210.

Michalak, P.S. and Gilkeson, L.A. (1994). *Rodale's Successful Organic Gardening; Controlling Pests and Diseases.* Rodale Press, Emmaus PA.

Michon, G. and Mary, F. (1994). Conversion of traditional village gardens and new economic strategies of rural households in the area of Bogor, Indonesia. *Agroforestry Systems 25:*31-58.

Milius, S. (2003). Sweet lurkers. *Science News 164:*374.

Miller, D.R., Rosenberg, N.J., and Bagley, W.T. (1973). Soybean water use in the shelter of a slat-fence windbreak. *Agricultural Meteorology 11:*405-418.

Mohd Ali, A.R. and Raja Bariza, R.S. (2001). Intercropping rattan with rubber and other crops. *Unasylva 52*(205):9-11.

Molles, M.C. (1999). *Ecology: Concepts and Applications.* McGraw Hill, New York.

Mollison, B. (1992). *Permaculture: A Designers' Manual.* Tagari Publications, Tyalgua, Australia.

Morales, H., Perfecto, I., and Ferguson, B. (2001). Traditional fertilization and its effect on corn insect populations in the Guatemalan highlands. *Agriculture, Ecosystems and Environment 84:*145-155.

Morna, C.L., Gikaru, G., and Ajayi, F. (1990). Farmers fight pests the natural way. *African Farmer 5*(November):28-33.

Morris, R.A. and Garrity, D.P. (1993). Resource capture and utilization in intercropping: Water. *Field Crops Research 34:*303-317.

Mountjoy, D.C. and Gliessman, S.R. (1988). Traditional management of a hillside agroecosystem in Tlaxcala Mexico: An ecologically based maintenance system. *American Journal of Alternative Agriculture 3*(1):3-10.

Moynihan, E.B. (1979). *Paradise As a Garden in Persia and Mughad India.* George Braziller, New York.

Muschler, R. (1999). *Árboles en Cafetales.* Materiales de Enseñanza No. 45. CATIE, Turrialba, Costa Rica.

Muschler, R.G. (2001). Shade improves coffee quality in a sub-optimal coffee-zone of Costa Rica. *Agroforestry Systems 85:*131-139.

Mutsaers, H.J.W. (1991). *Opportunities for Second Cropping in Southwest Nigeria.* RCMP Research Monographs No. 4. IITA, Ibadan, Nigeria.

Nabham, G.P. and Sheridan, T.E. (1977). Living fencerows of the Rio San Miguel, Sonora, Mexico: Traditional technology for floodplain management. *Human Ecology 5*(2):97-111.

Naeem, S., Thompson, L.J., Lawler, S.P., Lawton, J.H., and Woodfin, R.M. (1994). Declining biodiversity can alter the performance of ecosystems. *Nature 368:* 734-735.

Nair, P.K.R., Buresh, R.J., Mugendi, D.N., and Latt, C.R. (1999). Nutrient cycling in tropical agroforestry systems: Myths and science. In Buck, L.E., Lassoie, J.P., and Fernandes, E.C.M. (Eds.), *Agroforestry in Sustainable Agricultural Systems* (pp. 1-11). CRC Press, New York.

Nandal, D.P.S. and Bisla, S.S. (1995). Growing oats under trees for more and better fodder. *Agroforestry Today 7*(1):12.

Nasser, R., Velásquez, C., Velasco, C., Ruíz, J., Sánchez, E., Castillo, A.M., and Radulovich, R. (1994). Huertos caseros: Un actividad productiva con amplia participación de la mujer. In Radulovich, R. (Ed.), *Technologías Productivas para Sistemas Agrosilvopecuarios de Ladera con Sequía Estacional* (pp. 151-185). Serie Técnica No. 222. CATIE, Turrialba, Costa Rica.

Nebel, G. (2001). Sustainable land-use in Peruvian flood plain forests: Options, planning and implementation. *Forest Ecology and Management 150*:187-198.

Nevo, Y.D. (1991). *Pagans and Herders: A Re-Examination of the Negev Runoff Cultivation System of the Byzantine and Early Arab Period.* Achva Press, Jerusalem.

Newmann, K. (2003). New Guinea: The cradle of agriculture. *Science 301:*180-181.

Njoku, J.C., Okpara, D.A., and Ikeorgu, J.E.G. (2003). Response of sweet potato to duration of mucuna fallow. *Tropical Science 42:*44-47.

NRC (1981). *Alternative Agriculture.* National Resource Council, National Academic Press, Washington, DC.

NRC (1987). *Lost Crops of the Incas.* National Research Council, National Academy Press, Washington, DC.

NRC (1996). *Lost Crops of Africa:* Volume 1, *Grains.* Board on Science and Technology for International Development, Natural Resource Council, National Academy Press, Washington, DC.

Nyakanda, C., Mashingaidze, A.B., and Zhakata, E. (1998). Early growth of pigeonpea under open-pit planting and restricted weeding in semi-arid Zimbabwe. *Agroforestry Systems 41:*267-276.

Nyong, A.O. and Kanaroglou, P.S. (1999). Domestic water use in rural semiarid Africa: A case study of Katarko village in northeastern Nigeria. *Human Ecology 27*(4):537-546.

O'Brien, C. (2002). Bees buzz elephants. *The New Scientist 176:*8.

O'Conner, R.J. and Shrubb, M. (1986). *Farming and Birds.* Cambridge University Press, Cambridge, UK.

Ohte, N., Koba, K., Yoshikawa, A., Sugimoto, A., Matsuo, N., Kabeya, N., and Wang, L. (2003). Water utilization of natural and planted trees in the semiarid desert of Inner Mongolia, China. *Ecological Applications 13*(2):337-351.

Okigbo, B.N. and Greenland, D.J. (1976). Intercropping systems in tropical Africa. In Papendick, R.I., Sanchez, P.A., and Triplett, G.B. (Eds.), *Multiple Cropping* (pp. 63-101). ASA No. 27. American Society of Agronomy, Madison, WI.

Ong, C.K. (1991). Interactions of light, water and nutrients in agroforestry systems. In Avery, M.E., Cannell, M.G.R., and Ong, C.K. (Eds.), *Biophysical Research for Asian Agroforestry* (pp. 107-121). Winrock International and Oxford and IBH Publishing Co., New Delhi.

Ong, C. (1994). Alley cropping ecological pie in the sky? *Agroforestry Today 6*(3):8-10.

Ong, C.K. (1995). The "dark side" of intercropping: manipulation of soil resources. In: Sinofuet, H. and Cruz, P. (Eds.), *Ecophysiology of Tropical Intercropping.* Institute National de la Recherche Agronomique, Paris.

Ong, C.K. and Leakey, R.R.B. (1999). Why tree-crop interactions in agroforestry appear at odds with tree-crop interactions in tropical savannas. *Agroforestry Systems 45*(1-3):109-129.

Padoch, C. and de Jong, W. (1987) Traditional agroforestry practices on native and ribereno farmers in the lowland Peruvian Amazon. In Gholz, H.L. (Ed.) *Agroforestry: Realities, Possibilities and Potential* (pp. 179-194). Martinus Nijhoff Publishers, Amsterdam, the Netherlands.

Pain, S. (2003). Last of the lions. *The New Scientist 179:*36-40.

Palm, C.A. (1995). Contribution of agroforestry trees to nutrient requirements of intercropped plants. *Agroforestry Systems 30:*105-124.

Paoletti, M.G. (2001). Biodiversity in agroecosystems and bioindicators of economic health. In Shiyomi, M. and Koizumi, H. (Eds.), *Structure and Function in Agroecosysetm Design and Management* (pp. 11-44). CRC Press, New York.

Paris, Q. (1992). The von Liebig hypothesis. *American Journal of Agricultural Economics 74*:1019-1028.

Pearce, F. (2005). Return to Eden. *The New Scientist 185:*35-37.

Pearson, T.G., (Ed.) (1917). *Birds of America.* Garden City Books, Garden City, NJ.

Pennisi, E. (2003). Fungi shield new host plants from heat and drought. *Science 301*(12 September-5639):1466.

Pennisi, E. (2004). The secret life of fungi. *Science 304:*1620-1622.

Perales, R. H., Brush, S.B., and Qualset, C.O. (2003). Dynamic management of maize landraces in central Mexico. *Economic Botany 57*(1):21-34.

Percival, J. (1922). *The Wheat Plant.* E.P. Dutton and Co., New York.

Perfecto, I. and Armbrecht, I. (2003). The coffee agroecosystem in the neotropics: Combining ecological and economic goals. In Vandermeer, J.H. (Ed.), *Tropical Agroecosystems* (pp. 159-194). CRC Press, Boca Raton, FL.

Perfecto, I., Rice, R.A., Greenberg, R., and Van der Voort, M.E. (1996). Shade coffee: A disappearing refuge for biodiversity. *BioScience 46*(8):598-608.

Peri, P.L. and Bloomberg, M. (2002). Windbreaks in Southern Patagonia, Argentina: A review of research on growth models, windspeed reduction, and effects on crops. *Agroforestry Systems 56:*129-144.

Perry, C.H., Miller, R.C., and Brooks, K.N. (2001). Impacts of short-rotation hybrid poplar plantation on regional water yield. *Forest Ecology and Management 143:*143-151.

Peterken, G.F. (1993). Long-term floristic development of woodland on former agricultural lands in Lincolnshire, England. In Watkins, C. (Ed.), *Ecological Effects of Afforestation: Studies in the History and Ecology of Afforestation in Western Europe* (pp. 31-43). CABI, New York.

Peterken, G.F. (2001). Ecological effects of introduced tree species in Britain. *Forest Ecology and Management 141:*31-42.

Peters, W.J. and Neuenschwander, L.F. (1988). *Slash and Burn: Farming in the Third World.* University of Idaho Press, Moscow, ID.

Peterson, C.J. and Squiers, E.R. (1995). Competition and succession in an aspen-white-pine forest. *Journal of Ecology 83:*449-457.

Peterson, J.T. (1981). Game, farming, and inter-ethnic relations in northeastern Luzon, Philippines. *Human Ecology 9*(1):1-22.

Philbrick, H. and Philbrick, J. (1974). *The Bug Book: Harmless Insect Control.* Story Communications, Pownal, VT.

Picone, C. (2003). Managing mycorrhizae for sustainable agriculture in the tropics. In Vandermeer, J. (Ed.), *Tropical Agroecosystems* (pp. 245-262). CRC Press, Boca Raton, FL.

Pinedo-Vasdquez, M., Pareder del Aquila, P., Romero, R., Rios, M., and Pinedo-Panduro, M. (2003). Peru. In Brookfield, H., Parsons, H., and Brookfield, M. (Eds.), *Agrobiodiversity: Learning from Farmers Across the World* (pp. 232-248). United Nations University, Tokyo.

Platt, J.O., Caldwell, J.S., and Kok, L.T. (1999). Effect of buckwheat as a flowering border on population of cucumber beetle and their natural enemies in cucumber and squash. *Crop Protection 18*(5): 305-313.

Plucknett, D.L. and Smith, N.J.H. (1986). Historical perspectives on multiple cropping. In Francis, C.A. (Ed.), *Multiple Cropping Systems* (pp. 20-39). Macmillan Publishing Co., New York.

Pollack, A.W. (2004). No foolproof way is seen to contain altered genes. *The New York Times,* January 21, P. A10.

Pollan, M. (2003). The (agri)cultural contradictions of obesity. *The New York Times Magazine,* October, 12, pp. 41-48.

Porteous, J.D. (1996). *Environmental Aesthetics: Ideas, Politics and Planning.* Routledge, London.

Postel, S. (1999). *Pillar of Sand: Can the Irrigation Miracle Last?* W.W. Norton and Co., New York.

Pratap, N., Singh, R.K., Sindwal, N.S., and Joshie, P. (1998). Agroforestry for soil and water conservation in the western Himalayan Valley Region of India: I. Runoff, soil, and nutrient loss. *Agroforestry Systems 39:*175-189.

Price, C. (1995). Economic evaluation of financial and non-financial costs and benefits in agroforestry development and the value of sustainability. *Agroforestry Systems 30:*75-86.

Pringle, H. (1998). The original blended economies. *Science 282:*1447.

Prinsley, R.T. (1992). The role of trees in sustainable agriculture an overview. *Agroforestry Systems 20:*87-115.

Rabin, M. and Thaler, R.H. (2001). Anomalies: Risk adversion. *Journal of Economic Perspective 15*(1):219-232.

Rachie, M.D. (1983). Intercropping tree legumes with annual crops. In Huxley, P.A. (Ed.), *Plant Research in Agroforestry* (pp. 103-116). ICRAF, Nairobi, Kenya.

Racine, M., Boursier-Mougenot, E.J.-P., and Binet, F. (1987). *The Gardens of Provence and the French Riviera.* MIT Press, Cambridge, MA.

Rackham, O. (1998). Savanna in Europe. In Kirby, K.J. and Watkins, C. (Eds.), *The Ecological Basis of European Forests* (pp. 1-24). CABI, New York.

Raintree, J.B. (1983). *Guidelines for Agroforestry Diagnosis and Design: Draft for Comment.* Working Paper No. 6. ICRAF, Nairobi, Kenya.

Raintree, J.B. (1990). Theory and practice of agroforestry diagnosis and design. In MacDicken, K.G. and Vergara, N.T. (Eds.), *Agroforestry, Classification and Management* (pp. 58-97). John Wiley and Sons, New York.

Raloff, J. (2005). Food colorings: Pigments make fruits and veggies healthful. *Science News 167:*27-29.

Ranganathan, R., Fafchamps, M., and Walker, T.S. (1991). Evaluating biological productivity in intercropping systems with production possibility curves. *Agricultural Systems 36:*137-157.

Rao, M.R. (1986). Cereals in multiple cropping. In Francis, C.A. (Ed.), *Multiple Cropping Systems* (pp. 96-132). Macmillian Publishing Co., New York.

Rao, M.R., Sharma, M.M., and Ong, C.K. (1991). A tree/crop interface design and its use for evaluating the potential of hedgerow intercropping. *Agroforestry Systems 13:*143-152.

Raynor, W. (1992). Economic analysis of indigenous agroforestry: A case study on Pohnpei Island, Federated State of Micronesia. In Sullivan, G.H., Huke, S.M., and Fox, J.M. (Eds.), *Financial and Economic Analysis of Agroforestry Systems:*

Proceedings of a Workshop held in Honolulu, Hawaii, USA, July 1991 (pp. 243-258). Nitrogen Fixing Tree Association, Paia, HI.

Redford, K.H., Klein, B., and Murcia, C. (1992). Incorporation of game animals into small-scale agroforestry systems in the neotropics. In Redford, K.H. and Padoch, C. (Eds.), *Conservation of Neotropical Forests* (pp. 334-346). Columbia University Press, New York.

Redman, R.S., Sheehan, K.B., Stout, R.G., Rodriguez, R.J., and Henson, J.M. (2002). Thermotolerance generated by plant/fungal symbiosis. *Science 295:*1581.

Reilly, A. (1993). *Gardening Naturally.* Better Homes and Gardens Books, Des Moines, IA.

Reynolds, H.L., Packer, A., Bever, J.D., and Clay, K. (2003). Grass roots ecology: Plant-microbe-soil interactions as drivers of plant community structure and dynamics. *Ecology 84*(9):2281-2291.

Reynolds, L. (1991). Livestock in agroforestry: A farming systems approach. In Avery, M.E., Cannell, M.G.R., and Ong, C.K. (Eds.), *Biological Research for Asian Agroforestry* (pp. 233-256). Winrock International and Oxford and IBH Publishing Co., New Delhi.

Rezaul Haq, A.H.M., Ghosal, T.K., and Ghosh, P. (2004). Cultivating wetlands in Bangladesh. *Leisa 20*(4):18-20.

Rhodes, R.E. and Nazarea, V.D. (1999). Local management of biodiversity in traditional agroecosystems. In Collins, W.A. and Qualset, C.O. (Eds.), *Biodiversity in Agroecosystems* (pp. 215-236). CRC Press, Boca Raton, FL.

Rice, E.L. (1995). *Biological Control of Weeds and Plant Diseases.* University of Oklahoma Press, Norman, OK.

Rice, R.A. and Greenburg, R. (2000). Cocoa cultivation and the conservation of biological diversity. *Ambio 29*(3):167-173.

Rigby, D. and Cáceres, D. (2001). Organic farming and the sustainability of agricultural systems. *Agricultural Systems 68:*21-40.

Rindos, D. (1984). *The Origins of Agriculture: An Evolutionary Perspective.* Academic Press, New York.

Risch, S.J. and Carrol, C.R. (1982). Effects of a keystone predaceous ant, *Solenopsis geminata,* on arthropods in a tropical agroecosystem. *Ecology 63*(6):1979-1983.

Robinson, J. (2005). Silvopasture and the eastern wild turkey. *Agroforestry Notes 28*(February):1-4.

Ruc, J. (1990). Compounds from plants that regulate or participate in disease resistance. In Chadwick, D.J. and March J. (Eds.), *Bioactive Compounds from Plants* (pp. 213-224). John Wiley and Sons, New York.

Ruhigwa, B.A., Gichura, W.P., Spencer, D.S.C., and Swennen, R. (1994). Economic analysis of cut-and-carry and alley cropping systems of mulch production for plantains in south-eastern Nigeria. *Agroforestry Systems 26:*313-138.

Salam, A.M., Mohanakumaran, N., Jayachandran, B.K., Mammen, M.K., Sreekumar, D., and Babu, K.S. (1991). Thirty-one tree species support black pepper vines. *Agroforestry Today 3*(4):16.

Sanchez, P.A. (1995). Science in agroforestry. *Agroforestry Systems 30:*5-55.

Sandström, K. (1998). Can forests "provide" water: Widespread myth or scientific reality. *Ambio 27*(2):132-138.

Schenck, C.A. (1904). *Forest Utilization, Mensuration and Sylviculture.* Biltmore, Asheville, NC.

Scheu, S. and Parkinson, D. (1994). Effects of invasion of an aspen forest (Canada) by *Dendrobaena octaedra* (Lumbricidae) on plant growth. *Ecology 75*(8):2348-2361.

Schlich, W. (1910). *Schlich's Manual of Forestry,* Volume II, Fourth Edition. Bradbusy, Agnew and Co., London.

Schroth, G. (1995). Tree root characteristics: A criteria for species selection and system design in agroforestry. *Agroforestry Systems 30:*125-143.

Schroth, G. (1999). A review of belowground interactions in agroforestry, focusing on mechanisms and management options. *Agroforestry Systems 43:*5-34.

Schroth, G., Krass, V., Gasparotto, L., Duarte-Aguilar, J.A., and Vohland, K. (2000). Pests and diseases in agroforestry systems of the humid tropics. *Agroforestry Systems 50:*199-241.

Schulte, M. (1996). *Technología Agrícola Altoandina.* Swiss Agency for Development and Cooperation, Berne.

Schultz, B.B., Phillips, C., Rosset, P., and Vandermeer J. (1982). An experiment in intercropping tomatoes and cucumber in southern Michigan. *Scientia Horticulturae 18:*1-8.

Setälä, H. (2000). Reciprocal interactions between Scots pine and soil web structure in the presence and absence of ectomycorrhiza. *Oecologia 125:*109-118.

Shepherd, K.D. Ndufa, J.K., Ohlssong, E., Sjögren, H., and Swinkels, R. (1997). Adoption potential of hedgerow alley cropping in maize-based cropping systems in the highlands of western Kenya: 1. Background and agronomic evaluation. *Experimental Agriculture 33:*197-209.

Shiva, V. (1991). *The Violence of the Green Revolution: Third World Agriculture, Ecology and Politics.* Zed Books, London.

Shrair, A.J. (2000). Agricultural intensity and its measure in frontier regions. *Agroforestry Systems 49:*301-318.

Shuster, W.D. and Edwards, C.A. (2003). Interactions between tillage and earthworms in agroecosystems. In El Titi, A. (Ed.), *Soil Tillage in Agroecosystems* (pp. 229-252). CRC Press, Boca Raton.

Sillitoe, P. (1996). *A Place Against Time: Land and Environment in the Papua New Guinea Highlands.* Harwood Academic Publishers, Reading, UK.

Simpson, I.G. and Simpson, M.C. (1991). Systems of agricultural production in central Sudan and Khartoum Province. In Craig, G.M. (Ed.), *The Agriculture of the Sudan* (pp. 252-279). Oxford University Press, Oxford, UK.

Sink, M. (2004). Thirsty shrub is a target as the west fights drought. *The New York Times* May 15, p. A10.

Siwatibau, A. (1984). Traditional environmental practices in the South Pacific: A case study of Fiji. *Ambio 13*(5-6):365-368.

Small, M.F. (2003). How many fathers are best for a child. *Discover 24:*54-61.

Smith, M.E. and Francis, C.A. (1986). Breeding for multiple cropping systems. In Francis, C.A. (Ed.), *Multiple Cropping Systems* (pp. 219-249). Macmillan Publishing Co., London.

Smits, P.H. (1997). Insect pathogens: Their suitability as biopesticides. In *Microbial Insecticides: Novelty or Necessity?* (pp. 21-28). Proceedings of a Symposium organized by the British Crop Protection Council, University of Warwick, Coventry, UK, April 16-18.

Snyder, W.E. and Ives, A.R. (2001). Generalist predators disrupt biological control by a specialist parasitoid. *Ecology 82*(3):750-756.

Soemarwoto, O. (1987). Homegardens: A traditional agroforestry system with a promising future. In Steppler, H.A. and Nair, N.P.K. (Eds.), *Agroforestry: A Decade of Development* (pp. 156-170). ICRAF, Nairobi, Kenya.

Soluri, J. (2001). Altered landscapes and transformed livelihoods: Banana companies, Panama disease and rural communities on the north coast of Honduras, 1880-1950. In Flora, C. (Ed.), *Interactions Between Agroecosystems and Rural Communities* (pp. 15-30). CRC Press, New York.

Soorae, P. (2002). Jumping cholla—The ornamental that became a wildlife and livestock barrier. *Agroforestry Today 13*(1/2):25-27.

Sorgedrager, J.K., Flores, G., Herlant, P., and Schlaifer, M. (1991). *Sistemas Agroforestales Traditionales en Bolivia.* Projecto Desarrollo Forestales Los Andes, La Paz, Bolivia.

Soto-Pinto, L., Perfecto, I., Castillo-Hermandez, J., and Caballero-Nieto, J. (2000). Shade effect on coffee production at the northern Tzeltal zone of the state of Chiapas, Mexico. *Agriculture, Ecosystems and Environment 80*:61-69.

Stakman, E.C., Bradfield, R., and Mangelsdorf, P.C. (1967). *Campaigns Against Hunger.* Harvard University Press, Cambridge, MA.

Stanton, M. and Young, T. (1999). Thorny relationships. *Natural History 108*(9):28-31.

Starcher A.M. (1995). *Good Bugs for Your Garden.* Algonquin Books, Chapel Hill, NC.

Stark, M., Mercado, A., and Garrity, D. (2000). Natural vegetative strip: Farmers innovation gains popularity. *Agroforestry Today 12*(1):32-35.

Statälä, H. and Huhta, V. (1991). Soil fauna increase *Betula pendula* growth: Laboratory experiments with coniferous forest floor. *Ecology 72*(2):665-671.

Staver, C. (1999). Managing ground cover heterogeneity in coffee (*Coffea arabica* L.) under managed tree shade: From replicated plots to farmer practice. In Buck, L.E., Lassoie, J.P., and Fernandes E.C.M. (Eds.), *Agroforestry in Sustainable Agricultural Systems* (pp. 67-68). CRC Press, New York.

Steenbergh, W.F. and Lowe, C.H. (1969). Critical factors during the first year of life of the saguaro *(Cereus giganteus)* at Saguaro Monument, Arizona. *Ecology 50*:823-834.

Stevens, C. and Liebault, J. (1616). *The Countrey Farme.* Adam Flip, London.

Stockle, C.O., Papendick, R.I., Saxton, K.E., Cambell, G.S., and van Evert, F.K. (1994). A framework for evaluating the sustainability of agricultural production. *American Journal of Alternative Agricultural Production 9*:45-50.

Stocks, A. (1983). Candoshi and Cocamilla swiddens in eastern Peru. *Human Ecology 11*(1):69-84.

Stokstad, E. (2003). "Pristine" forests teemed with people. *Science 301:*1645-1646.

Strange, C. (1998). Windbreaks, snow and water. *Inside Agroforestry* (Winter):2.

Stratton, L.C., Goldstein, G. and Meinzer, F.C. (2000). Temporal and spatial partitioning of water resources among eight woody species in a Hawaiian dry forest. *Oecologia 124:*309-317.

Stringer, W.C. and Alverson, D.R. (1994). Use of fire as a management tool in alfalfa production ecosystems. In Cambell, K.L., Graham, W.D., and Bottcher, A.B. (Eds.), *Environmentally Sound Agriculture* (pp. 492-493). Proceedings of the Second Conference, April 20-22, Orlando, FL.

Stuhr, K. (2001). Alley cropping system buffers water table. *Inside Agroforestry* (Spring):1-2.

Stuhr, K. (2004). Pollinators: Wildlife we can't afford to live without. *Inside Agroforestry* (Winter/Spring):4.

Styger, E., Kakotoarimanana, J.E.M., Rabevohitra, R., and Fernandes, E.C.M. (1999). Indigenous fruits of Madagascar: Potential components of agroforestry systems in improving human nutrition and restore biological diversity. *Agroforestry Systems 46:*289-310.

Suckling, D.M. and Karg, G. (1998). Pheromones and other semichemicals. In Rechcigl, J.E. and Recheigl, N.A. (Eds.), *Biological and Biotechnological Control of Insect Pests* (pp. 63-99). Lewis Publishers, Boca Raton, FL.

Sun, D. and Dickinson, G.R. (1994). A case study of shelterbelt effect on potato *(Solanum tuberosum)* yield on the Atherton Tablelands in tropical north Australia. *Agroforestry Systems 25:*141-151.

Tarawali, G., Manyong, V.M., Carsky, R.J., Vissoh, P.V., Osei-Bonsu, P., and Galiba, M. (1999). Adoption of improved fallows in West Africa: Lessons from mucuna and stylo case studies. *Agroforestry Systems 47:*93-122.

Teasdale, J.R. and Rosecrance, R.C. (2003). Mechanical versus herbicidal strategies for killing a hairy vetch cover crop and controlling weeds. *American Journal of Alternative Agriculture 18*(2):95-102.

Teklehaimanot, Z., Millat-E-Mustafa, M., and Hall, J.B. (2001). Farmers' ecological knowledge about homegarden production systems in Bangladesh. *Forests, Trees and Livelihoods 11:*67-79.

Tenney, S. (1811). *Papers for 1811 Communicated to the Massachusetts Society for Promoting Agriculture.* Munroe and French, [Boston] MA.

Thurston, H.D. (1997). *Slash/Mulch Systems: Sustainable Methods for Tropical Agriculture.* Westview Press, Colorado, 196 pp.

Tilman, D., Knops, J., Wedin, D., Reich, P., Ritchie, M., and Siemann, E. (1997). The influence of functional diversity and composition on ecosystem processes. *Science 277:*1300-1306.

Tolme, P. (2004). Made in the shade. *Audubon 106*(July/August): 56-64.

Torquebiau, E. (1992). Are tropical agroforestry home gardens sustainable? *Agriculture, Ecosystems and Environment 41:*189-207.

Trawick, P. (2002). Trickle-down theory Andean style. *Natural History 111:*61-65.

Turner, N.J., Davidson-Hunt, I.J., and O'Flaherty, M.O. (2003). Living on the edge: Ecological and cultural edges as a source of diversity for social-ecological resilience. *Human Ecology 31*(3):439-461.

Valdéz, M.C. and Abastidas, I. (1993). *Agroforestería y Conservación de Suelos.* Corporación Hondureña de desarrollo Forestal, Honduras.

Van den Eynden, V., Cuera, E., and Cabrera, O. (2003). Wild food from Southern Ecuador. *Economic Botany 57*(4):576-603.

Vandermeer, J. and Perfecto, I. (2005). The future of farming and conservation. *Science 308:*1257.

Vandermeer, J. (1989). *The Ecology of Intercropping.* Cambridge University Press, Cambridge, UK.

Vandermeer, J., Perfecto, I., Nuñez, G.I. Phillpott, S., and Ballinas, A.G. (2002). Ants (*Azteca* spp.) as potential biological control agents in shade coffee in Chiapas, Mexico. *Agroforestry Systems 56*(3):270-276.

van der Valk, H.C., Niassy, A., and Bèye, A.B. (1999). Does grasshopper control create grasshopper problem? Monitoring side- effects of fenitrothion applications in the western Sahel. *Crop Protection 18*(2):139-149.

van Noordwijk, M. (1996). Mulch and shade model for optimum alley-cropping design depending on soil fertility. In Ong, C.K. and Huxley, P. (Eds.), *Tree-Crop Interactions: A Physiological Approach* (pp. 51-72). CAB International, Wallington, UK.

van Noordwijk, M., Hairiah, K., Sitompul, S.M., and Syekhfani, M.S. (1992). Rotational hedgerow intercropping + *Peltophorum pterocarpum* = new hope for weed-infested soils. *Agroforestry Today 4*(4):4-6.

Vasey, D.E. (1992). *An Ecological History of Agriculture 10,000 BC–AD 10,000.* Iowa State University Press, Ames, IO.

Velasco, A., Ibrahim, M., Kass, D., Jiménez, F., and Rivas Platero, G. (1999). Concentraciones de fósforo en suelas bajo sistema silvopastoril de *Acacia mangium* con *Brachiaria humidicola. Agroforestería en las Americas 6*(23):45-47.

Vera, F.W.M. (2000). *Grazing Ecology and Forest History.* CABI Publishing, Dordrecht, the Netherlands.

Versteeg, M.N., Amadji, F., Eteka, A., Gogan, A., and Koudokpon, V. (1998). Farmers' adoptability of Mucuna fallowing and agroforestry technologies in the coastal savanna of Benin. *Agricultural Systems 56:*269-287.

Vickery, J., Carter, N., and Fuller, R.J. (2002). The potential value of managed field margins as forging habitats for farmland birds in the UK. *Agriculture, Ecosystems and Environment 89:*41-52.

Vieyra-Odilon, L. and Vibrams, H. (2001). Weeds as crops: The value of maize weeds in the valley of Toluca, Mexico. *Economic Botany 55*(3):426-443.

Villachica, H., Silva, J.E., Peres, J.R., and Magno C. da Rocha, C. (1990). Sustainable agricultural systems in the humid tropics of South America. In Edwards, C.A., Lal, R., Madden, P., Miller, R.H., and House, G. (Eds.), *Sustainable Agricultural Systems* (pp. 391-437). St. Lucie Press, Delray Beach, FL.

Víquez, E., Prado, A., Oñoro, P., and Solabo, R. (1994). Caracterización del huerto mixto tropical "la asunción," Masatepe, Nicaragua. *Agroforestería en las Américas 1*(2):5-9.

Vitousek, P.M., Mooney, H.A., Lubshenco, J., and Melillo, J.M. (1997). Human domination of earth's ecosystems. *Science 277:*494-499.

von Weizsäcker, P. (1995). The yield of weeds. *Leisa 11*(3):6-7.

Waddell, E. (1975). How the Enga cope with frost: Responses to climate perturbation in the Central Highlands of New Guinea. *Human Ecology 3*(4):249-273.

Wale, M., Abate, T., and Tesfaye A. (1996). Getting to know the enemy—Mesoplatys beetles and sesbania. *Agroforestry Today 8*(3):17-18.

Wallace, D. (1997). Give the foodplain a break a waterbreak. *Inside Agroforestry* (Fall):1.

Wallace, D. (Ed.) (1980). *Getting the Most from Your Garden.* Rodale Press, Emmaus, PA.

Walle, R.J. and Sims, B.G. (1998). Natural terrace formation through vegetative barriers on hillside farms in Honduras. *American Journal of Alternative Agriculture 13*(2):79-82.

Walter, G.H. (2003). *Insect Pest Management and Ecological Research.* Cambridge University Press, Cambridge, UK.

Wang, H. (1994). Tea and trees: A good blend from China. *Agroforestry Today 6*(1):6-8.

Wang, Q. and Shogren, J.F. (1992). Characteristics of the crop paulownia system in China. *Agriculture, Ecosystems and Environment 39:*145-152.

Waring, R.H. and Pitman, G.B. (1985). Modifying lodgepole pine stands to change susceptibility to mountain pine beetle attack. *Ecology 66*(3):884-897.

Weaver, J.C. (1919). *The Ecological Relations of Roots.* Publication 286. Carnegie Institute of Washington, Washington, DC.

Weaver, J.C. (1920). *Root Development in Grassland Formulation.* Publication 292. Carnegie Institute of Washington, Washington, DC.

Weaver, W.W. (1997). *Heirloom Vegetable Gardening.* Henry Holt and Company, New York.

Weber, F.R. (1986). *Reforestation in Arid Lands.* Volunteers in Technical Assistance, Arlington, VA.

Weinzierl, R.A. (1998). Botanical insecticides: Soaps and oils. In Rechcigl, J.E. and Rechcigl, N.A. (Eds.), *Biological and Biotechnological Control of Insect Pests* (pp. 101-121). Lewis Publishers, Boca Raton, FL.

Wells, G. (2002). Biotechnical streambank protection. *Agroforestry Notes 23*(March):1-4.

Wezel, A. and Bender, S. (2003). Plant species diversity of homegardens of Cuba and its significance for household food supply. *Agroforestry Systems 57:*39-49.

Wiersum, K.F. (1984). Surface erosion under various tropical agroforestry systems. In O'Loughlin, C.L. and Pearce, A.J. (Eds.), *Symposium on Effects of Forest Land-Use on Erosion and Slope Stability* (pp. 231-339). East-West Center, Honolulu, HI.

Wight, B. (2003). Productive conservation: growing specialty forest products in windbreaks. *Inside Agroforestry* (Spring):4.

Wight, B. and Stuhr, K. (2002). Windbreaks: An agroforestry practice. *Agroforestry Notes* 25(March):1-4.

Wijesinghe, D.K. and Hutchings, M.J. (1999). The effects of environmental heterogeneity on the performance of *Glechoma hederacea*: The interactions between patch contrast and patch scale. *Journal of Ecology 87:*860-872.

Williams, T. (2002). America's largest weed. *Audubon* 104(January/February):24-31.

Williams, T. (2004). Drunk on ethanol. *Audubon* 105(July/August):26-35.

Willis, K.J., Gillson, L. and Brncic, T.M. (2004). How "virgin" are virgin forests. *Science 304:*402-403.

Wilson, J. (1994). *Landscaping with Herbs.* Houghton Mifflin Company, New York.

Wilson, T.D., Brook, R.M., and Tomlinson, H.F. (1998). Interactions between néré *(Parkia biglobosa)* and under-planted sorghum in a parkland system in Burkina Faso. *Experimental Agriculture 34:*85-98.

Wilson, W.M. (2003). Cassava (*Manihot esculenta* Crantz), cyanogenic potential, and predation in northwestern Amazonia: The Tukanoan perspective. *Human Ecology* 31(3):403-416.

Wohlgemuth, T., Bürgi, M., Scheidegger, C., and Schütz, M. (2002). Dominance reduction of species through disturbance—A proposal for central European forests. *Forest Ecology and Management 166*(1-3):1-15.

Wojtkowski, P.A. (1993). Toward an understanding of tropical home gardens. *Agroforestry Systems 24:*215-222.

Wojtkowski, P.A. (1998). *The Theory and Practice of Agroforestry Design.* Science Publishers, Enfield, NH.

Wojtkowski, P.A. (2002). *Agroecological Perspectives in Agronomy, Forestry, and Agroforestry.* Science Publishers, Enfield, NH.

Wojtkowski, P.A. (2004). *Landscape Agroecology.* Food Products Press, Binghamton, NY.

Wojtkowski, P.A. and Cubbage, F.W. (1991). A bordered matrix approach to the bioeconomic modeling of agroforestry systems. *Agroforestry Systems 14:*81-97.

Wojtkowski, P.A., Brister, G.H., and Cubbage, F.W. (1988). Using multiple objective programming to evaluate multi-participant agroforestry systems. *Agroforestry Systems 7:*185-195.

Wojtkowski, P.A., Jordan, C.F., and Cubbage, F.W. (1991). Bioeconomic modeling in agroforestry: A rubber-cacoa example. *Agroforestry Systems 14:*163-177.

Woodward, A. and Coppock, D.L. (1995). Role of plant defense in the utilization of native browse in southern Ethiopia. *Agroforestry Systems 32:*147-161.

Woolsey, T.S. and Greeley, W.B. (1920). *Studies in French Forestry.* Wiley and Sons, New York.

Worlidge, J. (1669). *Systema Agriculturæ.* F.W. Gent, London.

Worlidge, J. (1689). *The Second Part of Systema Agriculturæ or the Mystery of Husbandry.* F.W. Gent, London.

Worsham, A.D. (1989). Current and potential techniques using allelopathy as an aid in weed management. In Chou, C.H. and Waller, G.R. (Eds.), *Phytochemical Ecology: Allelochemical, Mycotoxins and Insect Pheromones and Allomones* (pp. 275-292). Institute of Botany, Taipei, Taiwan.

Yin, R. and He, Q. (1997). The spatial and temporal effects of paulownia intercropping. *Agroforestry Systems 37:*91-109.

Yoon, C.K. (2000). Simple method found to increase crop yield vastly. *The New York Times,* August 22, pp. D1-D2.

Yoon, C.K. (2004). To the rescue: Fungi that invade also protect leaves. *The New York Times,* February 24, p. D3.

Young, A. (1989a). *Agroforestry for Soil Conservation.* CAB International, Cambridge, UK.

Young, A. (1989b). The environmental basis of agroforestry. In Reifsnyder, W.E. and Darnhofer, T.O. (Eds.), *Meteorology and Agroforestry* (pp. 29-48). ICRAF, Nairobi, Kenya.

Zaldivar, M.E., Rocha, O.J., Castro, E., and Barrantes, R. (2002). Species diversity of edible plants grown in homegardens of Chibchan Amerindians from Costa Rica. *Human Ecology 30*(3):301-316.

Zerega, N.J.C. (2003). The Breadfruit Trail. *Natural History 112:*47-51.

Zhang, F. (1996). Influences of shelterbelts in Chifeng, Inner Mongolia. *Unasylva 47*(185):11-15.

Zihlmann, U., Weisskopf, P., Dubois, D., and Daellenbach, S. (2000). Mineral N-content in a loess soil under organic and integrated cultivation of potatoes. In Alföldi, T., Lockeretz, W., and Niggli, U. (Eds.), *The World Goes Organic, Proceedings of the 13th IFOAM Scientific Conference* (p. 171). August 28-31, Basel, Switzerland.

Zimmerer, K.S. (1999). Overlapping patchwork of mountain agriculture in Peru and Bolivia toward a regional-global landscape model. *Human Ecology 27*(1): 135-165.

Zingore, S., Mafongoya, P., Nyamugafata, P., and Giller, K.E. (2003). Nitrogen mineralization and maize yields following application of tree prunings to a sandy soil in Zimbabwe. *Agroforestry Systems 59:*199-211.

Zobel, B.J., Van Wyk, G., and Stahl, P. (1987). *Growing Exotic Forests.* John Wiley and Sons, New York.

Zohary, D. (2004). Unconscious selection and the evolution of domesticated plants. *Economic Botany 58*(1):5-10.

Zewdu, T. (2004). Grain and straw yield of barley as influenced by undersowing time of annual forage legumes and fertilization. *Tropical Science 44:*85-88.

Index

Introduction to Agroecology: Principles and Practices

doi:10.1300/5553_21